EUREL International Conference

THE DETECTION OF ABANDONED LAND MINES: A HUMANITARIAN IMPERATIVE SEEKING A TECHNICAL SOLUTION

7-9th October, 1996

Organised by

the Convention of National Societies of Electrical Engineers of Europe (EUREL) under the sponsorship of the Institution of Electrical Engineers with the support of the

Joint Research Centre of the European Commission

in association with the

Chinese Institute of Electronics
IEEE - Antennas and Propogation Society Chapter
Aerospace and Electronic Systems Society
Laser and Electroptics Society
Magnetics Society
UK and Republic of Ireland Section
Institute of Physics
IEE of Japan
USA National Academy of Sciences
Remote Sensing Society
South African Institute of Electrical Engineers
SPIE

Venue
EICC, Edinburgh, UK

Author Disclaimer

"While the author and the publisher believe that the information and guidance given in this work are correct, all parties must rely upon their own skill and judgment when making use of it. Neither the author nor the publisher assume any liability to anyone for any loss or damage caused by any error or omission in the work, whether such error or omission is the result of negligence or any other cause. Any and all such liability is disclaimed."

Copyright and Copying

Conditions of Acceptance of Advertisements

The Institution reserves the right to refuse to insert any advertisements (even if ordered and paid for) and/or to make alterations necessary to maintain its standards.

It is not guaranteed that any advertisement will be placed in any specified position or on any specified page unless agreement has been entered into and the agreed surcharge paid.

Every effort will be made to avoid errors but no responsibility will be accepted for any mistakes that may arise in the course of publication of any advertisement. These mistakes may include non-insertion, insertions other than those ordered and errors and omissions within the advertisement.

Notice to cancel any advertisement must be received in writing ten days prior to its next scheduled appearance, otherwise a charge will be made.

No responsibility will be accepted for repetative errors unless the advertiser's correction has been accepted in respect of that error.

No responsibility will be accepted for loss or damage alleged to arise from errors within advertisement copy, non-appearance of an advertisement or delay in forwarding box number replies.

Advertisers are required to ensure that the content of advertisements conforms with all legislation currently in force affecting such matters. They shall further indemnify the publisher in respect of any claims, costs and expenses that may arise from anything contained within the advertisement and published on their behalf by the Institution of Electrical Engineers.

The placing of an order or contract for insertion of an advertisement in any newspaper or journal published by the Institution of Electrical Engineers whether in writing or by verbal or telephone instructions will be deemed an acceptance of each and all the above conditions.

Published by the Institution of Electrical Engineers, London.

ISBN 0 85296 669 5 ISSN 0537-9989.

Printed in Great Britain by Omega Print and Design.

ORGANISING COMMITTEE

Professor R Voles, Consultant (Chairman)
Professor F Alder, UMIST
Mr D Barron, Pilkington Thorn Optronics
Brigadier (Rtd) P M Blagden
Mr D Daniels, ERA Technology Ltd
Professor H Griffiths, UCL

CORRESPONDING MEMBERS

Dr P Foster, Microwave and Antenna Systems, UK
Mr C Hamon, SAGEM SA (formerly SAT), France
Dr B Janzon, FOA Dept of Weapons and Protection, Sweden
Dr W Keydel, DLR, Germany
Professor D Longstaff, University of Queensland, Australia
Professor J-D Nicoud, Swiss Federal Institute of Technology, Switzerland
Dr A Sieber, Joint Research Centre, European Commission, Italy
Mr M Weathersby, International Society for Optical Engineering, USA
Professor Zong Sha, Chinese Institute of Electronics, Peopleís Republic of China

Contents

The Institution of Electrical Engineers is not, as a body, responsible for the opinions expressed by individual authors or speakers

Contents

Page No

Contents

Page No

Contents

Page No

Contents

Contents

List of Authors

List of Authors

PRACTICAL PROBLEMS IN DEMINING AND THEIR SOLUTIONS

K. Eblagh

UNOCHA Mine Clearance Programme Afghanistan

INTRODUCTION

Afghanistan is a country where some ten million abandoned land mines contaminate over 530 square kilometers of otherwise productive land threatening the every day life of twenty million Afghans. Like most clearance teams throughout the world, Afghan deminers are confronted with a wide variety of mines (over fifty different mines produced by more than ten countries) planted to deny combatants the use of residential areas, irrigation systems, agricultural and grazing lands and roads. Mines were laid by all combatants with little regard to recording mapping or marking of this legacy of conflict.

The variations in mine type, terrain and military application produces a complex combination of challenges. Challenges that generate the need for a "Tool Bag" approach to the clearance problem. No one solution will meet the needs of every situation.

The technologies applied to the mine clearance problem in Afghanistan include:

Manual Clearance Systems. Teams of thirty deminers working in two man clearance parties use metal detectors to identify metal fragments. Detector signatures are investigated by prodding or probing to uncover the metal located. Probing is done with a variety of equipment, the most effective being a bayonet. This technology has not changed since World War Two.

Dogs. Trained to indicate on the presence of explosive, dogs are used to check ground using tightly controlled search patterns. Supporting deminers investigate areas indicated by dogs using the same techniques applied to manual demining .

Mechanical Clearance Systems. Two mechanical systems have been employed in Afghanistan. Flail machines have been employed in the clearance of roads, grazing land and agricultural land. The machines work on the principle of using force to detonate or destroy mines. The concept is quite simple, hammers attached to rotating chains beat the ground in the path of the machine, all mines buried to a depth of 30 cm should be cleared.

The second mechanical clearance system is an armored light industrial backhoe. Traditional mud brick construction and the extensive use of mines in residential areas in Afghanistan developed a fairly unique demining challenge; the need to remove, in some cases up to 2 meters, of earth covering mined areas. Whilst mines covered by 2 meters of earth would not be detonated by people working over them, they will be uncovered by people returning to reconstruct their homes. The backhoe is used to assist in the excavation of material which would otherwise be removed using traditional manual demining techniques.

THE PROBLEMS

Battlefield Scrap

The majority of mine clearance is relies on metal detector technology either as the prime source for locating mines or as a check on other technologies, e.g. dog indications are investigated using metal detectors and probes. The majority of mined areas include large quantities of scrap metal producing a high "false alarm" rate in the clearance process. This false alarm rate can be as high as 1000:1, slowing the clearance process as every signature must be treated as a mine until its cause is confirmed and dealt with.

Metal signature investigation the most dangerous phase of the clearance process. 80% of all clearance accidents occur during the investigation of metal signatures. Most of the mines involved (PMN and PMN2) have relatively large quantities of metal and yet they pose a significant threat to deminers throughout the world.

Currently false alarm rates can be reduced through the use of dogs to locate explosives rather than metal.

What is needed is a device capable of reducing the false alarm rate to limits closer to 20:1. If this is achieved then all signatures could be dealt with by remote explosive means. This could reduce mine clearance accidents by up to 80% and have a significant multiplying effect on productivity.

First generation equipment of this type could provide the deminer with a visual image of shape and size of the metal signature. The deminer can then discard those images that do not conform to the mines prevalent in the area being searched and destroy those confirmed to be or highly likely to be mines.

Second generation equipment could be programmed to make those decisions for the deminer providing an output that may indicate the type and depth of the mine.

Shaped charges could also be developed to provide a range of tools need to deal with various types of mines without the need for probing and digging to uncover the mine in preparation for destruction.

Ferrite Soils

Ferrite soils continue to pose problems for the deminer. Most detectors/locators are unable to cope with these conditions. The machines capable of doing so are extremely expensive.

A major problem confronted in Afghanistan is the high metallic signatures given by kiln fired bricks in most of the mined residential areas. These areas do not act in the same way as ferrite soils as the metal content is not evenly distributed in the target areas.

Currently each signature is treated as a mine, which is extremely difficult, time consuming and very frustrating.

This problem could be addressed by the development of a machine capable of detecting explosive or plastics in the soil.

Hard Soils, Gravel and Roads:

Taking the present limited tools and equipment into consideration the excavation of hard/gravel surface, particularly roads, is an extremely difficult aspect of demining. It is both physically demanding and time consuming work.

The most efficient system available in the field today is the mine detection dog. The dogs ability to reduce the false alarm rate of produced by metal detectors reduces the need for deminers to use their bayonets and trowels to dig up kilometers of road investigating metal signatures.

Deeply Buried Mines:

Humanitarian mine clearance demands that areas are cleared of all mines. This aim generates the need for all land to be cleared to original or prewar levels. This need produces some very difficult clearance tasks particularly in residential areas and in the clearance of irrigation systems.

The war in Afghanistan included extensive use of mines in residential areas. During the conflict mines laid in these areas were covered by collapsed wall destroyed during the conflict or simply affected by erosion. Once covered by more than 500 mm of soil the mines were ineffective, the combatants then relaid the minefield producing a lasagna effect of mines and soil. In some cases there are three layers of minefield in the one area. Another similar problem has developed in irrigation system where combatants attempted to deny the use of irrigation systems as trenches and protected infiltration routes through the use of mines. This method of denial was ineffective and many of the irrigation systems were then filled in. The both situation we are confronted with the task of locating AP mines that are buried up to 2 m deep.

Metal detectors, bayonets and trowels have been used to excavate hundreds of thousands of cubic meters of soil to locate mines in these locations.

Recently the UNOCHA mine clearance programme has completed a very successful trial on the use of backhoe in the clearance of these areas. The backhoe is used to excavate much of the overburden and deminers conduct a final check of the ground using tradition methods. This has increased productivity of a demining team by 300%.

Production and supply of sensitive high power mine detectors which could detect mines to greater depths could also assist in the clearance of these areas. This equipment should also be capable of providing the deminer with information that could allow discrimination between scrap metal and mines reducing the extremely high false alarm rate confronted in residential areas.

Existence of Demobilized Vehicles in Minefields

Sometimes mines are placed around/under destroyed vehicles.

Clearance of such areas is extremely difficult and needs great caution because on one hand mine detectors are unable to locate mines laid beside these vehicles, and on the other hand moving the vehicles is also dangerous and has critical consequences.

Currently we have two alternatives. One is to simply prod and dig a path to the vehicle or metal obstacle, secure a line to the object and winch it into a cleared area and then continue with traditional mine clearance techniques. The second is to use dogs to check/clear an access route to the obstacle.

The first method is extremely slow and dangerous. The second is affective but requires highly trained personnel and support systems.

The only way out of this difficulty would be inventing detectors capable of focussing on specific target areas or able to provide a visual output of their signature. Detectors able to indicate the presence of explosives rather than metals would be the ideal solution.

Vegetation in minefields

Occasionally dense bushes with thick stems in minefield create a major obstacle to the clearance process. This problem would further intensify when the bushes are fresh and could not be burnt easily.

At present the mine clearance personnel first use defoliants to kill off vegetation then spray fuel onto vegetation and burn the bushes. Alternatively the vegetation is cut using knives or clippers in an extremely slow and dangerous process.

Two solutions could be developed to assist with this problem. The development of an armored, vehicle mounted slasher that could cut vegetation away. Secondly systems that could allow the easy projection of chemicals and fuel over the mined area to speed up the burning process.

Disposal of Bombs and Large Unexploded Ordnance

Unexploded bombs and ordnance are an enormous threat to the life and property of residents of most of the residential areas affected by armed conflict. It is especially true when some individuals tragically try to dismantle these devices when collecting scrap metal. The collection of scrap metal in post war areas is a common and lucrative occupation with often fatal consequences. For example eight persons who were attempting to defuse a bomb were killed in the garden of their house in Kabul. They were killed when the bomb exploded.

Solution

For removal of unexploded bombs and large shells the transport of which is feasible and safe we have adopted special measures as by using special techniques they are primarily defused, moved to sites quite far from the city and then destroyed.

If unexploded ordnance cannot be moved then it is destroyed in-situ. This is achieved by constructing a protective surround to encompass the item and then detonating the item by electrical initiation.

CONCLUSION

It is slowly being accepted that there is no panacea for the world wide landmine problem. What is clear, is that the detection of many mines based on their ferrous content is extremely difficult in many situations. False signatures from a variety of non-mine ferrous items on a battle area create many cul-de-sacs to the task of finding and dealing with landmines.

A possible solution, would be to produce an electronic instrument, linked to a sniffer system, capable of being programmed with a selected array of odours,-similar to that being developed here in Britain,- able to produce a signature odourgram. This would go a long way to providing the men who risk their lives in the service of others with another tool and a better chance to do their job more effectively, in a difficult and dangerous environment.

Can it be done, and can it be done quickly.

PLANNING AND RUNNING MAJOR MINE CLEARANCE OPERATIONS

G S Lucas

BACTEC International Ltd, UK

The requirement for mine clearance operations is based on the significant number of countries that are contaminated with mines together with the large numbers of mines already laid in the ground. Over 64 countries are contaminated with mines and there are in excess of 100million mines in the ground. Every day casualties and fatalities are being caused by land mines. The clearance of land mines is therefore of international interest and is attracting a considerable amount of effort from the international community in terms of technical solutions and funding.

Many of the mines have been in the ground for a number of years - for example in Angola mines have been laid over the last 30 years. In some parts of northern Africa and Europe mines date back to the 1939/45 war while in more recent internal and international conflicts mines have been laid recently or are still being laid now. The majority of mine clearance operations require the security situation within the blighted country to be stable and for there to be the will to clear the mines.

Somalia typifies the situation where there is a known threat and people are being killed and injured by mines but the political situation is such that no formal mine clearance operation can take place with confidence that once cleared, the areas will not be re-mined by other factions within the country.

Whilst a mines clearance programme is being established, one area where an immediate contribution can be made to reducing injury and saving life is a well run, well prepared, relevant Mines Awareness Programme. A Mines Awareness Programme can be initiated immediately upon cessation of hostilities. The programme should be aimed to educate the at risk local population ranging from the children through to the adults and targeting those groups of people who are most at risk, ie farmers, the members of a family who collect firewood, water and food.

A well structured Mines Awareness programme should adopt a multi media approach using images and language appropriate to the local situation. Media research is necessary prior to its development to ensure that both the media and the message are appropriate for the target audience. Whilst television, radio and newspapers are appropriate for urban populations, in some developing countries poster campaigns and community gatherings are the most effective method of dissemination in rural areas. The use of parliamentary techniques to ensure ownership of Mines Awareness campaigns is essential for their success.

A mines awareness campaign can be made interesting by using various techniques. Whatever method is used to put across the message it must be appropriate, easily assimilated and relevant. In most conflict situations there are a significant number of internally displaced personnel, (IDPs) as well as external refugees. It is often these two groups of people that are at greatest risk when they return to their villages as they may be unaware of the situation in the area to which they wish to return.

Upon declaration of a ceasefire or upon the breakout of peace a number of individuals may well wish to return to their villages so they can resume their old way of life as farmers, businessmen etc and therefore they will require access to their villages over cleared roads. It may be possible initially to clear roads and tracks but often the verges and the area just off the road will still be mined. It is therefore necessary to establish in the mines awareness campaign the extent of the likely contamination and where mines and likely to be found and therefore the areas to avoid. Mines awareness is equally appropriate for NGO field staff, commercial organisations and international humanitarian organisations.

Consulting Engineers and contractors rebuilding roads, utilities, restoring the infrastructure are equally at risk and often more so that the local community. The ex-patriates are not fully aware of the extent of mining and where mines may be laid or, more importantly, the local history of an area. It should therefore be an automatic process for all ex-pat personnel to be fully briefed on the situation and given a mines awareness briefing prior to deploying from the major city, port or airport where they have entered the country.

In order to carry out a mine clearance programme it is necessary to embark on a mines survey. The principal aspect of the survey is to identify the threat in terms of the type of mines likely to be encountered, to find out whether mines are likely to be booby trapped, to establish whether there are any records of the mines and indeed whether or not any mine clearance operations have taken place which could affect the overall programme.

When considering the threat due regard must also be given to the possible contamination of minefields or mined areas with other ordnance items including sub munitions. There are seldom instances where mines or mine contaminated areas have not been fought over or have not been subjected to some form of bombardment by artillery or by aircraft, often including the use of sub munitions and scatterable mines. Whilst most types of mine are well documented and the render safe procedures and disposal techniques determined, the actual situation in which the mines are laid - underground, on the ground, beneath vegetation and the time in the ground will have a significant impact on the subsequent clearance operation.

It is possible to identify the threat to a limited extent purely by local information obtained from military, local governments and personnel who have been involved in the laying of mines or lifting of mines. However, in order to plan a programme it is necessary to conduct a survey to identify those areas which are mined or possibly mined. A mines survey is a key element in a successful mine clearance programme. The survey should involve experienced mine clearers who systematically target various parts of a country so that a complete picture of the mine contamination can be obtained. The overall picture will identify those areas which are mined, those areas which are suspected of being mined, those areas which are definitely not mined and those areas where it is highly improbable that any form of land mines has been laid.

In conducting a survey it is necessary to obtain information on the ground from local military or civilian authorities, NGO's, other organisations with detailed knowledge on the ground and individuals who have local knowledge. Where possible contact should be established with the village chiefs and the farming communities to establish which areas the locals will not use and why. It may just be that an area has been fought over but there is no positive evidence of any incidents, just suspicion, or there may be areas where incidents have occurred and therefore there is positive evidence of the areas being contaminated.

There are a number of areas which are likely targets for mining - these include water sources, paths, bridges, roads and other areas of military significance. It should also be remembered that in some incidents mines are laid purely as a terrorist device to deny the locals access to the ground and thus achieve a military objective. It is only by detailed investigation on the ground by experienced mine clearers and people who have an understanding of mine warfare that the overall picture can be obtained.

The media will often exaggerate the extent of the contamination in an area. While a detailed survey may indicate that 90% of the roads in a particular area are totally clear of mines and are used on a day to day basis, the reality of the situation may not always be apparent many miles away in the capital city or in the various governors offices where the whole area is perceived to be contaminated.

Having identified the threat and carried out a survey and therefore being able to determine those areas which are contaminated, those which are likely to be contaminated or suspect and those which are not contaminated it is possible to apply the priorities for clearance in those particular areas.

In Angola, for example there are areas where it is known there are large barrier mine fields but these are in the south of Angola in areas where there is a low population density and as the locals know the location of the minefields they represent a low priority for clearance. On the other hand areas near towns or main access routes and key elements of the infrastructure should be considered as high priority. In determining the priorities for clearance due recognition must be given to the various agencies involved in either a peace keeping operation (UN) local government as in the situation in Kuwait following the liberation after the Gulf war, or other areas as in former Yugoslavia where there is a combination of UN, NATO and the local government.

Each of the authorities will have different priorities for clearance. It will range from military objectives to economic and humanitarian objectives. The World Bank and EBRD may prioritise economic projects to restore the infrastructure, ie opening road and rail links and reconstructing power stations and utilities. The UNHCR and NGOs may have the return of refugees and internally displaced people as their priority. There is likely to be a conflict between the various parties and it is in this situation that the programme manager must be able to liaise with all the various agencies.

Ideally the programme manager will be the focal point for all mine clearance activities within the country and will have the active support of all the various agencies. However, this is an ideal and is particularly difficult to achieve with the international community wishing to help the national organisations and military units. It is in this situation that the Programme Manager becomes a politician and a great deal of his time and effort is spent in liaising with the various government agencies and establishing a rapport with each of the agencies to try and co-ordinate all the activities.

Once the "political will" has been established and there is a clearly defined role for the mines programme manager supported by the various agencies it is then possible to proceed to the next stage of the mines programme. This is to obtain the necessary resources so that the actual mine clearance operation can commence.

Mine clearance resources can be obtained by training local personnel in humanitarian mine clearance operation. There is a significant difference between humanitarian mine clearance and the mining activities often carried out by both government and irregular forces, including terrorists, who are often very ill equipped and will improvise. The standard of the work they produce is not of an acceptable standard for humanitarian mine clearance.

While these personnel may well have the necessary skills and experience in dealing with the mines and the various mechanisms they have to be re-educated to adopt a systematic approach to carrying out a mine clearance operation. It is necessary to vet the mine clearers. It is possible that some mine clearers may wish to undertake mine clearance operations purely because they know they will be paid and therefore mine clearance offers them some form of security. However, they may not be psychologically suitable for mine clearance operations and therefore despite their enthusiasm and interest they should not be permitted to undertake mine clearance.

The suitability of the potential humanitarian mine clearers must continue to be examined during the training process. In some instances it is worth recruiting civilians or people who have no previous mine clearance experience. In this way there is no need to overcome the habits they have learnt during operational mine clearance and mine laying where there may have been scant regard for safety.

Training is naturally a key element of any mines programme and therefore resources and the necessary support must be made available for the establishment of training schools complete with the correct equipment for the local situation and with experienced instructors.

It is necessary to equip the student mine clearers with personal protective equipment, locators, prodders, medical back up etc. and this needs to be organised in advance of the training so that the students, when they arrive for the training, can undertake the training with the proper equipment in the right environment. The student mine clearers should also have access to inert mines for training and recognition purposes so they can become familiar with handling mines and they can run through procedures in a dry training area with inert mines prior to being deployed onto a live minefield situation under the direct control and supervision of the instructors who have taken them through their training. It is essential that the rapport and confidence levels are built up between the instructors and the students so that when the newly graduated mine clearers are deployed in the field they will act in a safe and controlled manner.

The initial training should be for basic mine clearers but after some time in the field a number of the mine clearers will prove themselves to be capable of becoming team leaders and supervisors. It will be necessary at a later stage to run management courses to encourage the local mine clearers to take a more active part in the management of the overall mines programme. Ultimately it should be the aim of the programme manager that the whole mine clearance programme is run and organised by the local personnel. It may take some years for the whole mine clearance programme to be transferred firmly into the hands of the local authorities but it is essential that where possible the locals are put in a position of authority so that the impetus of the programme can be maintained for many years. It is likely in a country such as Mozambique that the programme will last for a number of years and therefore it is imperative that the local authorities assume responsibility for the whole programme including the training.

The training of local personnel as humanitarian mine clearers is one option to providing the mine clearance resource. As a short term expedient measure it is possible to contract commercial companies to undertake mine clearance with fully experienced, fully equipped teams who can be deployed at short notice to undertake high priority clearance. This is an expensive option. In the same way a number of NGOs have a mine clearance capability and they will often have an infrastructure within a country so they are able to recruit and train local personnel as mine clearers. This is a totally separate operation to that being undertaken by the Government in the Government training schools.

NGOs have worked very successfully in Afghanistan, Laos, Cambodia, Angola, Mozambique, Sudan, Yemen and many other countries. The NGOs provide a valuable contribution but the rate of progress may not be as great as that from commercial companies or indeed organisations using some form of mechanical assistance.

A great deal of emphasis is now being placed on the standard and level of clearance. Where the United Nations has influence on the award of mine clearance contracts, it has stipulated a clearance probability of 99.6%. This appears to have been generally accepted within the mine clearance community as an acceptable standard although every company aims to clear the ground
once hundred per cent.

Quality Control and Quality Assurance are therefore key factors in any mine clearance programme. Each organisation carrying out mine clearance should be responsible for its own internal QA. In addition to this a third party should ideally be selected to undertake Quality Control and Quality Assurance checking of mine clearance projects. The Quality Control measures should include spot checking of areas already cleared, checking of instruments to ensure they are functioning to the manufacturers specification and that routine quality checking is carried out by the contractor or mine clearance organisation.

Prior to the start of work every instrument should be checked against a typical target mine buried to the depth of the target mines to give the operator confidence that the equipment is working and also to ensure that the locators are able to locate the mines at the depth they are likely to be found at. Because Quality Assurance and Quality Control are such important aspects of any project it is necessary to define accurately in any contract how the quality parameters will be applied and in particular what constitutes a "find" and the action that needs to be taken in the event that an item of ordnance is found on a area that has been cleared.

In order to run a successful mine clearance operation it is necessary to form a structure and organisation which is recognised and has sufficient authority to be able to operate in a semi-autonomous mode. There is a need for a Programme Manager who must be an experienced mine clearer. He will require experienced staff to cover Mines Awareness, mine survey, training, Quality Control and safety as well as an operations team covering data preparation, mapping and data retrieval.

Research has shown that initially mine clearance programmes receive tremendous support and enthusiasm but that this may wane after a few years. To maintain momentum what is needed is impetus and interest from senior government officials supported by a proper budget. It is essential that whatever organisation is sponsoring the mine clearance programme it is not directly related to a Ministry of Defence. Donors and the international community are most reluctant to provide money which could possibly be re-directed within a Ministry of Defence towards non-mine clearance related operations. If it can be demonstrated that the allocation of funds is controlled by the mines programme manager under the guardianship of, say, the United Nations, this gives the sponsors and donors confidence that the money they are allocating is going to be used expeditiously and to the benefit of the country. It is also for this reason that command and control of all demining operations should be vested in the Mines Action Centre or an appropriate organisation set up specifically for the purpose of carrying out the mine clearance operation. Situations have arisen where some organisations working independently of any other authority have carried out extremely competent demining operations but have not been attacking the highest priority areas for clearance and in effect their plaudible endeavours have not achieved the maximum benefit for the country.

There is understandably a reluctance on behalf of some organisations to conform to a central organisation in that it implies bureaucracy and inefficiency. With a mine clearance programme it is the duty of the programme manager to ensure that this situation does not arise and that his organisation acts in an efficient and proper manner at all times to ensure effective deployment of resources.

Direct interface with government officials, including military at the right level, will ensure that the mine clearance programme is implemented. It is equally important that all records are centralised and that there is a centralised database so that should individuals, companies, government or any other organisation can have ready access to information which could help increase safety. Media and public information systems should be fully utilised so that as areas are cleared information can be disseminated to the local population so that farmers, businessman, local officials are aware that an area is being opened up and can therefore react in a positive manner to encourage agriculture and other programmes to take place.

It is again a function of the organisation to determine the most suitable equipment for use in mine clearance operations to ensure that there are standard operation procedures for dealing with certain terrain conditions and certain mines. It is not possible to write instructions or procedures covering every eventuality but there should be general procedures covering routine clearance operations and where there is a change of circumstances so a new procedure should be written by the team leaders on the ground and this should be disseminated so that all other teams have access to the same information. This is the advantage of having a focal point where all the information can be gathered and then disseminated as appropriate - the Mines Action Centre.

Within the international demining community there is a wealth of experience and this should be tapped to ensure that the staff in any mine clearance programme are the most experienced and most suitable staff that can be obtained for that particular programme. Although formal military training forms the basis for most mine clearance operators some have learnt through hands on experience but may not have the depth of knowledge or the experience in dealing with other items of ordnance. It is therefore necessary select the key personnel for a mines programme so that the personnel are able to perform the function required by the particular programme.

The selection of equipment may unfortunately be determined by budget restrictions and may not be determined on technical grounds. Procurement agencies working for international organisations may not fully understand the technical requirements of the mine clearance operator in terms of the locator he needs and therefore it is essential that at all stages the mine programme staff are directly involved in the selection of equipment. Operational requirements and technical requirements of the operators should outweigh commercial considerations.

This is particularly so in the case of locators being used for the detection of minimum metal mines where it is not necessarily the cheapest detector that gives the operator the greatest degree of safety. The operator's safety must be of paramount importance in selecting any personal protective equipment or mine detection/location equipment. The sensitivity of an instrument in locating minimum metal mines must take priority over commercial considerations. The detector with the greatest detecting range will be the most suitable piece of equipment from the mine detectors point of view and therefore this should be the prime consideration and not the cost. Whilst it is understood that cost is a consideration, safety cannot be compromised.

In summary a well run Mine Programme is the key to a successful mine clearance operation. A well coordinated Mines Awareness campaign will probably produce a significant reduction in injuries and fatalities. The effectiveness of a mine clearance operation will be dependent upon a properly defined organised and funded mine programme combined with the will for the programme to succeed.

IMPROVING THE TECHNOLOGY FOR HUMANITARIAN DEMINING

B S G Janzon

FOA - National Defence Research Establishment, Sweden

SUMMARY

• Land-mines may be the most toxic and widespread pollution facing mankind

• Land-mines are the weapons most consistently dangerous to non-combatants.

• They kill and maim soldiers and civilians alike, without preference

• Many of the civilians killed or injured will be children.

• Bombs can be aimed at military targets, bullets fired by soldiers at other soldiers.

• But, while land-mines can be employed for legitimate military purposes, they are by nature hidden and kill or wound whoever triggers them.

• They do not differentiate between tanks and relief trucks, between soldiers and playing children.

• They can remain deadly for decades

BACKGROUND

To-day, over 125 million uncleared landmines can be found in about 60 countries throughout the World. Many of these countries belong to the third world, and many are among the world's poorest countries. Their chances of ever solving their national mine problems by using their own resources can be judged to be virtually nil.

The following tables show estimated, world-wide pollution by land-mines in the ground (Janzon, 1995, Strada 1996) :

Continent/Zone	No. of mines [millions]	Countries affected
Africa	44	18
Middle East	26	9
East Asia	23	8
South Asia	25	4
Europe	7	13
Latin America	1	8
Total	≥126	60

Some of the most affected countries are

Country	No. of mines [millions]	Land-mines per km^2 [1/km^2]
Bosnia-Herzegovina	3	59
Cambodia	10	55
Croatia	2	36
Iraq	10	23
Egypt	23	23
Afghanistan	10	15
Angola	15	12
Eritrea	1	11
Iran	16	10
Yugoslavia	0.5	5
P.R. of China	10	1

Every year, about 26000 persons are estimated to be killed by landmines, or about 500 a week. The number of persons injured is judged to be of the same order of size. In the countries affected a land-mine injury, usually involving the loss of at least a foot or a leg, and often also abdominal injury, results in a high mortality. Surgical care and hospitalization is seldom readily at hand, and in the delay before treatment can occur many victims will succumb (Nechaev & al, 1995).

Mines are not selective, but affect soldiers and civilians alike. Soldiers can be trained to protect themselves, mine awareness for civilians is much more difficult to teach.

The ten situations most vulnerable to land-mines are everyday activities, such as (Rae McGrath, in Janzon, 1994) :

Hazard	Situation
1	Fuel collection/woodcutting
2	Grazing livestock,
3	Planting, harvesting, plowing
4	Land and ditch clearing
5	Walking on tracks
6	Fishing/collecting water
7	Playing
8	Scavenging for food and scrap metals
9	Driving, esp. overtaking on verges of road
10	Recovering dead or injured relations/friends from minefields

As may be realized from this list, children are especially vulnerable to land-mines. You cannot expect a small child to remember all precautions against mines at all times, when playing or doing everyday chores. It is thus no surprise that sometimes up to 80 % of persons killed or injured can be children.

Even if one person killed by land-mines is, indeed, one too many, the number of victims may seem surprisingly low to you, comparing, for instance the number of traffic deaths or victims of cancer or AIDS. However, the figures are quite uncertain since not all deaths and injuries will be reported. And, their effects on the societies affected are much more pronounced than these figures may let you believe.

• One main effect is denying return and repatriation to refugees, thereby permanenting a crisis situation which might otherwise have been possible to resolve.

• Agriculture cannot be resumed until all mines are cleared, thus starvation may ensue.

• Communications are blocked or hampered, obstructing the economical life and development of the country. In Mozambique, a 50 km stretch of main road had remained closed to all traffic for more than five years. When it was, finally, demined by South-African specialists, they

"Detection of abandoned land mines", 7-9 October 1996, Conference Publication No. 431

found only **two** land-mines, plus a few other ammunition items, on the road.

• The cost of demining is excessive to most countries affected. In Cambodia, it is estimated that the entire national economy could be consumed over the next five years, for demining purposes only.

It is obvious that there is a great need for new methods to handle the worldwide problems with land mines. The problem is further aggravated by the fact that, whereas only about 100.000 mines are cleared annually by the worldwide demining community, over 2 million are laid out. And there are a further 200 million estimated to be in storage, ready to be deployed at any moment!

Indeed, the land-mines may well be the world's no. 2 environmental problem, second only to that of water supply. And it is certainly much more hazardous!

This situation calls for new concepts which can speed up the current manual mine clearance process with the same or a higher reliability level. It may be necessary to increase the speed of the process by a factor of fifty to a hundred, without sacrificing safety.

Is the mine an efficient weapon?

Unfortunately, it is! Developments in land-mine technology have been numerous since the second world war. There are now novel types of mines, such as robotic mines, anti-helicopter mines, side attack mines, which are blast-hardened and electronically fuzed. Non-metallic mines have existed since the 1950's.

However, these high-tech mines do not constitute the main bulk of the problem, but that is rather the common, anti-personnel high-explosive contact mine, of the same type as those used during WW 2.

Some military experts claim that the high-tech mines do need these simple anti-personnel mines, or other devices, to protect them from being easily cleared. An exclusively anti-personnel mine-field is rather easily breached by manual clearance or by mechanical means, thus only hampering military operations. Area denial mining is also much less effective against a well trained and well equipped military force than against unprotected civilians.

Mine detection and clearance

There are some important concepts to be distinguished between:

Minebreaching - is a military operation designed to breach a way or lane through a land-mine field to allow passage. It is often done by blasting, e g a so-called mine-serpent, which is a hose thrown or pulled out by a thrower or rocket, and is filled with high explosive. When detonated it will trigger mines beneath it or in its immediate vicinity, leaving a lane suitable for passage of a soldier on foot. Wider lanes for vehicles can be cleared by mechanical means, such as rollers, flails or plows. They may be mounted on an armoured vehicle. A reinforced bulldozer can also be quite useful for this purpose.

Mine-clearance - denotes another military operation, where an area needed for passage or military operations is cleared of mines. The standard method is by mine-detector and then by mechanical prodding by a single soldier. When a mine is found it will be neutralized and removed, or detonated *in situ* .

Demining - mostly used as "humanitarian demining" - denotes a process where all mines, duds and other explosive devices are removed from an area, which is often extensive, and with a very high reliability. The term can also be used in connection with peace-keeping operations or similar.

Whereas a military operation may afford to take a risk of losing one or two soldiers or vehicles in a mine-field, a clearance reliability of about 99.9 % is usually required in the demining situation. Quality assurance is carried out in Afghanistan by the deminers forming a line and marching elbow to elbow through the fields which they have just cleared. Anything less may not convince the population that an area is safe!

Although mines have developed considerably, mine detection and clearance technology has evolved much less. Essentially the same techniques as developed during the second world war are still being used. The mine/metal-detector is still the main device used for detecting mines. When an echo is found the operator will have to yield his place to another specialist, the prodder, who performs mechanical prodding, sometimes just with a bayonet. His prodding will determine whether there is any mine or not. He will then carefully dig around it and uncover it. In humanitarian demining, the mine should then be destroyed without being moved, since this minimizes hazard to the deminers and eliminates the risk that the mine might later be stolen and re-used.

There has been development in mine detection and clearance, but only for the typical military mine-breaching situation, then often with mechanical means. These methods may give a clearance ratio of only up to 80 %. Furthermore, for instance flails have been shown to have the capacity to throw unexploded mines around them, up to a kilometer, thus re-contaminating already cleared areas.

The environmental conditions can be vastly different, from swamps and dried-out rice-paddies in Cambodia, to the Kuwaiti desert, to the rocks and mountains of Afghanistan and the jungles of Ruanda. Thus it cannot be expected that there will be one single solution to the demining problem, rather there should be several different and combined technologies and methods.

One of the major problems in demining is that on a former battle-ground, the soil will be severely contaminated with both metal and explosives. A fragment may yield the same response of the mine-detector as a mine. Each fragment will then have to be prodded and identified. Typically there may be 1000 fragments to one mine. After about 900 fragments with no trace of a mine, the deminer may tend to become impatient, and then the risk of accident will greatly increase. Indeed, demining is one of the most arduous, boring and monotonous tasks existing, often to be carried out under extreme, adverse and harsh environmental conditions.

Of course, totally non-metallic mines, and there are several varieties, a chinese design being the most common one, will not be detected by the mine-detector, and may lose you both a valuable peace of equipment (a mine detector may cost about USD 3000) and a trained deminer.

Like a dog can detect odor from drugs and other chemical substances, it can also be trained to detect the vapor from explosives. Dogs are excellent for use in demining, but unfortunately they are expensive to breed and train, and tire easily. One dog is usually not enough to assure that an area is not mined, you must use two or more to be sure. Their most efficient use may be in delineating the boundaries of minefields, or to declare areas safe.

In South Africa an innovative method of using dogs was designed. A mine-resistant vehicle collects air samples throughout a suspect area. The samples are brought to a center with several dogs and opened. The dogs will only need a few seconds to mark the samples where explosive vapours are present. The dogs and mine-detection equipment can then be brought out to further investigate the sectors where the positive samples were collected.

A land-mine, costing about USD 3 to manufacture typically costs USD 200-1000 to clear. And the cost per person injured may typically be about USD 5000 in a developing country, for treatment and rehabilitation. A person maimed by a landmine is not likely to have a good life afterwards. He/she will often have great difficulty to earn a living and may also, in some cultures, become a social outcast.

Indigeneous deminers will be preferred by most countries, since the cost of Western specialists may be about a hundred times higher. Very few countries affected - with Kuwait being one exception - can afford such resources. International cooperation can, though, be effectively provided in the form of coordination, specialist services, training assistance, equipment and funding. Still, an indigeneous deminer may have wages equivalent to that of a cabinet minister in his country. His income may support up to fifty members of family and kin.

International demining operations are run by many kinds of organisations, some excellent and efficient, some more enthusiastic to help than knowledgeable, some even in it for the profits and "excitement". Co-ordination may sometimes be poor or lacking. The UN has a demining coordination office within the Department of Humanitarian Affairs, but numerous other UN agencies are also involved in demining, and coordination resources are scarce. An international clearinghouse for demining information is urgently needed, but so far, the resources to establish such an organisation are insufficient.

Indeed, the main obstacles to demining are neither technical, nor a lack of available manpower, but rather of a political and economical nature. However, even all the world's combined resources would not suffice as long as the rate of mine-laying surpasses that of mine-clearance by a factor of over twenty.

International Law

In the 1981 UN Convention on the so-called particularly Inhumane Weapons, there are rules governing the use of land-mines. Some are that:

• Remotely delivered mines, such as artillery-fired or air-dropped ones, must have automatic neutralization devices, which destroy the mine or render it ineffective after a lapse of time.

• Remotely delivered land-mines may only be used against an area which is in itself a military objective, or which contains military objectives.

• The location of pre-planned land-mine fields must be recorded by means of maps and logs. These records must be preserved and must be handed over to the UN and to the adversary upon the cessation of active hostilities.

These rules may have some effect, but neutralization devices are seldom 100 % reliable. Most land-mines, which have been manually emplaced, will have no self-destruct or neutralizing mechanism. The second rule precludes the mining, from aircraft, of vast areas. However, some 40.000 km^2 in Afghanistan is certainly large enough, anyway. Maps and logs are generally of a poor quality. Can you expect soldiers, para-military or guerrilla fighters, who may not know how to read and write, to do a good job of recording land-mines?

In May 1996 a UN Review Conference of the 1981 Convention finished in Geneva, Switzerland. One of the main subjects was land-mines. Sweden and other nations had proposed a total ban on anti-personnel land mines, supported by Russia and others. The outcome was less, but some rules additional to those of 1981 were agreed upon (UN, 1996):

- Internal conflicts are now covered by the Convention
- Mines must contain at least 8 g of iron, to be detectable by the current types of mine-detectors
- Self-destruction/neutralization devices that will function within 30 days are required on all mines remotely delivered, or manually emplaced outside marked and monitored military areas.
- Such mines must also have a back-up self-deactivation device, effective within 120 days.

- The use of mines with a mechanism designed to trigger detonation of the munition by the presence of a mine-detector is prohibited
- States Parties undertake to introduce national penal sanctions for violations of the Convention.
- Export of non-detectable and other prohibited mines is prohibited
- The protection of UN forces and humanitarian missions is strengthened.

What technologies may be applicable?

In June 1994 FOA organised an International Technical Expert's Workshop in Stockholm. The purpose of that was to bring together experts with demining field experience, scientists and industrial engineers to exchange views and to try to formulate some basic requirements for new equipment for demining purposes. You may judge, yourself, how well we succeded by reading the report (Janzon, 1994).

One of the efforts during the Workshop was a listing of several candidate technologies for mine detection, and assessing them relative to various problems and environments.

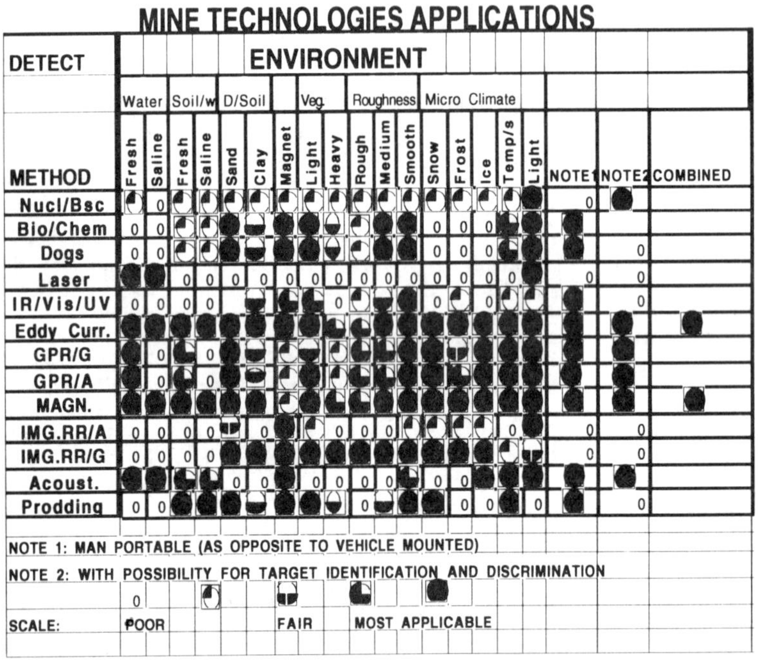

Figure 1. Judgement of the applicability of candidate technologies to various environments (Janzon, 1994)

The Challenge

In Sweden we have joined the knowledge from defence research and industry to form a new and more efficient concept for mine clearance. Our idea is to use new technology to get faster and more selective mine surveillance and clearance in different situations.

Our solutions are based on FOA's own research and also on discussions with experts from other countries. The program has four main areas:

• To construct a low cost, low weight hand-held mine detection device for detection of anti-personnel and anti-tank mines, with or without metal content, for use in open or overgrown areas. The equipment will have the capability to discriminate between a steel fragment and a land-mine, preferably also to identify the mine type.

• To construct a vehicle-borne mine surveillance and mine-clearance system for use on roads and tracks with a higher searching speed than that possible today, for detection of both covered and surface-laid anti-tank and anti-personnel mines, applying the same technologies as for the hand-held equipment.

• To develop airborne equipment for detection of mine-fields or mined areas, with a mapping capacity and a positioning accuracy to within 0.2 m.

• To develop a vehicle for rapid mechanical clearance of mines, surface-laid and buried, and with a very high clearance ratio. The equipment must be able to withstand anti-tank mine blasts with no more severe consequences than repairs possible to carry out in the field. It will be possible to operate remotely in cases of extreme hazard.

To make it possible to use more than one technology in a combination of sensor data by use of data fusion and advanced, real-time signal processing is a precondition of all the systems above.

TECHNOLOGIES

Based on work carried out in the Swedish defence research program and on experience from defence industries we have chosen the following technologies for our systems:

- Wideband Ground Penetration Radar (GPR, Impulse Radar)
- Metal Detector (Electromagnetic Induction)
- Magnetometers
- Biosensors
- IR and Visible Multispectral Sensors
- Nuclear Backscatter

FOA is also, together with the Swedish Armed Forces, which are also providing the funding, developing a mine and unexploded ordnance clearance database. This is soon due for field tests in Bosnia. We are also planning an experimental mine field facility in Sweden, permitting realistic but safe field tests of new detection and location equipment.

CONCLUDING REMARKS

The most imminent problems in demining are, in my personal opinion (however supported by many experts with practical experience in the field, Janzon, 1994):

• There is still an urgent need for a total ban on the use of anti-personnel, contact land-mines,

• a need for much increased funding of demining operations by the developed countries,

• a need for increased funding for victim assistance, and

• a need for strengthened centralised management of world-wide demining operations by the International Community. An organisation with a clear mandate for this purpose should be developed.

• It is absolutely necessary to develop new methods to detect mine-fields and individual mines, especially to discriminate between a fragment in the ground and a mine.

• There is an urgent need for methods and equipment to detect non-metallic mines

• There will be a continuing necessity for manual deming methods

• There is a great and urgent need for a mine and ordnance identification database system, readily accessible by all interested parties.

• There is also a need to develop mechanical or other area clearance methods, adapted to the demining situation, giving a clearance ratio approaching 100 %.

It appears clear that even the successful application of existing technologies, if they can be combined in fieldable and affordable systems, may contribute greatly to the solution of the demining problem.

There is certainly also scope for research, and the development and application of new technologies. It would certainly be beneficial if the following could be agreed upon:

• Research programs to test new technologies for demining

• Co-ordinated, internationally funded scientific programs

• U.N. (DHA) as the preferred coordinating body (but resources are still lacking)

• Common test facilities and a database, available to all involved in the programs.

Limited funding from Sweden's Ministry of Foreign Affairs, comprising a total of MSEK 20 (ab. Mecu 2) in 3 years, has now been made available to FOA. We, and cooperating Swedish industries, are prepared to do our utmost to present some viable and fieldable solutions.

Moreover, FOA would be quite willing to talk to anyone who would have an interest in contributing to better demining methods and equipment, either through research or through the provision of funding.

Many others are also quite active, however mostly connected with military programs and requirements, which are somewhat differing from the humanitarian needs. A research program was initiated in the US by the National Academy of Sciences, also sponsored by the Pentagon. The European Union is also tentatively planning for a MEcu 50 program (Mullins, 1995).

The challenge is ours: The ideas are certainly there, the knowledge and expertise are there, the research and development resources are there. Funding for developing much improved systems is, seen altogether, available. However, coordination and co-operation are largely lacking. The eagerness to get into this new, fascinating research field of demining seems too great to permit the sharing of ideas with others, or cooperation within joint projects. Meanwhile people are dying and human suffering prevails, world-wide.

If we join our resources, rather than for every country and organisation to try out its own path, the efforts would be immensely more efficient and the chances of reaching our humanitarian goals would increase much.

A slightly better world, and much reduced suffering for the Millions affected might be our reward!

REFERENCES

Janzon B (Ed.), 1994: Report FOA-R-94-00034-2.0-SE, ISSN 1004-9154, FOA, Stockholm, Sweden

Janzon B, 1995: Report FOA-D-95-00185-2.6-SE, FOA, Stockholm, Sweden

Mullins J, 1995: New Scientist, 4 May 1996, 32-37

Nechaev E A, Gritsanov A I, Fomin N F, *et al*, 1995: Mine Blast Trauma. Russian Ministry of Public Health and Medical Industry, Russian R.R. Vreden Research Institute of Traumatology, S:t Petersburg, Russia, 1995

Strada G, 1996: Sci. American, May 1996, 26-31.

UN Document CCW/Conf.I/14, 1 May 1996

A STATEMENT OF USER REQUIREMENT - MINE DETECTION

J A Craib

United Kingdom

INTRODUCTION

The problem facing the current generation of mine detectors is well known and needs little repetition here. The basic technology has been used since World War II and although there have been many improvements, they still remain little more than a sophisticated metal detector. If there is no metal in the landmine, or too little, a mine detector will not locate it. There are two other practical problems. Firstly, laterite soils - common in Africa - have a heavy ferrous content which creates a further difficulty in the reliable detection of mines by masking the signal from them. Some detectors can minimise this but it is not entirely effective. Secondly, landmines are often laid in areas in which battles have been fought and there can be a large amount of metal scrap in, or on, the ground. Some is very small and can be confused with the small amount of metal in what are usually called 'minimum-metal mines'. It is, of course, possible to manufacture a mine with no metal at all.

There are many minefields around the world which contain landmines with sufficient metal for reliable detection. Here, the present mine detectors can be used. On this basis, there will always be a role for these metal-seeking detectors and, in some situations an advanced detector will not be needed.

Nevertheless, there is growing concern that many minefields do exist which require an advanced detector in order to achieve the almost certain level of detection now needed. The concern is all the greater because the means to meet this need has not yet been found. Indeed, there is a wide debate within the scientific community about how best to meet the need. However, this paper will consider only what the user - the people who will use the equipment to find the mines - actually want.

WHAT DOES THE USER WANT?

Apart from agreeing that they want something better than that which is available now, the humanitarian user has rather sat back and waited for someone else to find the solution to his problems. For almost all other equipment the needs of the military and the humanitarian community are quite different. The need to reliably detect minimum-metal mines is common to both and the military are spending some quite large sums of money on it. The humanitarian community is, by contrast, spending nothing. The major work which is being undertaken to find solutions to humanitarian needs is all being undertaken by military research agencies under specific funding from their governments. There are many smaller agencies, including universities and private companies which are also looking at this problem but they can do very little without external funding, and that is not easy to find.

Even the users themselves have done little to help the scientific community and have even not stated what capabilities they would expect from any new equipment. Yet without such information, how does the scientist know whether what he might design will be acceptable to the user?

To overcome this, a small meeting was held over three days at Heathrow Airport in February this year at which a small group of users (humanitarian and military) met to agree a list of key equipments. In most cases, and certainly for the detection requirement, a full Statement of Requirement (SOR) was developed as a document. It is this discussion and the final document which forms the basis of this paper.

THE PROBLEMS THE USER FACES

It is not enough to say that conventional detectors do not entirely meet the modern need - it must be quantified so the scientist can eliminate possibilities which he might otherwise wish to consider but which might lead to the same problems.

Clearly the mines which need to be detected pose the greatest problem and from two particular perspectives. Firstly, and as already mentioned, the major difficulty is from those mines which contain too small an amount of metal to be reliably detected by existing mine detectors. Secondly, there are a few sophisticated mines which are sensitive to electro-magnetic radiation, so any new sensor technology must not actively initiate mines they are seeking. This latter problem may be one which is of more concern to the military engaged in a high-technology conflict than to the humanitarian community involved in the developing countries.

A second problem is the environment in which the mines are located. The best known problem is that of laterite soil with its heavy ferrous content (thus the red colour) which can mask the response coming from minimum-metal mines. But there are many other situations in which mines are found:

- Heavy vegetation
- Inhomogeneous soils
- Rubble from demolished buildings
- Under water or saturated soil
- Under snow
- Saline soil
- High humidity
- Very dusty conditions
- Metallic contamination in the ground

This latter needs some further explanation as it can cause considerable time-wasting by forcing an operator to investigate what turn out to be false alarms. The UN in Afghanistan have stated that for every mine they locate, they also locate over 1,000 metal fragments. This slows down the whole process of clearing ground. Many minefields are laid in areas in which battles have been fought, or around once defended military locations. Any battlefield will contain much metal scrap such as cartridge cases, fragments from exploded weapons (grenades, shells etc), barbed wire, and so on. In addition to the above, a former defended location will have all the detritus left by the occupants (food tins, items of military equipment, discarded tools etc). A lot of this will be visible and easily collected but much of it will be embedded in the ground, covered by shifting ground, or hidden by vegetation. Since a conventional mine detector cannot distinguish between a buried mine and buried scrap, great care is needed.

WHAT IS NEEDED?

Both the military and the humanitarian communities need a detector which is capable of finding all mines whether non-metallic, minimum-metal, or metallic. It is required as a replacement for the current conventional hand-held metal-only seeking detectors and should be contained in a single equipment.

There are some situations in which a vehicle mounted system would be acceptable but its use would be terrain limited. Only a hand-held system - to be used rather like present detectors - can cover all needs.

ECONOMIES OF SCALE

This might normally be discussed later in an equipment paper but it has important implications, particularly for the humanitarian community. Cost is a major factor in the development of any equipment and no less so here. Arguably, the military can afford to spend more on a new detector than their civilian counterparts. However, since the two requirements are likely to be very close, there is considerable merit in considering both together and thus benefiting from economies of scale. The sensor technologies and data processing will be common and most of the differences will be a matter of packaging or presentation of the data.

COST EFFECTIVENESS

No one expects the solution to the problems to be cheap (or, at least, as cheap as the present range of detectors) but if the cost is too high, sales will be reduced. The degree to which this might be true will be a balance between cost and the benefit which a new detector provides. As a minimum, there must be a clear and quantified improvement in detection performance compared to existing detectors. An equipment giving a marginal improvement is unlikely to succeed.

How this improvement can be measured will depend upon 2 factors;

> A near 100% guarantee that all mines within the detection envelope will be located, irrespective of terrain or environmental conditions.
>
> The ability to discriminate between mines and other surface and sub-surface objects, thus substantially reducing false alarms.

DETECTION EFFICIENCY

Ideally one might hope that any new equipment would be totally effective and would detect all mines in all circumstances and, indeed, this is the ultimate aim. Most regard it as unrealistic given that so many variables exist now and more may be identified later. The UN recognises that 100% is unrealistic at present and has devised a figure of 99.6% to represent the minimum effectiveness expected. In simple terms this means that in a minefield containing 1,000 mines, it is 'permissible' to miss 4 of them. The instinctive reaction to this is that missing these 4 is not acceptable but it raises an important issue.

In the very rare situation where the number of mines in a minefield is known, this instinctive reaction would be correct. The ground would most probably be checked again until the 4 were located. The difficulty is when no one has any idea of how many mines there are in an area, and this is the most common situation. In this case checking whether at least 99.6% of the mines have been found is not possible. From a quality assurance perspective the figure is probably meaningless. However, from a performance perspective it remains the minimum target to be achieved using any new equipment and will be tested either practically or through computer modelling, or some similar method acceptable to prospective purchasers.

PERFORMANCE REQUIRED

The following sections discuss the capabilities expected of any new equipment. They are not absolutely fixed and some compromise may be possible as a trade-off to reduce cost, complexity or timescales. Nevertheless, the details below are a very good indication of what is being sought by users.

Detection Capability

The equipment must be capable of detecting all mines (ferrous, non-ferrous, and metallic) on the surface and down to 200mm (ground surface to the top of the mine) in all types of soil. The orientation of the mine (at an angle, on its side, or upside down) must have no effect on the detection capability.

The detection performance will be tested against a range of mines which will be specified by a potential buyer but are likely to include:

VS50	TS50
TMA 1 to 5	PMA 1 to 3
T72A	Type 59
R2M1	Type 72 (AT)

The equipment must have an efficiency of at least 99.6%.

The false alarm rate must be as low as possible but a ratio of 5:1 will be acceptable.

It must not have to be in contact with the ground and should be capable of full operation at up to 50mm above the ground.

It should be in a self-contained unit and capable of being carried by one man.

Operation

As this is to be a hand-held equipment, the interface with the operator is critical and the following features are important. The features reflect the need of the humanitarian community rather than the military although, as said before, the variations are likely to be mainly packaging.

It must be designed for use by personnel whilst standing upright, squatting, or lying prone. One man should be able to carry and operate it for one continuous hour without excessive fatigue.

The equipment must be suitable for use by people, having a minimum of special training or technical education, in the following situations:

rural and urban areas.

remote, environmentally hostile areas in the developing world.

Any extending sensor head must weight no more than the equivalent of 2kg at the end of a 1.25m handle.

It must be able to detect mines in real-time so that any mine located by the equipment will be indicated to the operator before he stands on it.

Clearance should be possible at a speed which allows a 100m long path, one metre wide to be checked in no more than 15 minutes (assuming no mines are found).

Electronic Characteristics

Discrimination. The detector must discriminate against all ground types including metallic, water and/or salt saturated, at all times of day or night, without loss of sensitivity. Some limitations may be agreed provided they are clearly specified and acceptable to the user. The system should require the minimum of calibration or be automatic.

Controls and alarms. These are an important part of the interface with the operator and the following are considered the most important.

The number of manual controls is to be kept to a minimum. They must be simple to operate at any time of the day or night.

The equipment must include an automatic BITE to prove the system before each operation and to maintain user confidence.

There must be an audio alarm which activates when a mine is located.

All audio alarms must be easily distinguished from each other and be clearly heard above background noise without interfering with the operator's ability to hear instructions.

Electromagnetic Compatibility

The equipment should not initiate any known mine or associated device. Any limitation in this respect must be clearly defined and will be assessed against representative test mines. The user will determine whether any such limitation is acceptable.

Two detectors must be capable of working within 10m of each other without mutual interference and must not interfere with any other equipment.

Operating Conditions

Climatic. The equipment must be capable of being operated in temperatures between $-40^{0}C$ and $+55^{0}C$ and stored in temperatures between $-55^{0}C$ and $+85^{0}C$.

Contamination. The equipment must be capable of operating and being stored in dusty and humid environments without any reduction in performance.

Storage and Packaging. The equipment will normally be stored individually in stores. Each equipment should be placed in a suitable container for storage and transportation. The container should be as light and compact as possible and should carry the Complete Equipment Schedule (CES) including operating instructions and spare batteries.

Assembly and Preparation for Use. It should be possible to assemble the equipment from its container in no more than 3 minutes. Any 'warm-up' period should not significantly exceed and additional 2 minutes.

Operational Day. Equipment used by the humanitarian community is normally subjected to much greater use that the same equipment used by the military. The detector will be used almost continuously for 8 hours on the days when it is deployed. Over a year, it is probable that the equipment will be used on over 250 days. They will be subjected to rough treatment during transport over roads and tracks of a very poor standard.

Power. The equipment must be powered by one a range of commonly available batteries. Each battery charge must allow a minimum of at least one full days operation and, if possible, at least 40 hours. Recharging should not take longer than 20% of a fully functioning battery charge.

Reliability

It is essential that high levels of reliability are achieved in order to ensure adequate availability and acceptable life-cycle costs. The mean time between failures must be at least 1000 hours.

Maintainability

High levels of maintainability are required. The equipment will be used in remote areas of the developing world without easy access to the manufacturer. Modular components which have to be exchanged must be kept to a minimum. Humanitarian programmes, particularly those under the control of national staff, have little money to pay for expensive components. They can often repair equipment using relatively crude tools with remarkable success. However components which are 'potted' cannot successfully be repaired.

CONCLUSION

The user very much needs the skills of the research

scientist and the engineer to provide the quantum leap in detection capability that is required. Similarly the scientific community needs guidance from the user community on what is actually needed by the latter. There is little point in providing a solution which does not meet the problem. It is hoped that this paper will give some insight into what the user will require.

ADDED VALUE: THE REQUIREMENTS OF MINE CLEARERS AND DONORS FROM NEW MINE DETECTION TECHNOLOGIES

Robert Keeley

Head of United Nations Mine Action Centre (Croatia)

INTRODUCTION

Estimates suggest that there are some 6 million mines laid in the former Yugoslavia. These mines are mainly laid in belts along the former confrontation lines. The greatest contamination is estimated to cover about 8000 square kilometres throughout Bosnia and Croatia.

Mine clearance is an occupation that does contain an inherent element of risk, and some casualties are inevitable. Although many mines are damaged or even booby trapped, training and the use of the appropriate render - safe procedure (RSP) contribute to the process of minimizing the risk to mine clearers during the rendering safe process *once a mine has been located.* Indeed, although some casualties amongst mine clearers are caused by the handling of the mine as part of the RSP, they are often caused by mines not detected by the mine clearer. In other words, It is the mine that the mine clearer doesn't see that 'bites' him on the way to deal with the mine that he intends to disarm.

Furthermore, and to put this into perspective, most of the number of mine incidents in Yugoslavia involved people *not* involved in mine clearance operations. For example, of the 200-plus incidents involving UN personnel in the former Yugoslavia, only 16% were engaged in mine clearance drills. The others were people engaged on normal tasks. Most of these incidents occurred because the mines were not visible. Similarly, the majority of mine casualties within the civilian population occurred because the victims were unaware they were in a mined area.

In other words, it is the difficulty involved in the detection of mines and the identification of mined areas that present the greatest problems to mine clearance and bomb disposal operations, and limits our ability to warn civilian populations of the location of mined areas.

There are therefore 3 distinct areas where the risk from mines could be reduced by the application of new technologies. These are:

- The identification of areas of contamination;
- Detection of individual mines; and
- Rendering safe of mines.

AIM

The aim of this paper is to explain the requirements of mine clearers and donors from new mine detection technologies.

The utility of new technology solely for the rendering safe of mines or UXO will not be considered.

DEFINITIONS

Mine. A mine is any munition placed under, on or near the ground or other surface area and designed to be detonated or exploded by the presence, proximity or contact of a person or vehicle.

Mine clearance. For the purposes of this paper, mine clearance involves all processes involved in the removal and disposal of mines from a defined area. This may include an instance where the conduct of a mine detection operation allows an area to be declared as free of mines.

Mine detection. For the purposes of this paper, mine detection is defined as the ability to identify the presence of a mine and determine its location without causing its fuzing mechanism to function.

Render safe procedure (RSP). For the purposes of this paper, the term Render Safe Procedure (RSP) describes the action taken on a mine (or any item of unexploded ordnance) to separate the explosive from the action of the initiation mechanism.

Mechanical clearance device. For the purposes of this paper, the term 'mechanical clearance device' is used to describe an item of equipment that detonates, disrupts or destroys mines without manual approach by the equipment operator or other mine clearance personnel to within the danger area of the mine, except where protected by the armour of the clearance device.

REQUIREMENT

The requirement of humanitarian mine clearance is not merely to clear mines *per se*. Rather, it is to return land contaminated, *or believed contaminated*, by mines or other unexploded ordnance (UXO), to use by the civil population whilst minimizing the risk to the mine clearance team personnel.

It is reported that current mine clearance resources are only able to clear an equivalent of 1/35 of the mines still being laid around the world. Therefore, the utility of any mine clearance technology must be measured against its ability to assist in this process by improving this ratio. The value of such new technology will be measured by its ability to assist mine clearers and donors to maximize the return of donors from mine clearance programmes whilst improving the effectiveness and safety of mine clearance operations.

IDENTIFYING AREAS OF CONTAMINATION

The identifying of areas of contamination assists considerably in the clearance process. Firstly, it allows mine clearance planners to assess the extent of the problem, thus assisting in the resource allocation process; secondly, it allows mine clearers to maximize effort by concentrating resources in the areas of greatest contamination, rather than merely starting at one end of the affected region and working towards the other. Thirdly, given the long time that mine clearance takes, it allows the marking of areas of contamination thus advising the local population of the threat.

Level One Survey

The process of identifying areas of contamination in general terms is known as a 'level one survey'. This involves all assessment that can be carried out without actually engaging in the physical penetration of a suspected contaminated area. At the moment, a level one survey is usually carried out via a deductive process based on the following data:

- the historical record of the conflict, from which a prediction of the areas of greatest general risk can be made, based on general military intelligence.

- analysis of any minefield records left by the combatants, although experience shows that these records can never be regarded as complete or accurate; and

- reconnaissance reports (including correlation of mine incident information).

Given the established effectiveness of the analysis techniques described above, any new technologies would have to demonstrate their cost-effectiveness against these existing methods. For example, where considerable survey and mine record data exists, as in the former Yugoslavia, new survey technology will be hard pressed to compete with existing analytical methods *for general planning purposes*.

However, in 'green field sites' where mine clearers do not have access to such information, the ability to generate a 'level one' survey at short notice would be of considerable planning value.

Level Two Survey

The identification of the exact edge of mined areas is the main aim of 'level two' survey. For civilian clearance purposes it is becoming increasingly more common to place mine clearance in the context of the overall humanitarian or reconstruction process, rather than as an isolated project. In other words, the emphasis is placed on the need to find out how much of the target area is clear, not necessarily on identifying the location of the nearest mine.

While the 'state of the art' gives only an approximate location of mined areas, it is impossible to define and fence off the edge of a suspect mined area without actually clearing the area that lies immediately under and outside the fence, as a fence implies that the area on the outside of the fence is clear (indeed, mine casualties just outside fences are common).

Unfortunately, a general 'here be dragons' warning is of comparatively little use to farmers who want to know where the exact edge of the mined area is when it is on or next to their land.

Therefore, any process that could assist in the identification of the edge of mined areas - either by improving the accuracy or speed of reconnaissance, would assist in the maximization of mine clearance effort and the promulgation of timely and useful warnings.

For this purpose, it would not appear necessary at first glance to identify the location of each mine with 100% accuracy. However, if the 'level 2 survey' process is to identify the edge of mined areas accurately and dependably without the need for further reconnaissance, it needs to recognise the mines at the edge of the target as well as those in the central area of probability of the contaminated area.

A secondary requirement of level 2 survey is to identify the type of mine found in the mined area. This is often useful for military mine breaching operations where it might allow a faster breaching method to be used. However, the intelligence process conducted as part of initiating clearance work in the region, and the training package conducted by the mine clearers should mean that there is actually very little value added - for humanitarian mine clearance - by level 2 survey being able to identify the types of mines in the target area.

The potential utility of new survey technology

The ability to carry out level 1 and 2 surveys remotely would be useful where political and security conditions do not guarantee freedom of movement for humanitarian mine clearance teams, and is an area where there may also be significant 'cross over' between humanitarian and conventional military requirements.

However, any survey technology used to identify the limit of mined areas must ensure that it achieves the humanitarian safety standard for areas declared outside contamination, or else be sold with clearly understood caveats of its limitations. If survey technology offers too vague a definition of a mined area, then the existing analytical methodology will remain the most cost-effective; if the survey equipment retains too wide a safety margin then the mine clearer will continue to expend resources clearing areas that are not contaminated; if it retains too narrow a safety margin then clearance operations and civilian populations may incur casualties, thus reducing customer confidence in the technology.

MINE CLEARANCE OPERATIONS

The actual clearance of contaminated areas can be split into 2 distinct functions: detection of mines and their subsequent rendering-safe. Although some technologies (such as flails) combine these 2 functions into 1 process, it is generally the detection of mines as the precursor to rendering-safe that presents the greatest problems during the actual clearance operation.

Manual mine detection techniques

In order to understand where new technologies will be expected to offer their greatest potential, it is worth explaining how manual mine clearance tasks are currently carried out. The mined areas in and around the villages and towns of the former Yugoslavia offer what is perhaps the global 'worst case' of mine clearance tasks.

A typical task site. A typical mine clearance task site in Bosnia or Croatia will often include the ground immediately around a house or other building. Building plots in these villages are typically separated by metal-framed wire fences. The garden will include the underground utilities in iron pipes - gas, water and sewerage - and electricity and telephone wires usually pass into the house through overhead wires from wooden 'telegraph' poles running along the street. In areas where there has been direct or indirect weapons fire, these poles are down and the wires lie across the ground.

The land plots also contain the normal household scrap found in gardens, eg pieces of car, discarded nails, screws and children's toys.

As a result of military action, building rubble and shrapnel from explosions will scatter more metallic scrap into the gardens, and bushes and other crops in the gardens become overgrown if the house is no longer occupied or tended. Where the house is occupied or fought over, empty cartridge cases, discarded ammunition or unexploded ordnance may be found.

The utility of metallic mine detectors is therefore limited in such sites. This is exacerbated by the tiny amount of metal content in the Yugoslav mines which are the main threat in the region.

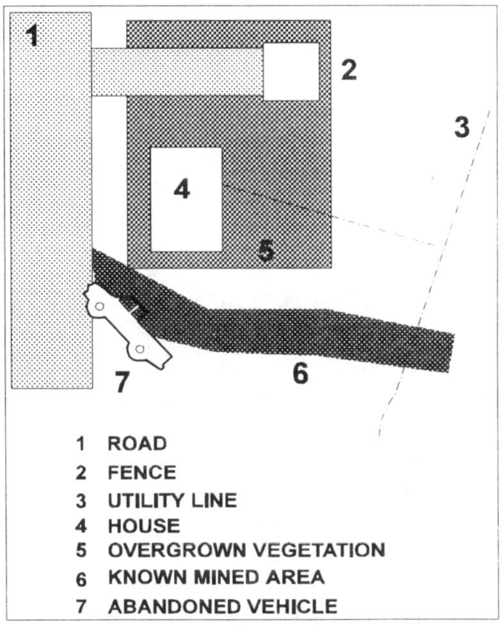

Manual mine clearance drills. As a result of the situation described above, the basic manual mine clearance drill recommended for use in the former Yugoslavia involves the mine clearer making his way along a lane of about one metre width. The first phase of the process includes a visual and a (gentle) hand search for surface mines and tripwires[1] just ahead of the mine

1. Tripwires are a common trigger for some types of mines and are often made from plastic similar to fishing line.

clearer. The second phase is a probing process, which involves the insertion of a lightweight mine probe or prodder into the ground, at an angle of 30°, to a depth of 100 mm, repeated every 30 mm across the width of the clearance lane. The operator uses the probe to feel for hard objects buried under the surface. The insertion angle of 30° is intended to minimize the risk of the operator inadvertently pressing on the top of the mine, and the spacing of 30 mm is intended to take account of the smallest mines found in the region.

An electronic mine detector able to detect Yugoslavian anti-personnel mines to a minimum depth of 100 mm in dry metal - free soil *may be used as part of this process as additional confirmation* before the mine clearer advances a further 30 mm along the clearance lane. However, given the degree of metallic contamination found in such locations it may be decided to dispense with the mine detector. Thus the use of minimum-metallic' mines[1] in the former Yugoslavia prevents mine clearers from utilizing electronic mine detectors as the primary means of mine detection. Any new detector technology will be required to allow reduction in the time spent prodding if it is to add value to the clearance operation.

Performance data. The painstaking nature of manual clearance techniques, combined with the discomfort provided by the protective clothing and further limitations on progress as a necessary part of maintaining overall site safety standards, means that an individual mine clearer may only achieve a clearance rate of 2 m^2/hr, while, allowing for rest periods, a team of 10 mine clearance personnel might only be expected to clear 40 or 50 m^2 *per day* in the typical task site described above.

Operating costs: exploding the '$1000 per mine' myth. It is therefore easy to accept that, at present, the greatest component of cost of mine clearance programmes is the *per diem* cost of mine clearance teams. Indeed, the cost of mine clearance is more accurately expressed as a cost per m^2 than as a cost 'per mine' (unfortunately, the press still refers to the soundbite-friendly statistic of '$1000 per mine' that was once achieved by dividing the annual cost of a mine clearance programme by the number of mines found that year). Because the aim is to clear ground, and the first stage of mine clearance is to detect and locate mines, much effort is expended on searching the ground *between the mines*. Mine clearers will therefore be seeking, from any new technology, the ability to reduce the amount of clearance effort currently expended on the ground between the mines.

1. Typically, Yugoslavian AP mines contain a small piece of metal foil (approximately 3-4 mm in diameter) as part of the fuze assembly.

Mechanical mine clearance techniques

Counter-mine history has already seen the development of mechanical mine clearance devices, including rollers, flails and mine ploughs. Although Mechanical mine clearance devices offer fast coverage, they require skilled operators, considerable maintenance support and are greatly affected by the type of terrain they encounter. Indeed such devices, which are primarily intended for use in military minefield breaching operations, are not regarded as providing the necessary level of clearance for humanitarian clearance tasks. Indeed, some mines are specifically designed to resist such devices.

However, although mechanical clearance devices may not, on their own, provide the necessary level of clearance, they may be used in support of manual clearance operations. For example, mechanical methods may be used as a proving method in areas where mines are suspected, as part of the quality assurance process after manual clearance operations, as part of a route clearance operation, or to clear vegetation before manual clearance.

Mine sensing dogs

It is possible to train dogs to react to the smell of explosives, thus indicating the presence of mines. The value of such dogs is that they can screen large areas very quickly , may be used in terrain unsuitable for mechanical devices , and can allow mine clearers to focus clearance operations on the areas where the dogs indicate the presence of mines. The dogs require skilled handlers who are able to interpret their signals and maximize their limited endurance. Dogs have proved worthwhile in several mine - contaminated regions around the world, However, field trials carried out in Croatia in 1994 were disappointing. The results of the trial suggest that the abilities of the dog teams may have been limited by comparatively low temperatures and high explosive contamination. Further work needs to be done to establish the operating parameters of mine sensing dogs in the former Yugoslavia.

STANDARDS

The football pitch test. Any mine clearance technique used for humanitarian purposes must effectively achieve a standard whereby those undertaking the operation would 'be prepared to let their own children play football on the ground concerned' when they have finished. The current generation of mechanical clearance devices (rollers, flails and sniffer dogs) are not regarded as passing this test, and are considered more suitable as means of detecting the edge of contaminated areas and as quality assurance tools. Therefore, at present, only manual mine detection and clearance techniques are considered as providing the necessary clearance standard suitable for use in areas known to be mined.

Residual problems: evolution of the 99% standard. Mine clearers will not *guarantee* that any cleared ground is '100%' clear, but will normally 'certify' that the ground is clear to 'the best of their knowledge and belief'. This is intended to cover the eventuality of mines or UXO subsequently turning up, perhaps as a result of flooding or earth slip moving mines from nearby contaminated areas. As a result, clearance standards are generally expressed as 'in excess of 99%'.
In Bosnia and Croatia we are therefore developing a process of quality assurance which will rely on a 3-stage assessment process, involving the following:

a. technical assessment of potential mine clearers at the project tendering phase, to identify potential weaknesses in the proposed methodology, while at the same time a clear scope of works included in the contract will clarify the exact expectation of the client from the mine clearer;

b. random inspections of task sites to ensure that the mine clearers are actually conducting mine clearance in accordance with the terms of the contract; and

c. final testing of a sample area of the task site, during which we expect to find *no mines or UXO within the area of the sample.*

If all of these conditions are met, then, although the mine clearer has to meet stringent QA standards at the time of handover, his liability for subsequent mine or UXO contamination found in the task site area is mitigated. The organization accepting the ground for use is instructed that there is a residual risk (ie 'less than 1%') of mines or UXO subsequently turning up.

THE POTENTIAL OF NEW TECHNOLOGIES

The potential of new mine detection technologies would be realised when equipment that is able to detect the location of the latest generation of land mines will allow the mine clearers to:

- survey contaminated ground faster and more accurately, in order to improve the mine clearance planning process; or

- reduce the unit cost of clearance (per m^2), ie to allow mine clearance teams to cover more ground per day

At present, a mine clearance team manager directed to clear a football pitch where mines are known to be present will divide up his team to operate in clearance lanes, and, in order to meet humanitarian standards, will probe the entire pitch at 30 mm intervals. Mine detection technology offered to the mine clearance manager will have to offer him unambiguous and dependable advice, such as 'there are no mines in the area west of the centre spot' or 'there are 15 mines here, and their locations are....'. This information will have to be offered in such a way that the mine clearer is confident that he can therefore reduce (or speed up) the amount of physical clearance that he has to undertake, *whilst still subsequently being able to certify the ground as suitable for handover.* In other words, being told there are 'probably' no mines at one end, or that there are 'between 12 and 18 mines present' *adds no value to the mine clearance process, as the team will still have to conduct normal manual clearance drills.*

CONCLUSIONS

The use of manual mine clearance techniques has been developed to meet the threat from minimum-metallic mines, whilst the use of mechanical mine clearance techniques is being refined to recognise the advantages and limitations of such equipment against the latest generations of mines in an humanitarian environment.

Whilst enhancement to mine survey capability would be of some value, the major difficulty remains the limited ability to reliably detect minimum metallic mines without use of mine probing drills or mechanical 'mine proving' operations.

Any new detection equipment that can promote operator confidence to the extent that mine clearers can dispense with (or speed up) manual clearance techniques whilst increasing the surface area that can be covered each day, will be regarded as adding value and will be a cost-effective investment.

Any new detection equipment that does not add value (ie does not allow mine clearers to dispense with or speed up existing survey or detection techniques) will be regarded as expensive and irrelevant and will not find a place in the developing and increasingly competitive world of humanitarian mine clearance.

In short, if they cannot add value to the existing mine clearance processes, the smart pieces of kit will stay on the shelf.

THE COUNTERMINE CHALLENGE: AN EVOLVING REQUIREMENT IN A CHANGING ENVIRONMENT

Colonel A. R. R. McAslan

Deputy Director Operational Requirements (Land Systems), UK Ministry of Defence

This paper expresses the personal views of the author and does not necessarily represent official UK Government policy.

INTRODUCTION

The threat posed by mines is extensive and diverse. Advances in technology have led to increased lethality through better fusing, improved sensing, variable modes of attack and enhanced explosive efficiency. For nations who cannot afford such sophisticated weapon systems there is a plentiful supply of simple but effective anti-tank (AT) and anti-personnel (AP) mines available on the world market. Strict rules govern the siting and guarding of minefields to prevent the unnecessary suffering of non-combatants, however these rules are frequently ignored by poorly trained armies and armed factions.

The British Army's countermine requirement has evolved in recent years in response to NATO's changing military doctrine, which has moved away from attrition through the linear static defence of forward positions to a manoeuvrist approach. Such an approach requires an ability to counter the enemy's obstacles (including mines) quickly and effectively across the width and depth of the battlefield. This has involved a major change to our equipment, organisations and training priorities.

We have also had to respond to the growing mine threat to our troops deployed on Peacekeeping Operations (PKO) around the world - six British soldiers have been killed and 30 seriously wounded by mines in Bosnia-Herzegovina (B-H) since July 1993. Our approach to this threat has been entirely consistent with the increased priority given to humanitarian activity by the UN, NATO and the European Union.

The aim of this paper is to summarise the evolving UK military countermine requirement in response to the changing security environment

OPERATIONAL SPECTRUM

Within the spectrum of military activity, two distinct types of mine clearance are envisaged. The first is during *warfighting*, that is the capability to conduct operations to achieve a military objective. The second is the capability to conduct *operations other than war*, including post-conflict battle area clearance (BAC). Whilst the overall approach to countering the mine threat for both these categories is the same, there are two distinguishing criteria: the acceptability of risk and the timescale.

In *warfighting* the predominant mine threat is likely to arise from a combination of mechanically laid pattern mines, scatterable AT mines, both ground and air delivered, together with an increasing use of area defence weapons and off-route mines. The use of these systems is broadly commensurate with a sophisticated and well resourced enemy, although advanced systems are increasingly being fielded by less capable armies, for example the TMRP-6 mine in B-H (see Figure 1). It is deemed essential to have the equipment to counter this threat rapidly, and the acceptance of an appropriate level of risk is implicit.

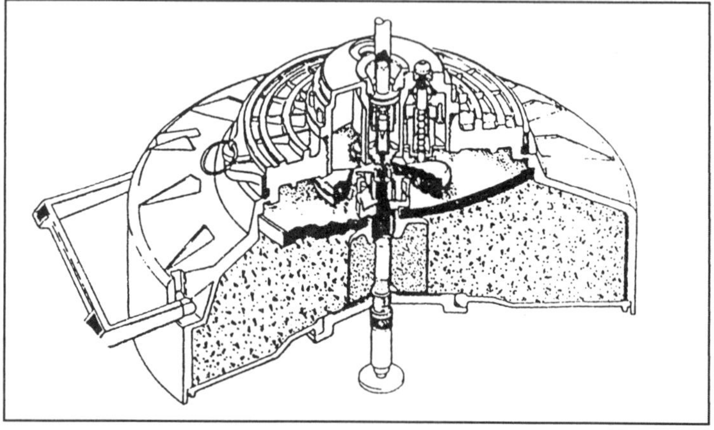

Figure 1: The TMRP-6 mine with a lethal EFP (explosively formed projectile) plate which penetrates the belly armour of all vehicles. These mines are deployed in large numbers in Bosnia-Herzegovina.

In *operations other than war* the mine threat is likely to focus more on nuisance and ambush mining, with the emphasis on AP mines, booby traps and sabotage devices. Large pattern minefields will be uncommon. AT mines may be used, particularly by poorly equipped forces as the main counter to armoured vehicles. Regardless of the type of mine, they will be used in a much less predictable manner. Targets in depth may include lightly armoured logistic and administrative vehicles. In contrast to warfighting, the degree of acceptable risk in countermine operations will be extremely low, but the time constraints will be less exacting.

POST-CONFLICT BAC

The aim of post-conflict BAC is to restore a country to normal by removing the threat from land contaminated by mines, submunitions, unexploded ordnance, ammunition, missile fuels, weapons and other hazardous debris. It is an invaluable contribution to positive peace. The British Army has two regular regiments dedicated to the task of explosive ordnance disposal (EOD): 33 Engr Regt (EOD) which is trained to conduct BAC and deal with all forms of unexploded ordnance, and 11 EOD Regt RLC which has responsibility for the clearance of Land Service Ammunition other than that found during BAC. The RN and RAF also have a capability to conduct EOD: the former for RN property and vessels, and the latter for airfields and crashed aircraft.

In addition to these specialist units, combat engineers from engineer regiments and assault pioneers from infantry battalions are trained to carry out mine and booby trap clearance.

The organisation, equipment, procedures and training for EOD are aimed to achieve 100% clearance with minimal risk to life. The standards equate to those required for humanitarian mine clearance, although the scale of the problem is considerably less.

British troops in B-H have a range of countermine equipment. Primacy for BAC in B-H, as elsewhere, rests with the Royal Engineer EOD units although the armoured engineers have Engineer tanks with mine ploughs and scatterable mine clearance equipments. Appliqué kits have been developed by DRA Chertsey to enable these tanks to be controlled remotely, although they are not permanently fitted. Recently we have extended our capability with three sets of South African Chubby mine clearance systems, three Aardvark JSFU flails and six Alvis mine protected vehicles. This significant enhancement is designed to reduce the ongoing AT and AP threat, and to meet the refined IFOR mission - which gives greater emphasis to peacekeeping through the reconstruction of the civilian infrastructure rather than satisfying purely military objectives. In order of priority, this new equipment is required to: enable the rapid rescue and recovery of troops and civilians from a mined area when there is an immediate threat of death or injury; to prove routes through areas which may be mined when access is deemed to be operationally essential; and to prove routes to utilities that serve military bases including power supply, water mains and pumping stations, telephone exchanges, sewerage and disposal sites. This latter task will also greatly benefit civilians living in the vicinity of military bases.

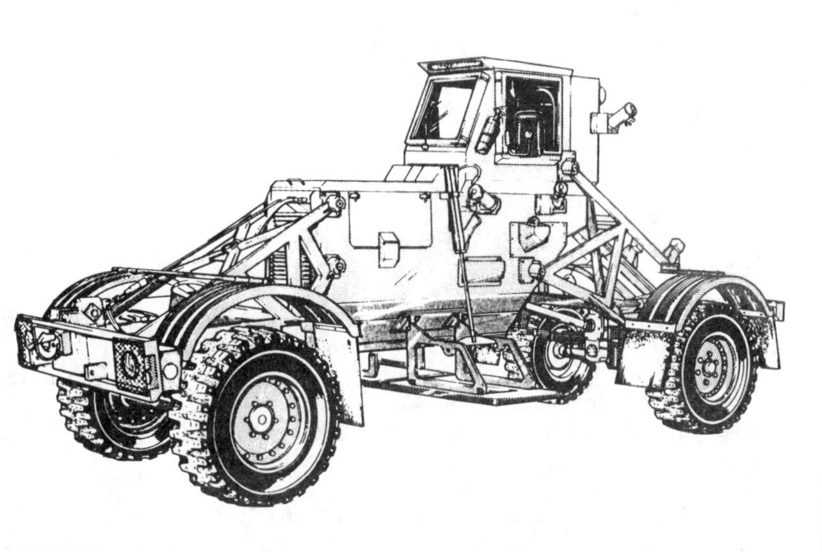

Figure 2: The South African RSD Mine Detection Vehicle which forms part of the Chubby mine clearance system. The British Army has recently deployed three complete systems to Bosnia-Herzogovina. The vehicles are fitted with special armour to defeat the TMRP-6 mines.

EVOLVING UK POLICY

The current UK position on the involvement of British Forces is clear and unambiguous, as it reflects the extant UN mine clearance policy. This is based on the fundamental principle that primary responsibility for clearance lies with the former warring factions. In B-H this activity is co-ordinated by IFOR and will, in due course, be assumed by the new Bosnian government.

Although it is currently possible in B-H to encourage the former warring factions to honour their demining obligations, this is principally due to the presence of sixty thousand IFOR troops. This will not always be the case in B-H and elsewhere, and the extent to which the UN may have to bear responsibility will depend on the capacity and willingness of the former warring factions.

The international response to the humanitarian issue is ongoing and individual countries and organisations (including NGOs) have separate views and agendas, although a broad consensus appears to be developing. Four recent activities, in particular, are influencing the UK's evolving BAC policy and equipment requirements:

- The UN conference on the *Global Landmine Crisis* held at Geneva in July 1995.
- NATO's Defence Research Group report of March 1996 on the suitability of technology to humanitarian mine clearance.
- The UN sponsored conference on mine clearance technology, procedures and standards held in Copenhagen in July 1996.
- The EU statement to the UN Security Council on 15th August 1996 which confirmed and clarified the EU's commitment to humanitarian mine clearance.

COUNTERMINE EQUIPMENT

Our current countermine equipment strategy identifies four capability areas:

- ***Detection***, involving both the long range detection of minefields, and the short range location and marking of individual mines.
- ***Breaching***, which is the traditional armoured or combat engineer operation to create lanes through minefields in order to allow passage of our own combat forces. Our current capability is based on the mineplough and explosive mine clearance hose.
- The need for ***survivability*** is becoming increasingly important, in response to the growing political significance of casualties, but also because survivability against mines is now technically feasible.
- And mine ***clearance***, that is the clearance of the area in the immediate vicinity of our own forces, routes, logistic installations and the like, and also the subsequent BAC.

Along with other NATO nations, the British Army is placing more emphasis on remote minefield detection. Two major projects are funded: a close-in mine detection and neutralisation system known as MINDER which will be fitted to specialist Sapper vehicles, and REMIDS which will be mounted on unmanned aerial vehicles and will enable us to detect minefields out to 100 kms. Both equipments will be phased in over a decade using three field standards (FS): FS 1 will detect surface laid mines, FS 2 will detect buried mines, and FS 3 will detect area and off-route mines. By matching the equipment development to the availability of suitable technology we will reduce the risk, and hence the cost, of the project.

We also have a funded programme to field an advanced hand-held mine detector which will be capable of detecting non-metallic mines. The equipment will need to be lightweight, reliable, and with a low false-alarm rate. It must also have the total confidence of the user. The programme involves many technological challenges and we do not believe it will be possible to field such a system before 2000, at the earliest.

MINDER, REMIDS and the advanced hand held mine detector will be optimised for warfighting operations, but will have wider utility including BAC. There is a need, however, to develop equipment which is optimised for BAC. Most of our current BAC systems have evolved from EOD operations and, although extremely effective, are slow and inappropriate for the decontamination of large areas. An approach based on *processing* the land may be more appropriate, using equipment such as Ground Clearance International's *Unisift*. This equipment can sift soil to a depth of 50cm operating at up to 0.5 hectares per hour. AT mines will be disposed of using normal EOD techniques; all other sifted material, including AP mines, will be crushed and rendered safe. Clearly, such an approach could not deal with all BAC, but it may be appropriate for decontaminating agricultural land.

RESEARCH

Research into countermine has increased significantly over the past three years, and is expected to remain high for the next five years at least. The challenges include sensor technologies, information processing and data compression, communications, command and control, operator interface, ruggedness and reliability. Much of this work will be carried out by industry and academia, but will continue to be assessed and integrated at DRA Chertsey.

It has recently been decided to establish at DRA Chertsey a separate focal point for BAC. This area of work complements the research being conducted in support of operational mine clearance and is being funded by the MoD, although extramural contracts for humanitarian mine clearance are being pursued.

QUALITY STANDARDS

The standards and agreed processes for operational mine clearance and BAC have evolved incrementally and informally. A minimum required level of clearance for humanitarian mine clearance of ".... better than 99.6%, and ideally 100%" has become the de facto international standard. While such a standard is admirable it has created a series of problems, not least the practical assessment of percentage clearance In certain circumstances 100% is unnecessary, unachievable or unaffordable.

At the International Conference on Mine Clearance Technologies held in Copenhagen in July 1996 an alternative requirement was proposed:

> *"...... The area should be cleared of mines to a level/standard which is agreed to be appropriate to the residual/planned use of the land, and which is achievable in terms of the resources and time available. The (contractor) must achieve at least 99.6% of the agreed level/standard of mine clearance."*

This definition has yet to be endorsed, but it represents a significant change to BAC and humanitarian mine clearance as it acknowledges the need to match the level of clearance to the residual and planned use of the land. In many cases the target will still by set at 100%, and the requirement will be to achieve 99.6% of that target. But in other cases, for example upland used principally by sheep or goats, a lower target may be deemed to be appropriate. By lowering the target it will be possible to make greater use of mechanical demining processes. This will increase the rate of mine clearance and thus reduce the overall problem.

The Copenhagen conference also considered the need to establish a set of universal quality standards, perhaps based on the ISO9000 series. ISO9000 has many attractions as it will create a more professional mine clearance culture and it will demand consistent standards. Not least, it will produce an agreed lexicon for BAC. A working group is due to be established by the UN to assess the relevance, scope and implications of adopting ISO9000.

We fully support both these UN initiatives. Indeed, we are already conducting a study into the application of quality standards and procedures to operational and military BAC. The results of our work will be made available to the UN and other agencies for humanitarian demining.

CONCLUSIONS

The British Army has invested heavily in countermine equipment and continues to give the highest priority to research into more effective operational mine clearance and post-conflict BAC. The military requirement of BAC and humanitarian demining are converging in both form and quality.

The strategic goals of the military and humanitarian communities are complementary, and we thus have a joint responsibility to ensure that our equipment, procedures, processes and standards are consistent and holistic.

A GUIDE TO SENSOR DESIGN FOR LAND MINE DETECTION

K.Langer

South Africa

Perhaps the most important factor in the design and development of a land mine detector is the initial direction of research to determine which technology is the most suitable for this purpose.

It must be borne in mind that the ultimate product, the land mine detector, must have an extremely high reliability in land mine detection, be able to work reliably under harsh field conditions and simple enough to be operated by non-scientific personnel. It must be able to detect a large variety of different types of mines, metallic and plastic, buried in the ground or scattered on the surface.

The detector equipment must be safe and practical to work with in the field, taking into consideration that a number of mines are booby-trapped, often with trip wires above the ground or other devilish devices below the surface. It also must be reasonably priced.

At this conference you will hear about the state of the art of different technologies such as ground penetrating radar, electro-optical, electro-magnetic and other technologies.

All these systems have proved that they are basically suitable for land mine detection and yet no single method seems to be meeting all the criteria needed to satisfy all requirements for the purpose of detecting land mines.

What is the reason ?

It would appear that this is largely due to the fact that these technologies are probing secondary characteristics of land mines, rather than the heart of a land mine : the explosive material !

There is therefore an alternative way open to find a solution to the problem : To investigate and develop a detector which can identify the presence of a land mine directly by detecting the explosive material itself.

Following this thought then comes the recognition that a suitable detection device already does exist and has proved to be highly successful : the nose of a dog !

THE SENSE OF SMELL

It is a significant fact that " smelling out " land mines by dogs, (mostly Alsatians or German Shepherd dogs as they are called), is at present the most reliable and effective method in detecting land mines.

It may at first seem almost incredible that the explosive chemicals which are contained in a land mine do emit a vapour sufficiently strong to be detected by a dog, in spite of the mine being buried in the ground and probably lying there for a long time.

This demonstrates the outstanding sensitivity of a dog's nose since only relatively few volatile molecules are released by a land mine but even such small amounts are sufficient to reveal its presence without doubt.

It is believed that dogs are able to detect odours in a concentration of about 5 parts per billion. In practical terms this amounts to the fact that a trained dog can identify the " smell " of a land mine at a distance of about 5 metre, even if the mine is buried underground.

Smell is a very powerful means of sensing , especially in the animal world.

To give another example of this : when a female of a certain type of moth releases a minute amount of a specific pheromone, the two antennae of a male moth may pick up the scent at a considerable distance away. These antennae, which look like tiny feathers, are only about 5 millimetres long and protrude from both sides of the moth's head. They are covered with smell sensitive receptors. These organs are so extremely sensitive that they can detect and identify females of their own species over distances of several kilometres.

Animals, including man and dog, pick up smells through their nose. Even the average human nose is remarkably sensitive.

Some years ago, the National Geographic Magazine, in co-operation with the Monel Chemical Senses Center of Philadelphia put out a survey among their readers concerning smell. For this purpose they included in their questionnaire some scratch-and-sniff panels. The amazing thing was that they only used 27 grams of a particular odorant and divided this quantity in micro capsules on nearly 11 million copies of the magazine. This extreme dilution proved to be completely adequate for the purpose.

Dogs have approximately 20 to 40 times more olfactory cells than a human being and this probably accounts for their greater scent sensitivity.

A PRACTICAL EXAMPLE

A specific system which a South African company has developed relies entirely on trained dogs as primary land mine detectors.

A suspected mine field is first surveyed by a mechanical sniffing-device which collects concentrated explosive vapours in sampling tubes from the air near the soil and from vegetation. Plants absorb explosive vapours and so dense vegetation actually helps in detecting land mines by this method. The sample tubes are marked with the area code from which the vapours were collected and are then sent to a base camp or a special dog centre. The concentrated vapours inside the sample tube remain there for quite a long time. At the dog centre, specially trained dogs examine each tube and indicate which are positive by lying down. This procedure is repeated several times with different dogs to increase the accuracy. The area's from which negative samples were collected are then regarded as mine-free. The confidence level of this operation is claimed to be better than 99 %

During the second phase of the de-mining method, free-running dogs are utilised in the sectors which were identified as positive. These dogs have been trained to lie down immediately as soon as they smell explosive vapour. Each danger area has now been reduced to approximate ten metre diameter and this limited area is further searched using conventional methods.

The utilisation of dogs and the method used has proved highly successful. Thousands of land mines in Angola and Mozambique have already been detected and cleared by this method.

However, the training of suitable dogs is a matter of great expertise and patience. An additional difficult task is the handling of dogs in all sorts of climates and different operating conditions. Then there are other problems. To mention just a few : dogs, like human beings, get tired and can only operate efficiently for a limited time. They need supervision and require logistic support. Added to these realities is the unfortunate fact that although the system is very effective, it is also rather slow and costly.

If the operating principle of the olfactory sensors in dogs or other animals could be translated into technological terms, then the search for a suitable land mine detector would be over. But is it possible ?

The answer is yes !

MAN-MADE ODOUR SENSORS

Man has already produced excellent technical duplicates of the hearing-, seeing- and touch senses. But the knowledge to build a land mine odour detector as good as an Alsatian's nose seems to have sofar eluded him. The reason is simple:

Up to now there was not sufficient impetus to warrant the research and development costs of such specialised apparatus. This situation has now dramatically changed with the need for finding a solution to the land mine disaster. Odour detection has sofar proved to be a highly reliable method of detection and this method is not only restricted to land mine detection but could encompass other explosive devices as well. In addition, this technology can be adapted for the detection of drugs, which also form a colossal, global problem. Furthermore, there are many potential commercial, industrial and other fields where an odour detector could be applied in.

COMMERCIAL OLFACTORY SENSORS

Research in olfactory sensors, which form part of the family of biosensors, has developed rapidly during the last years. The field of biosensors has become a specialised branch of science with a number of dedicated journals on the subject appearing regularly.

A number of reliable vapour sensors are already available on the world market.

I will have to leave the description of various suitable biosensors to other speakers at this conference.

However I would like to highlight one sensor type which shows outstanding potential for land mine detector use at present : this is the conducting polymer sensor. This type of sensor has a high degree of sensitivity and other desirable features and has been in operation for some years now by a British company for the purpose of aroma analysis and quality control of industrial and agricultural products. In this function it produces excellent results and in addition, the long term sample-to-sample reproducibility, which must be an essential feature of a reliable detector system has proven to be very good.

It would certainly be possible to develop this type of sensor further for land mine detection if additional funds could be found for a specialised research project.

Various solid state- and other biosensors are at present under development at a number of universities world wide and close attention should be given to these institutions. An excellent way to do this is via Internet.

However, it would appear that sofar all the presently available olfactory sensors are based on secondary detection characteristics, such as the conductivity of special materials when brought into contact with vapours. Other types of sensors exploit different detection principles such as the changing electrical charge density at the surface of an electrode or the altered optical characteristic of a sensing element.

Although many of these scientific principles hold the promise of suitability for land mine detection, there is one primary detection principle, which has totally proven itself and that is again.... the biological smell organ.

THE PRINCIPLE OF BIOLOGICAL SMELL DETECTION

In considering the design of a vapour detector based on the animal olfactory system, it may be useful to look very briefly how smell is detected by biological sensors :

This principle is not as complicated as is generally imagined:

Very basically speaking, a number of odour receptor cells are brought, through sniffing, into contact with volatile molecules, which constitute " the smell ".

The odour molecules are dissolved by mucus with which the odour receptors are normally surrounded and a chemical reaction takes place. This causes the sensors to send out signals to the brain via special nerves, called neurons. The process by which information is transmitted from sensors to the brain, via these neurons, is quite well known and could be reproduced by electrical signals.

The brain is the differentiating instrument which analyses and classifies the type of odour and this function could be duplicated by a computer in conjunction with an artificial neural network.

The precise principle of operation and the physical structure of the odour sensors remain therefore the only relatively unknown areas. However a wealth of knowledge on the subject of olfaction has been published and is waiting to be gathered and utilised.

CONCLUSION

I hope that this very short introduction to sensor design for land mine detection, utilising olfactory principles, has shown an alternative way to achieving a solution to end the devastating land mine problem.

What is needed to approach the problem from this angle, is to set up an organisation or association, consisting of analytical researchers, biologists, biochemists, electronic- and mechanical engineers and computer experts. The analytical researchers mentioned here are scientific people who research published articles and literature covering the field of olfaction. They are in contact with libraries, universities and other institutions and commercial companies and people all over the world. These researchers filter and analyse the information gathered and consolidate the resulting data which is then communicated to the other workers.

Such a correlative and multi-discipline organisation, concentrating on developing a land mine detector based on olfactory technology, should be able to produce excellent and relatively quick achievements since it aims to detect the explosive material of a land mine directly.

However, as stated before, no single technology can at this stage be relied upon completely to guarantee a 100% reliability rate in the detection of land mines. Hence it is essential to combine any technology chosen with at least one other, different, technology and fuse these methods into a single, integrated detection system.

The presentation of various suitable technologies at this conference is therefore extremely valuable and helpful.

BOFORS SCHNAUZER - A BIOSENSOR FOR DETECTION OF EXPLOSIVES

Sten-Anders Brink
Bofors Applied Technologies AB
Sweden

We all know of the communication ability animals have and many times the human race seems inferiour compared with animals because we have lost our natural instinct as we are far too urbanized. What would, for example, the French farmer do without his pigs when searching truffles? And what would the result be without dogs when people are lost during avalanches or earthquakes. During the last World War, in a small town in Germany, a goose flew up and down the street making a lot of noise, waking up all the inhabitants and only minutes afterwards a heavy bombing raid hit the road. And you will surely remember the human dream of becoming a bird or at least be able to fly as one.

And therefore, it is our greatest wish to work out a dog's nose but with unlimited endurance our modern technique is able to achieve and to use our knowledge for the good cause of saving life and land for future generations.

The UN requirement for mine detection system are stringent: it needs to be capable of locating 99,6 % of all mines, plastic or metal, in any varieties of soil and under any conditions of vegetation. Its output signal should give an indication of the size and shape of the object being detected and it should be usable by local deminers.

The detecting system need to be both vehicle mounted and manpack. No single sensor meets the requirements. An efficient sensor system will probable use a combination of different technologies brought together through a single signal processing system in order to be the answer to the problem.

Research into different technologies is taking place in many countries. Unfortunately, there is overlap, duplication and potential waste in this work. To get the best results from the scattered recources available, they need to be coordinated. Such coordination should start at national level, but international collaboration should be promoted using United Nations or any international humanitarian organization as a forum or catalyst.

With this in mind, Bofors made an initiative to take a leading position by introducing the Bofors Concept of Demining. The Bofors Concept of Demining comprises concepts for a total system incorporating different technologies that will be the tools for:

- search and detection of minefields
- detection, location and identification of individual mines
- destruction of mines
- verification of the work carried out.

The crucial item on this agenda is detection, location and identification of individual mines. The overall plan is to build a multisensor system which will be based on in combination:

- biosensor detectors
- ground penetrating radar
- metal detectors.

A multisensor system must at least consist of two supplementary sensors. It seems that the biosensor, in combination with a metal detector is the most cost effective combination.

By incorporating also a GPR you will be close to the ultimate system in terms of enhancing the detection probability, reducing the false alarm rate and increasing the speed of detection. By combining the technologies we will from each sensor receive different information that can be processed in an advanced signal processing and information handling unit. This unit will be an artificial neuron network and will furnish the operator with the necessary information.

Within Bofors we have concentrated our activities to research and development of the biosensor, leaving to other companies the responsibility to develop the other types of sensors that will be incorporated in the system. By doing so we will avoid duplicating work and also concentrate the efforts to have systems working in the field within a reasonable time.

Bofors Weapon Systems are at the same time working in the area of destroying mines. For this purpose a mine clearance vehicle has been developed.

I mentioned that we are leaving the development of other types of sensors to other companies. In order to join the efforts made on the national level we have with the full blessing of the Swedish Government formed a Joint Swedish Demining Concept. In this respect we combine the skills and effort of almost the whole Swedish Defence Industrial complex. The parties involved have agreed upon co-operation, exchange of information and a reasonable workshare.

As I mentioned, Bofors Applied Technologies has responded to the challenge of developing the detection device based on biosensor technology.

By demining is meant that an area will be safe to use by civilians after the demining operation. No mines or unexploded ammunition are allowed to remain in the entire area. The U.N. requires that 99.6 percent of all mines, both AP and AT, as well as unexploded ammunition, will be detected and located.

To meet U.N. requirements a multisensor system or a combination of sensors is needed. No sensor exists today that by itself is capable of detecting and locating all mines and unexploded ammunition. The biosensor system will be an essential part of a such systems. The reason why the biosensor needs support from other sensors is that in its first version it is not able to locate an individual mine. The biosensor is cabable to locate a mine in an accuracy in the region of a few meters. So in a multisensor system the biosensor will have, at least, two important advantages, such as:

* The biosensor system will be of great value to verify that an area is cleared from mines and unexploded ammunition.

* The biosensor system will be of great value to eliminate false alarms in a multisensor system.

The most common method today of detecting and locating mines is by means of a person probing through the mine-field, inch by inch. This method is considered to be very effective as regards finding all mines and unexploded ammunition, but it is extremely time-consuming and dangerous. The U.N. estimates that for every 1000 mines detected, one deminer is killed or injured.

The biosensor system will be a very effective device for the deminer as regards:

* His safety will increase enormously.

* He can work faster than today. He only needs to probe a small area to exactly localize an indicated mine.

The method which currently is also considered to be most effective is the use of dogs. It is a well-known fact that dogs can detect drugs and explosives, therefore they are used all over the world for drug- and mine detection.

But the use of dogs has disadvantages such as:

1. A dog trained for mine detection is very expensive because it takes between 6 months to a year of training before the dog can be used in a minefield.

2. It is hard to teach a dog to avoid tripwires.

3. Dogs tire quickly, after half an hour it needs several hours of rest. During a period of 24 hours a dog can only be used for two half-hourly working periods. Which means one hour in 24 hours.

4. Even the most well-trained dog is still an animal, and it can ignore smells that are not as interesting - from the dog's point of view!

It is, as mentioned, an old dream that one day modern science will be able to develop an artificial dog-nose having the same sensitivity as a real dog-nose.

There are several devices for the detection of smells available on the market today. They are usually called "sniffers" but there is no sniffer on the market today which has a sensitivity close enough to the sensitivity of a dog's nose.

Dogs are believed to be able to detect down to one picogram. But nobody knows for sure, because even the most sophisticated laboratory equipment is not able to match the sensitivity of a real dog-nose.

The technique we have chosen to realize this old dream is to use the latest in micro-electronics, micro-mechanics and in biotechnology. I will come back to how the biosensor itself works. But first I will point out some important factors that characterize the system.

Compared to normal sniffers we are very selective. The sensor only reacts to one substance. This is a great benefit as it makes the system difficult to fool. A dog smells many different smells simultaneously (whether it wants to or not). Consequently it learns to recognize a pattern of smells. But the trainer wants to teach the dog to recognize only one particular odour. Even during very sophisticated training procedures it is almost impossible to prevent dogs from learning to recognize a specific pattern of smells. It does happen that dogs, trained for drug detection, do not recognize the smell of a particular drug, but they do recognize the type of plastics used in the boxes the drug was kept in during the training. This shows that it is possible to fool a dog if the smell pattern is not the pattern it was trained to recognize.

A dog's level of intelligence enables it to become more or less skilled in avoiding being fooled. Normal sniffers have the same problem but they don't have the dog's intelligence to deal with this problem. By using neuron-net artificial intelligence created by modern science it could be possible to copy the dogs ability to avoid being fooled by changed smell patterns. In the system we do not have this problem at all as the sensor only reacts to one specific vapour.

We have a system that is very difficult to fool without the need for a lot of signal processing.

The biosensor system is much, much more sensitive than other "sniffers". It will have a sensitivity of less then 10 picograms. Which is 100-1000 times less compared to the ability of current sniffers.

The biosensor system has a very short response time, therefore it is possible for the deminer to walk slowly through the minefield.

The biosensor system works as follows:

All mines containing TNT, and having a plastic, bakelite, rubber or wood casing, leak TNT vapour into the ground around the mine. Even mines with a metal casing leak, due to bad sealing but they are easy to locate by a metal detector.

Within a few months the area around the mine will be so contaminated with TNT that it starts to leak TNT into the air above the mine. We will collect TNT from the air above the mine or from the surface of the ground.

We will suck up a volume of that air. The concentration of TNT will be enriched by a filter system. The filter will be heated up after the air has passed through the filter.

When the filter has been heated up, TNT will evaporate and be collected in the next sub-system which we call the trap. The TNT, which so far has been in gas form, will then be trapped in a small volume of water, or an organic solvent.

So what we have achieved so far, is that we have concentrated the TNT vapour from a volume of air into a tiny volume of solvent. This small volume of water is then presented to the biosensor itself.

The biosensor combines micro-electronics and micro-mecanics together with the latest in biotechnology.

A quartz crystal oscillates at its own-frequency, approximately 8 to 9 Mhz. Such a system is very sensitive to changes in the mass of the resonator.

Antibodies which are sensitive to TNT are attached on the surface of the oscillator.

If the small volume of solvent contains TNT, some of the antibodies will release from the surface of the oscillator. The own-frequency of the oscillator will then change due to the change in mass.

The change in own-frequency is then detected, and the change of frequency will be in the range of 5 to 25 Hz. If the frequency has changed, the operator will be warned immediately. The system has a sensitivity of less than 10 picogram.

The interface between the equipment and the operator will be very simple. We have in mind three indicating lights. A green light to indicate that the system is running, and no TNT has been detected.

A yellow light to indicate that something has caused a change in the own-frequency of the oscillator. If it is due to TNT, a red light will be lit in a few seconds. If the reason for the change in own-frequency is due to other reasons, the green light will soon be lit again. This arrangement allows a person with little or no training or even education, to use the biosensor system.

The project is divided into three phases:

The first phase started a year ago and will be completed with a prototype. The prototype is planned to be ready in December this year.

The prototype will show that we are able to detect less then 10 picograms of TNT within a reasonable response time, hopefully within 15 to 20 seconds. This is already a very short response time for a system that uses antibodies. We will be able to reduce the response time even more during early 1997.

At the end of the second phase, that means 1999, The biosensor system will be ready for the market.

So what we do is that we first launch the biosensor system on the market and than we move the focus on to the multisensor system.

The third phase is the design and manufacture of a complete multisensor system able to fulfil the requirements of the UN.

We are currently concentrating our efforts on detection of TNT. The use of biotechnology and antibodies makes it easy to adapt the sensor to other explosives of volatile substances such as different types of drugs etc.

This fact opens a lot of interesting applications of the system and thus we anticipate a substantial market. The demand for safety at airports, postal stations etc is really endless and so is the task to facilitate duty officer's daily routine when handling containers, lorries and luggage of all kinds.

But still the greatest and most serious scourge in postwar time are mines maiming lots of innocent people, children. Vaste areas of land cannot be used without great danger and civilian loss. This cannot continue and we are many, I am sure, who feel committed in this task and we have to use our best techniques, our best scientists and economic recourses to find the best demining solution in a short time.

A DEMINING TECHNOLOGY PROJECT

J.D. Nicoud

EPFL, CH-1015 Lausanne, Switzerland

INTRODUCTION

Applying technology to humanitarian needs is a stimulating objective. Detecting and removing antipersonnel mines seems to be a perfect application for robots. Indeed it is, but a good understanding of the problem and a careful analysis must filter the goals of the projects in order to avoid deceptions.

On one side, it is important to have as many projects as possible being started, even naive and redundant, in order to increase the probability of getting results and showing the industry that there is a market for demining. On the other side, the problem is difficult, lifes will quickly be lost if the work is not done with the highest degree of security, and financial partners will be discouraged if the promised results are not reached in a significant way.

Teams willing to really do something useful for a demining campaign are faced, in addition to the many scientific problems to be solved, with the difficult task of changing the habits of conservative users. Since any new technology may cause accidents, tests must be done carefully. But prices will come down only if new technologies are introduced and widely distributed.

FINANCING

For an academic research group, getting adequate financial support may seem difficult :

- The military have been the usual provider, but you need to get accepted and you have to work their way.
- Fundamental research money is good for paying PhD students. They will work on some innovative and narrow topic, not on demining. They need 3 years, the typical duration of a thesis, to get their results.
- Industrial money is scarce. The potential market will take long to bootstrap. Demining teams dream of the perfect sensor, but they need to see it before they order it. Companies proposing a demining sensor need to sell before developing the next improved generation.
- Humanitarian money is insufficient and difficult to divert into technological developments, since its objective is to bring a direct help to the population.

But the society´s awareness toward demining actions is increasing. Private money is available, and once a group is known by adequate local publicity, money is more easy to get, at least in quantities adequate for conducting research projects. Starting small is the best way to understand the problem, adapt to real needs and build up the required highly motivated team. Bootstrapping important projects takes anyway a lot of time and efforts.

A PLAN FOR ACTION

Except for the metal detectors used by the demining teams, research for detecting and neutralizing mines is mostly under the control of the military and has made little progress for the last 50 years. It is time now to encourage the engineers, who have developed nice control applications, personal computers and robots, to be active in this field, and ask for appropriate funding. Coordination of this research must be made at the international level (EC, UN). Start-ups and existing industries must take over the manufacturing of the sensors and robots for a market of several billion dollars, if the objective is to clean the planet in 20 years, and not in 1000 years as given when dividing the number of mines (100 Mio) by the number of presently yearly removed mines (100´000). The present average cost for removing a mine is far too high ($ 1000) and is now increasing, since the methods are not changing.

Augmenting humanitarian money, without working on the technology, will reduce the duration of the clearing, and hence the number of innocent civilian victims (10´000 per year), but it will not reduce the number of injured deminers (1 for 2000 mines). A new technology will reduce this number, and pay for itself, due to the increased efficiency. Let us invest enough on the technology to generate the dynamism of the first days of personal computers. Demining sensors and robots are also mostly electronics and software.

Let us assume, as proposed by Nicoud (1) that for the next 3 years, the budget for demining doubles every year, and 25% of it is devoted to research, development and support to the industries providing demining equipments (figure 1). Support to industries is required to help them making the initial investment, and keep the price of the technology as low as possible. Even if the demining teams obtain more money, they may not be tempted to buy new equipment they have to validate, if it is more expensive.

With new sensors and an increasing production, the cost for removing a mine should decrease by a factor 3 to 5 in 10 years, 10 to 20 in 20 years. Computer technology has experienced a 10 times larger factor over the last 20 years.

SENSORS

The demining problem mostly lies with the sensors. Present methods use metal detectors, which give many false alarms, demanding a careful prodding of the suspect area. Ground penetrating radar (GPR) is accepted as a valid technology, see the detailed description by Daniels (3). Systematic tests have been made for example by Fritzsche (5).

Microwaves, thermal imaging, odor sensors, biosensors, microcalorimetry, ion mobility spectrometry, X-ray tomography, as surveyed by Gros (4), are still in a research phase and may never be applicable to humanitarian demining. Research in these fields has been mostly conducted with military money and objectives. It is time to start industrial oriented researches in some of these areas, particularly on very sensitive odor sensors. But if the objective is to bring shortly an improved solution to demining teams, one have to rely exclusively on technologies which are already well developed, light in weight and reasonably priced, that is GPR and improved metal detectors.

Year	Investment	Demining teams	R&D	Industries	Cost per mine	Mines removed
1996	100 M$	100 M$	0	0	1000 $	100 000
1997	200 M$	150 M$	50 M$	0	1000 $	150 000
1998	400 M$	300 M$	80 M$	20 M$	800 $	375 000
1999	800 M$	600 M$	100 M$	100 M$	600 $	1 million
2000	1 B$	800 M$	100 M$	100 M$	400 $	2 million
2001	1 B$	900 M$	50 M$	50 M$	300 $	3 million
2002..9	idem	idem	idem	idem	decrease	increase
2010	1 B$	900 M$	50 M$	50 M$	90 $	10 million
2011..16	idem	idem	idem	idem	idem	idem

Fig. 1 Scenario for removing all mines before year 2020

THE EPFL DeTeC PROJECT

Private money in 1994 allowed us to work on a "demining robot". Our speciality is to develop autonomous mobile robots and we started with a very naive understanding of the problem. It was fun inventing the Pemex and demonstrating a prototype that proves the cost-effectiveness of its simple design and its excellent motion abilities. But this robot, like all the other proposed ones, is of no practical use for demining since it now uses a primitive metal detector. Moreover, the autonomous navigation aspect is not yet solved, and we have stopped for some time working on this: it is too difficult to be cost competitive, since human deminers in Cambodia or Angola are paid $ 200 to $ 1000 per year, and they have no life insurance. Who will buy a robot in these countries? One can hope that soon the differences in human life value will decrease, and robots will soon be required, as it is already the case in former Yugoslavia where 3 million mines have been laid.

The mine sensor system is now the top priority for humanitarian demining. The project at the EPF Lausanne is to develop a man portable system combining a ground penetrating radar (GPR) and a metal detector (MD). We are experimenting several approaches. The problem indeed cannot be solved with all its generality with 4 persons in 2 years. Hence we will limit ourselves to "simple" demining situations: flat field, standard soviet mines. The efficiency of the proposed solution will be precisely measured. If the results are positive, it should encourage demining teams to ask for advanced solutions, other teams to develop improved systems and providers of technology to invest in this market.

TEST SYSTEM

Extensive tests in a "sand box" are required to get varied data under well known conditions, allowing for software development. Many situations have to be simulated and a large amount of data has to be acquired.

A sand box must be large enough so that the measures are not influenced by the surrounding structures. Our sand box consists of two adjacent containers with an internal volume of 3.2 x 3.2 x 1.2 metres, built with concrete blocks. One is filled with clean sand, the other with arable land earth. A cartesian positionning system has been built to move the sensor under test over the ground. For the time being, we are not controlling the vertical motion; a spring adjust the pressure of the sensor when it touches the ground. The cost of this installation was about $15'000 plus an equivalent amount in salaries for technicians and mechanics.

The stepping motors control box receives displacement orders from a serial line. Data acquired from a GPR or a MD (see next sections) is stored on a PC's disk and transferred later to some server. Most of these files are available on Internet, see http://diwww.epfl.ch/w3lami/detec/gprsoft.html.

More realistic tests will be carried later in the open. The cartesian positionning system is easy to dismantle and carry. It just needs 4 support points for installation and can operate with the PC from a small power generator.

Fig. 2 The EPFL test system

GPR SENSOR

As said before, the most mature new technology is the Ground Penetrating Radar one, where a small antenna (20 cm in diameter) scans the ground and measure the echo to build an image of the 20cm top layer of the soil. The information from the GPR will be combined with the data given by the metal detector, adapted in cooperation with a specialized manufacturer. A scanning movement of the sensor may also be used to provide a more precise indication on the nature of the object, and to guarantee both a good level of confidence when rejecting an alarm and a low number of false alarms.

Work is in progress now with the SPRScan GPR we bought from ERA Ltd. We had to choose among several manufacturers and we continue to update our knowledge about existing GPRs, see (http://diwww.epfl.ch/detec/gprhard.html).

We use the Reflex software and MatLab for file processing. Several other presentations in this conference will detail the hardware and software aspects of GPRs.

For the final implementation, the algorithms will have to be optimized, and implemented on some DSP. We have experience with the 56001, 320C40, 320C80, but we will appreciate the help of any team having a reasonably priced solution with an adequate volume and power consumption.

INDUCTION COIL SENSORS

An interesting work has been performed at DASA-Dornier by Borgwardt (2). An imaging induction coil sensor based on a rotating antenna displays metallic scrap and mines in the upper soil. We have used a Foerster Minex 2000SL metal detector which generate two continuaous wave frequencies (2 kHz and 20 kHz). We intercept the four signals available at the output of the pre-amplifier. Phase differences make the recognition of magnetic and non-magnetic metals easy. Tests and software development will be made using our sand box. For a man-carried sensor, a precise (and not yet defined) positioning mechanism will be required. The operator will see on a screen or from a set of lamps where he has to continue the scanning to improve the recognition process.

THE PEMEX ROBOT

The Pemex (Anti-personnel Mine Explorer) has been proposed by Nicoud (6) as a low cost solution for carrying a mine sensor and exploring automatically an area. The Pemex is light enough not to make the mine explode and can explore a complete area, reporting on a central point the critical spots. It can put visual marks, or deposit hollow charges with a radio-controlled detonator, allowing for all mines being detonated simultaneously. Such a scenario still implies a lot of implementation efforts and the resulting product will be expensive, until produced in large quantities; due to the low cost of human deminiers, we are now emphasizing the work on the sensors.

SYSTEM VALIDATION

For any new demining technology, the number of mines found, the ease of use by locally trained operators, the reliability, the repairability, the number of false alarms, and the operating cost have to be be reported and checked against standard manual procedures. To give the project a fair chance of success with present technologies, the kind of terrain and the type of mines must be specified. We will limit ourselves to well known kind of fields and mines. Even if a solution applies reliably to 5% of the mine fields only, about 5 million mines could be removed using this procedure. Other projects will progressively address tougher mine fields as well as more dangerous or hard-to-detect mines.

The problem of neutralizing the mines is not of concern here, since several techniques are well mastered by demining teams and contribute only to a small part of present demining costs.

CERTIFICATION OF THE FIELD

When a demining team has finished its work, a soccer game takes place on the field, to prove to the population that it is safe. Engineer developing sensors should be ready to do the same. It is however possible that the demining teams are not confident with a new technology they do not understand. A mechanical device stamping or rolling on the ground in order to make the possibly still active mines explode is required. Mechanical solutions are claimed to be only 90% reliable. A soccer game too! Multiplication of probabilities will allow to reach the required confidence level.

The Dervish and the Mattock systems proposed by S.H. Salter of Edinburgh University are simple and efficient solutions that still need to be experimented. A special vehicle, e.g. one based on a land cultivator, could be developed to stamp the ground and demonstrate the terrain's safety. This vehicle should be left to the population recovering the land for its own ploughing and for carrying goods.

CONCLUSION

The scientific community should do more research on demining. If it is easy to find money for a small group at the national level, an important international project with the active participation of the industry is still missing. Technology can lower the cost for removing a mine. If this cost is getting low enough, the military advantage of mines will disappear, and mines will be more easily forbidden.

We would like to thank John Walker for its initial support and our present supporting organization: the Fondation Pro Victimis in Geneva, the Swiss Foreign Office and the EPFL. The 4 people working now on the DeTeC project are doing a great job and deserves my thanks and congratulations.

Fig. 3 Pemex-B

References

1. Nicoud, J.D., 1996, "Mine Clearance: not only a problem for the military any more", ISMCR'96, Brussels, May 9-10
2. Borgwardt, C., 1995, "ODIS - Ordnance Detection and Identification System", WAPM'95 Workshop on Antipersonnel Mine Detection and Removal, Lausanne, June 30-July 1st
3. Daniels, D. J., "Introduction to Subsurface Radar", IEE Proceedings, Vol 135, No 4, 1988, pp 278-320
4. Gros, B., Bruschini, C., 1996, "Sensor technologies for the detection of AP mines: a survey of current research and system developments", ISMCR'96, Brussels, May 9-10
5. Fritzsche, M., 1995, "Detection of Buried Landmines using GPR and Metal Detector. First Results and Field Experiments", WAPM'95 Workshop on Antipersonnel Mine Detection and Removal, Lausanne, June 30-July 1st
6. Nicoud, J.D., Mächler, Ph., 1996, "Robots for Anti-Personnel Mine Search", Control Engineering Pratice, 4(4), April

A SUSPENDED DEVICE FOR HUMANITARIAN DEMINING

J. P. Trevelyan
The University of Western Australia, Australia

SUMMARY

A device suspended from cables could provide safe access to a minefield for scanning with sensors or to detonate undetectable mines with a mechanical probe. Such a device could provide an economic and semi-automatic means for humanitarian demining, and has the potential for some local manufacture. This paper examines the concept and economic performance issues. We expect to build a prototype by the end of 1996.

DEMINING PROBLEMS

Apart from deaths and injuries to people, the major effect of landmines is to deny access to land and its resources, causing deprivation among the affected populations. Regardless of any efforts to stop the further use of landmines, the safe restoration of productive land is an urgent issue in affected countries according to sources such as McGrath (1, 2), CMAC (3) and the UN (4). If demining costs can be reduced to levels which are comparable with the economic potential of cleared land, financial institutions might provide long term finance for expanded and faster demining work.

Demining operations measure their work in terms of a few square kilometres per year. The rate of demining must be increased to make a major impact on countries with landmine problems.

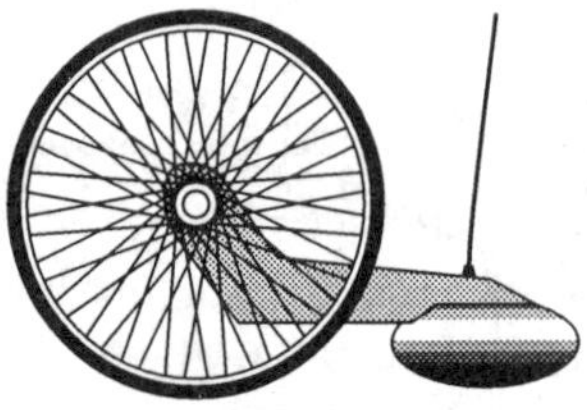

(a) Lightweight sensor robot (Pemex)

b) Remotely operated armoured robot

Figure 1. Two recently proposed concepts for robotic mine clearance.

Improving demining technology to reduce costs without compromising effectiveness presents a formidable technical challenge. The high risk of explosions, the highly variable terrain, absence of infrastructure and the operating environment are all factors which tend to rule out the use of complex technical solutions such as robots or remotely operated devices (5).

DEMINING ROBOTS

Figure 1 shows two types of device which have been proposed recently. The "mine-rat" is a small light-weight autonomous vehicle which carries a sensor to detect landmines (6 ,7). It is light enough not to trigger surface or buried mines. An alternative is a larger and heavier vehicle which can withstand anti-personnel mine explosions (8). Both have serious disadvantages.

The "mine-rat" can still disturb trip-wires and cannot be "hardened" against accidental explosions. Also, some surface mines (such as the Russian butterfly mine) can be triggered by successive applications of lower force loads, so a "mine-rat" would be at risk from these devices.

The heavier vehicle would almost certainly be susceptible to anti-tank mines.

Both vehicles, however, require a reliable means for navigating themselves across the terrain. While GPS[1] can provide a position reference, it does not solve the problem of locating and identifying obstacles, and selecting an appropriate path to follow towards a desired position. So both vehicles require "human" drivers (in addition to GPS for position measurement and recording). This, in turn, necessitates continuous manual operation from a "drivers station" in a support vehicle nearby, much like the more familiar bomb disposal robots. While both vehicles can work on sloping or rough terrain, both can be blocked by relatively small obstacles.

So far, robotics research has not come up with low cost control systems which could allow such robots to operate automatically for extended periods. Hence they are unlikely to provide cost-effective solutions.

Both these devices attempt to access the minefield by traversing the ground, and expose themselves to severe control difficulties and explosion hazards as a result. If we approach the minefield from above, instead, we can almost completely eliminate these problems.

A SUSPENDED ROBOT

Figure 2 shows another approach, though it does not look much like a robot. Here a tool carrier is suspended

[1] Global Positioning System - using satellites for accurate navigation to an accuracy down to a few centimetres.

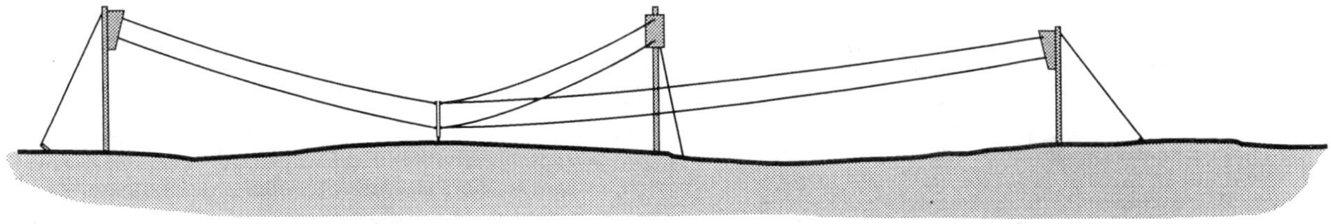

Figure 2 Suspended robot for demining.

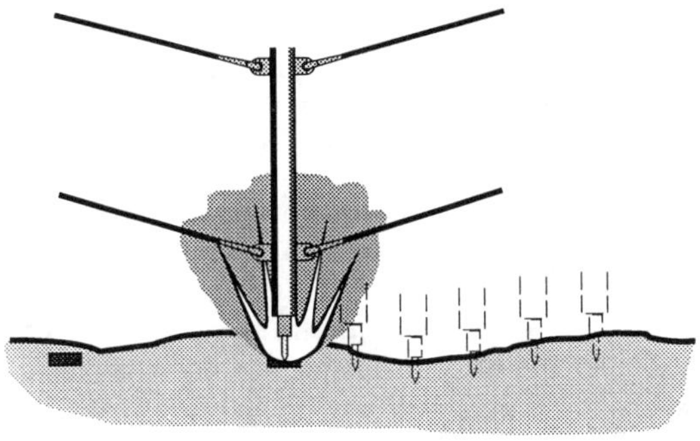

Figure 3. Ground pounding to detonate mines

above the minefield by a system of steel cables controlled by winches on three or more poles or towers at the corners of the minefield.

The tool carrier can moved to a chosen position and orientation between the poles under fully or semi-automatic control. It could carry one or more of:

a sensor package for mapping the minefield,

a rotary cutter and gripper for removing unwanted vegetation or clearing pieces of metal, or

a ground "pounding" device to detonate mines. This could be designed with an expendable foam plastic foot which would absorb the mine blast.

This device could significantly improve existing detection techniques because the location of the detector is known for each measurement. Compared with conventional hand-held detectors, much more sophisticated signal processing should be feasible, and it may be possible to reduce false alarm rates on metallic mines and non-metallic mines with metal fuses.

The risk of damage from explosions can be minimized because the more "delicate" motion control system is mounted on the poles and can readily be protected from dust or flying metal and rock fragments. Steel cables are also robust and can readily be replaced if necessary.

This device will work over any terrain where temporary poles can be erected. Therefore, it should work over hilly ground and waterways as easily as open agricultural or urban land.

This device requires access to sites for the poles which may need back stays for support. Therefore, if the minefield extends further than the maximum allowable span, a path must be cleared to provide access to erect the poles carrying the winches. In an urban environment, taller buildings or power poles could provide mountings for the winches. In practice, a minefield is first explored with 1 to 2 metre wide test lanes to determine the density and types of mines laid, and these lanes would provide sites for erecting poles.

The winches need to be well controlled to provide accurate positioning. A computer can do this with suitable electric drive motors. Some damping is needed to help eliminate unwanted "bouncing" or swaying in the suspension cables.

As shown in figure 4 a suspended robot can access an area within the triangle defined by the nearest 3 winch mounting poles. To retain reasonable control, all cables must have some tension. Therefore, the probe or sensor can only operate some distance inside the triangle. 4 poles provide significantly greater working area than 3.

An obstacle such as a tree restricts access to part of the operating area extending away from the two nearest poles. If the tree branches are well clear of the ground, it may be possible to bring the device close to the trunk in which the "shadow" (inaccessible) zones are much smaller.

The shadow zones are obtained by remembering that the probe needs clear lines to three poles. For each combination of three poles, the probe must be within the inner triangle shown in figure 4.

In a forest, it may be wiser to use the tree as supports for the winches and operate over much smaller areas between trees.

In practice, this type of device would be operated in a semi-automatic manner. An operator would manually move the tool carrier (using a remote joystick type device) to mark the outline of a small work area (say 20 to 50 square metres). Then the control system take over and automatically move the tool carrier in a regular scanning pattern within this defined work area for sensor mapping, or for pounding the ground (much slower). Obstacle clearance would require continuous manual control.

OPERATING COSTS AND FINANCE

How much would a device like this cost to build and operate?

Apart from the computer to control it and optional sensing devices, the major cost would be the winch units and the associated controls. Using typical costs for conventional robots as a guide, each winch device would probably cost about US$10,000 after allowing for environmental

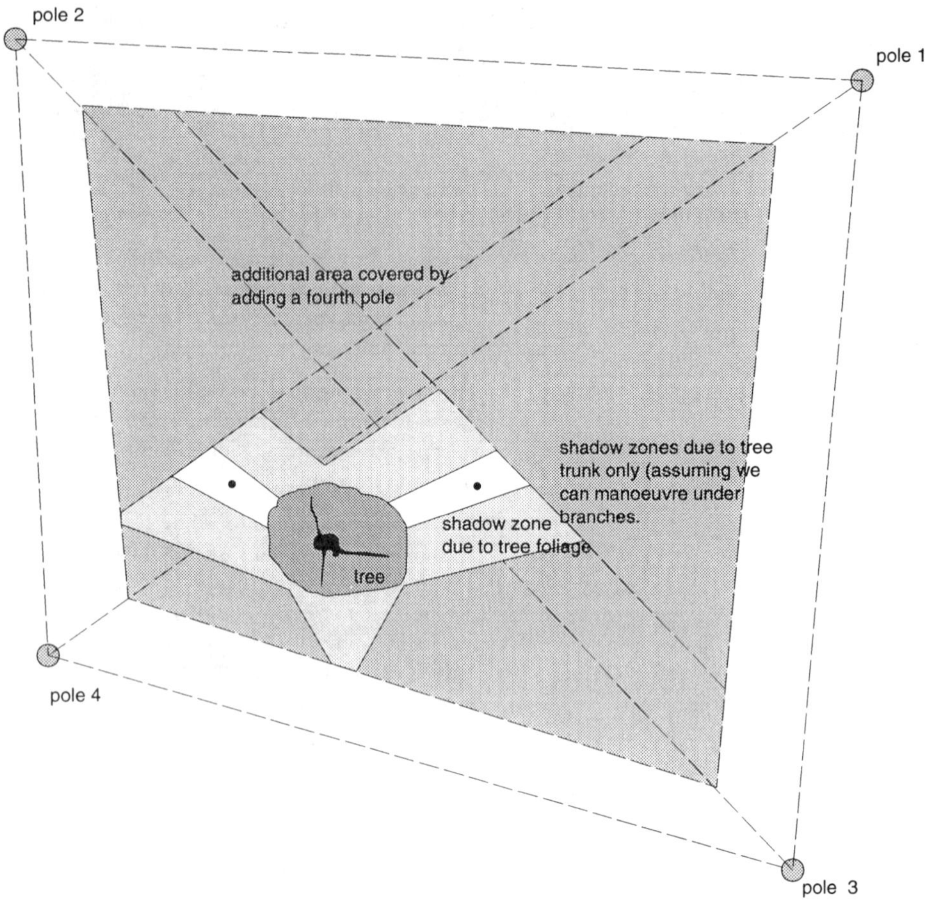

Figure 4. Working area for suspended tool carrier (refer to text)

protection[2]. Each winch would handle two cables in tandem, and so four winches would provide a basic system. Assuming that the system design and software could be written off by appropriate research agency funding, it would cost about $60,000 to $80,000 to build a field operable system. This would be capable of mechanically probing for mines at a rate of about 1 position every second.

I assume that supporting labour, transport and electric power (from small mobile generators) is already available—these costs have not been included. Maintenance support and training will be covered by a fixed percentage of the capital cost—assumed to be 15%. Further, we will assume that the equipment will last 5 years and has to be completely replaced after then. Therefore, the operating costs can be approximated as 35% of the capital cost.

Assuming we need to probe the ground 400 times per square metre (5 cm between probe points), the probing rate would be about 8 square metres per hour.

A 50 by 50 metre work area (2,500 sq. metres) would therefore take about 2 weeks to clear this way. Given appropriate programming, the system could operate automatically for extended periods (up to a day) with minimal supervision. Each time a mine explosion occurs, a visual check would be needed for damage to the cables, though this could be detected by appropriate sensors. Based on a capital cost of $80,000, the annual operating cost would be about $28,000.

If the minefield originally contained 50 mines, and the mines were detonated this way, the demining cost would be about $1200, and the 50 mines would have been eliminated at a cost of about $25 each (about $5,000 per hectare)

A sensor such as surface penetrating radar has been suggested as one means of detecting buried objects such as mines (see 9 - 13). Assuming that we can scan 50 square metres of ground per hour, it would take 50 hours to scan the work area. Assuming that the false alarm rate is 50:1, and that possible mine locations about 20 cm square are probed to trigger an explosion, we would need to probe 2500 sites (for 50 mines) taking about three minutes each. This would take about 120 hours. Thus the time required to clear the minefield would be reduced to about 1 week. Assuming that the sensor technology doubles the cost (i.e.

[2] A typical mid-range industrial robot system costs about $50,000 including the controller. Of this, the joint motors (6 per robot) cost about $5,000 each with their controller circuits).

the sensor and software costs $80,000), the operating cost for 1 hectare would be about $2500 and the mines would have been eliminated for about $12 each.

What is the economic potential of demined land?

It is not easy to compare land values in countries affected by conflicts. Assuming that the land being recovered is comparable to productive agricultural land in SE Asia, we can compare clearance costs with rice production. Typical rice land produces between 1 and 8 tonnes per hectare (14). (Cambodia seems to have the lowest productivity—Encyclopaedia of the Third World (15) quotes 1.2 tonne/hectare for Cambodia - one of the lowest in the world.) Rice is worth between $500 and $1000 per tonne (16). Urban land usually has more value — typical urban fringe land in developing countries is worth $50,000 per hectare. Therefore, we can see that the demining costs predicted here are comparable, within reason, of the economic potential of the land.

ACKNOWLEDGEMENTS

The author would like to thank all the people who have contributed to this paper, including D. Macey and C. Trevelyan for background research and checking, R. Bird (DSTO Melbourne) for outline technical information on Australian military work and published technical reviews, Dr. A. Churches (UNSW) for background and encouragement, and the Australian Army School of Military Engineering for technical briefing and operational experience from Afghanistan and Cambodia. Also to Mines Advisory Group, WA Branch of International Red Cross, and the Swiss Federal Institute of Technology Microprocessors and Interfaces Laboratory (EPFL-LAMI), Lausanne, Switzerland for supplying their publications.

REFERENCES

Copies of all reference material available from the author on request.

1. McGrath, R. (1994). *Landmines, Legacy of Conflict: A Manual for Development Workers*, Oxfam, Oxford, UK.

2. McGrath, R. (1995). *1994/95 Report on the Activities of the Mines Advisory Group*, 54a Main Street, Cockermouth, Cumbria CA13 9LU England.

3. CMAC (1995). *Facts and figures: Progress report 1992-1995.* CMAC, Phnom Penh.

3. Red Cross (1995) *Landmines must be stopped: Chapter VI Mine Clearance.* Special Brochure, International Committee of Red Cross, Geneva, Switzerland.

4. United Nations (1996) UN Database on Landmines, available from World Wide Web at http://www.un.org /Depts/Landmine/

5. Anonymous (1995) *International Meeting on Mine Clearance, Background Paper No. 5.* For discussion by experts, Panel E "New Technologies in Mine and Minefield Detection and Mine Clearance", available from author.

6. Nicoud, J .D. (1995) *Proceedings of Workshop on Anti-personnel Mine Detection and Removal WAPM '95*, Swiss Federal Institute of Technology Microprocessors and Interfaces Laboratory (EPFL-LAMI), Lausanne, Switzerland.

7. Nicoud, J. D. and Mächler, P. (1995) *Pemex-B: a low cost robot for searching anti-personnel mines.* In WAPM '95 (above)

8. Velez, G. and Thomas, H. (1995) *Requirements for robotics in explosive ordnance disposal operations in tropical and desert areas.* In WAPM '95 (above)

9. Azevedo, A. G., Gavel, D. T., Mast, J. E., and Warhus, J. P. *Landmine detection and imaging using micropower impulse radar (MIR).* In WAPM '95 (above)

10. Chignell, R. J. *Ground probing radar - a technology ready for mine detection equipment development.* In WAPM '95 (above)

11. Daniels, D. and Dittmer, J. *Design challenges for ultra-wideband radar for AP mine detection.* In WAPM '95 (above)

12. Fritzsche, M. and Trinkhaus, P. *Detection of buried landmines using ground penetrating radar and a metal detector: first results of field experiments.* In WAPM '95 (above)

13. Garriott, G., Land, D., Nillies, J., Vance, G, and Sherbondy, K. *A multi-sensor system for demining.* In WAPM '95 (above)

14. Khor, M. (1994). Economic Globalization and Eroding International Cooperation: Implications for Agricultural and Rural Sustainability, Third World Resurgence Series No. 67.

15. World Bank (1996). World Development Report 1994, Oxford University Press (in press).

16. Public Ledger No. 71960 (1996) from http://www.netbenefit.co.uk/ledger/index.html

APPLICATION OF SYSTEM IDENTIFICATION AND NEURAL NETWORKS TO CLASSIFICATION OF LAND MINES

J. W. Brooks and M. W. Maier

The University of Alabama in Huntsville, Alabama, USA

INTRODUCTION

The ability to discriminate or classify land mines is critical; false classifications in the form of "false positives" may result in unnecessary remediation efforts. On the other hand, "false negatives" could lead to disastrous consequences by leaving people vulnerable to hazardous ordnance which has not been remediated. Often, all detections are classified as hazardous, and that can lead to unnecessary remediation efforts.

The detection and classification of unexploded ordnance (UXO), including land mines, is a critical technology issue in the United States.The US Congress passed House Resolution (H.R.) 2401 and H.R. 3116 mandating the establishment of advanced technology programs to address the issues of UXO detection, classification and remediation. From August 1993 to December 1993, 29 different systems were used in tests at the Jefferson Proving Grounds (JPG) in an effort to characterize their ability to detect and classify broad categories of UXO, including land mines (1). Detailed evaluations of the performance of the demonstrators at JPG are published in 1995 (2), and reveal a very high percentage of both "false positives" and "false negatives" for all sensor types. (Both publications are available through the Defense Technical Information Center, DTIC, 8725 John J. Kingman Road, Suite 0944, Ft. Belvoir, VA, USA, 22060-6218)

Methods of classifying objects in the past have included various high resolution radar (HRR) approaches, including Synthetic Aperture Radar (SAR), whereby range profiles of various targets are compared to a library of profiles. Another methodology includes various resonance-based approaches, whereby the Singularity Expansion Method (SEM) (Baum(3)) parameters are estimated by some method, often a Prony or Maximum Likelihood approach. The natural resonance approach is very appealing two reasons: Firstly, SEM theory states that the natural resonances are invariant with target aspect angle and polarization. For application to land mines, this is appealing since the mines are probably oriented randomly. Secondly, one need store only a limited number ($O(10^2)$) of identification parameters in a library for each object, as opposed to the ($O(10^9)$ or so parameters for each object using the HRR approach (4).

It is for these reasons that we have chosen to investigate the natural resonance approach, as applicable to an impulse ground penetrating radar (GPR). We have restricted the scope of the paper to a subset of conducting bodies of revolution (BOR). In particular, we seek to decide if the radar return from a buried BOR is either a "Cylinder Class" or a "Cone Class," first described in Brooks (5). From the perspective of land mine classification, the cylinder could represent a mine, and the cone a non-hazardous object. The methods described herein are applicable to a wideband incident waveform (or stepped frequency waveform) which excites natural resonance modes of the BOR. Exact resonance parameters of canonical BOR have only recently been calculated (Vechinski (6)), and those results are used in this paper.

BACKGROUND

In 1971 Baum (3) postulated that the solution for electromagnetic interaction of currents due to an incident field, and the resulting scattered fields, could be formulated in terms of the singularities (poles) of the current distribution in the complex plane (The Singularity Expansion Method, or SEM). A key result of this postulate is that the pole locations, and resulting summation of exponentially damped sinusoids in the time domain, would be invariant with target orientation relative to the radar (Mains and Moffatt (6)). In the years since 1971, numerous studies have been conducted in SEM parameter estimation including methods based on Prony, pencil-of-function, linear and nonlinear least squares, etc. A recent paper by Baum (7) provides an excellent summary of research to date. That paper also describes methods known variously as extinction-pulse (E-pulse) or kill-pulse (K-pulse) techniques, whereby a linear time-domain filter is synthesized which, when convolved with the target responses to which they are matched, annihilate pre-selected natural frequency content of those responses (Rothwell, *et. al.*(9) and Chen, *et.al.*(10)). These methods have been verified in a laboratory setting, but even though much work has proceeded over the years, an optimal synthesis technique has not been discovered and there remain several problems with pulse generation, etc. (7)

Recently, Dowling, *et al.* (11) applied a linearly constrained total least-squares (LCTLS)-Prony method in recovering the SEM parameters from a conducting sphere, but no attempt was made at classifying two targets based on their estimated SEM parameters.

SEM may be considered to be a hybrid between time-domain and frequency domain methods in which the scattering process is separated into parameters such as natural complex resonances, modal currents and coupling coefficients. Once the complex resonances or poles are determined, the transient response may be constructed by a series of exponentially decaying sinusoids which are related to each pole. A major advantage of SEM is that it predicts the natural poles of the scatterer which are theoretically aspect independent.

It was shown in (6) that the current distribution on a body may be represented by

$$(s) = \mathbf{Z}^{-1}(s)\mathbf{V}(s \qquad (1)$$

where $\mathbf{Z}(s)$ is a generalized systems impedance matrix in the Laplace domain, $\mathbf{J}(s)$ be the unknown current vector and $\mathbf{V}(s)$ be the excitation vector. Then, solving for $\mathbf{J}(s)$, the inverse of the impedance matrix may be expanded into a series of singularities as(12)

$$\mathbf{J}(s) = \sum_i \left(\frac{\eta_i(s)\overline{\mathbf{M}}_i(s)}{s - s_i} \right) + \overline{\mathbf{W}}(s) \qquad (2)$$

"Detection of abandoned land mines", 7-9 October 1996, Conference Publication No. 431 © IEE 1996

where $\overline{\mathbf{M}}_i(s)$ are the natural modes, $\eta_i(s)$ are the coupling coefficients, the poles s_i represent the natural resonances (complex), and $\overline{\mathbf{W}}(s)$ is an entire function. The coupling coefficients are the magnitudes of the natural oscillations given by the complex frequencies at the s_i and are aspect dependent.

Thus, the "natural resonances" of a body can be determined by examining the poles of the transfer function which relates the input (incident pulse) with the output response (reflected pulse). The problem reduces to one of system identification. In theory, distinct targets have distinct sets of natural resonances which may be used as identification features. The concepts of "early time" and "late time" are fundamental in SEM analysis. "Early time," in a systems context, is equivalent to the forced response, and "late time" is equivalent to the unforced, or natural response. In principle, the resonant currents on the body are not fully developed until passage of the incident pulse beyond the body (Figure 1). Prior to this ("early time"), the pulse is interacting with the individual scatterers on the body and the currents are changing with time in a highly complex manner, so we can not, in a strict sense, speak of "natural resonances" until the "late time." However, the reflected pulse amplitude in the "late time" is considerably smaller than during the "early time." As shown in Van Blaricum (12), using the "early time" response as well as the "late time" response leads to time-varying complex coefficients and is the subject of current research. Attempts have been made (Morgan, *et.al.*(13)) to show that the early time response can be modeled using the Kumaresan-Tufts (K-T) algorithm and the late time by the Cadzow-Solomon Algorithm, but this was done using synthesized pulses consisting of damped exponentials. Generally, the late-time response is of a much lower magnitude than the early time response, and is thus more vulnerable to noise effects.

THE TARGET SET

We consider two bodies of revolution (BOR), for which the output responses were calculated using the methods in (6). A conducting cone and cylinder are illuminated by an incident impulsive waveform at seven aspect angles, from 0° to 90°, in 15° increments, and at both $\varphi - \varphi$ (Electric field vector perpendicular to the BOR axis of revolution) and $\theta - \theta$ (Electric field vector in the plane of the axis of revolution) polarizations. The length-to-base ratio for both is 1.0, and the incident pulse width is approximately 6.7 nsec. Figure 2 shows the true complex (*s*-domain) pole locations for the cone and cylinder targets from (6). Note that the pole clusters have a distinctly different pattern, particularly for real frequencies less than about 50 MHz.

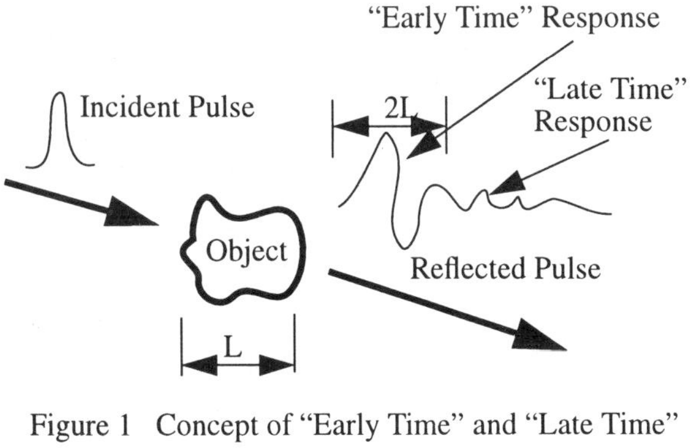

Figure 1 Concept of "Early Time" and "Late Time"

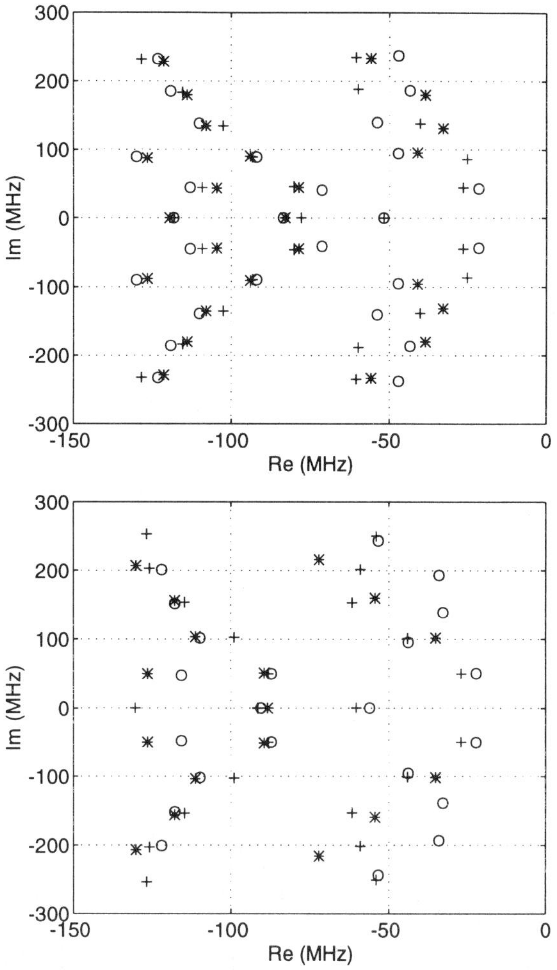

Figure 2 True Cylinder (Upper) and Cone (Lower) Natural Poles, Mode 0 (o), Mode 1 (+), Mode 2 (*)

RESULTS

Figure 3 and Figure 4 are representative radar returns from the cone and cylinder, respectively, at the indicated polarization.The Autoregressive with Exogenous Input (ARX) and optimal Instrumental Variables (IV) methods (Ljung(14)) were applied to the data in an attempt to replicate the pole positions in Figure 2. Additional identification methods such as the Autoregressive Moving Average with Exogenous Input (ARMAX) (14) were also tried, but the results were generally not as good while consuming considerably more computational resources. Parameter estimates were derived for each BOR response. Practical limitations of nearly singular matrices and computer roundoff error limited the maximum model order to be less than about 40.

Figure 5 is an example of 20th order model fits to the cone response with an aspect angle of 30° and $\Theta - \Theta$ polarization. Note the significantly improved fit provided by the IV method. Other measures of performance, such as the Akaike Information Theoretic Criterion (AIC) (Fukunaga(15)) provided adequate results, but the IV method consistently led to a more visually satisfying result, i.e., a "good fit" corresponds to a "good model".

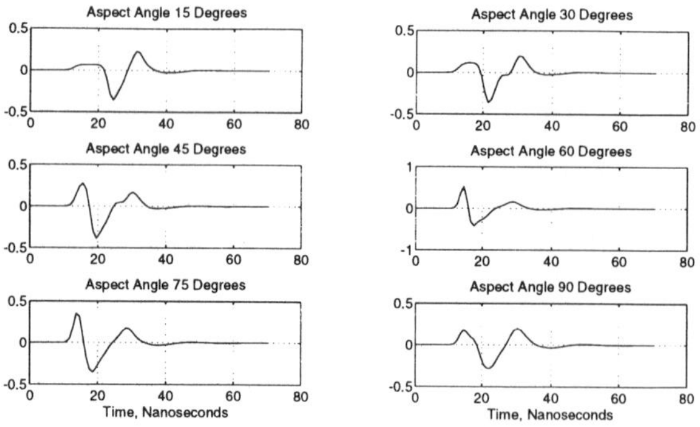

Figure 3. Cone Response to Input Pulse, θ – θ Polarization

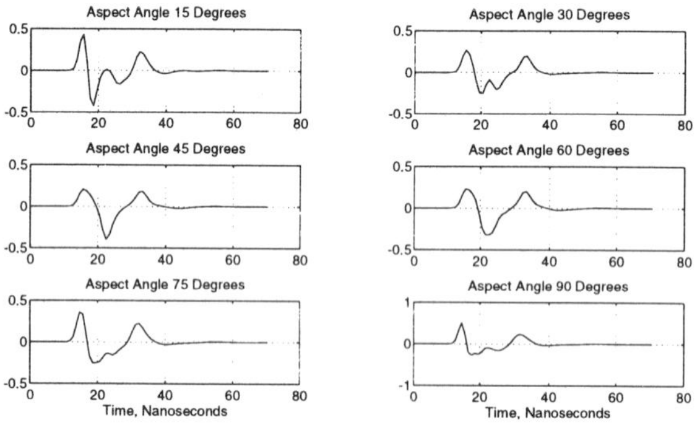

Figure 4. Cylinder Response to Input Pulse, θ – θ Polarization

Figure 6 and Figure 7 show the composite results of the pole estimation by each method for the cylinder and cone targets. The true Mode 1 and Mode 2 poles are plotted as "o," and the estimated poles are denoted by "+." The estimated poles represent the composite for all aspect angles and the two polarizations. No distinction between the modes is made, as a particular mode may be excited more than another at some aspect angles. Mode 0 poles were excluded, as that mode is very weakly excited under any circumstance (6). In general, Mode 1 is excited more than either Mode 0 or Mode 2.

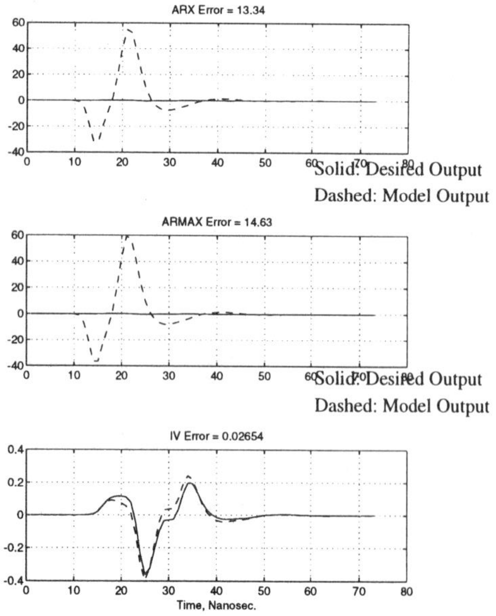

Figure 5 20th Order Models Applied to Cone, Aspect Angle 30°

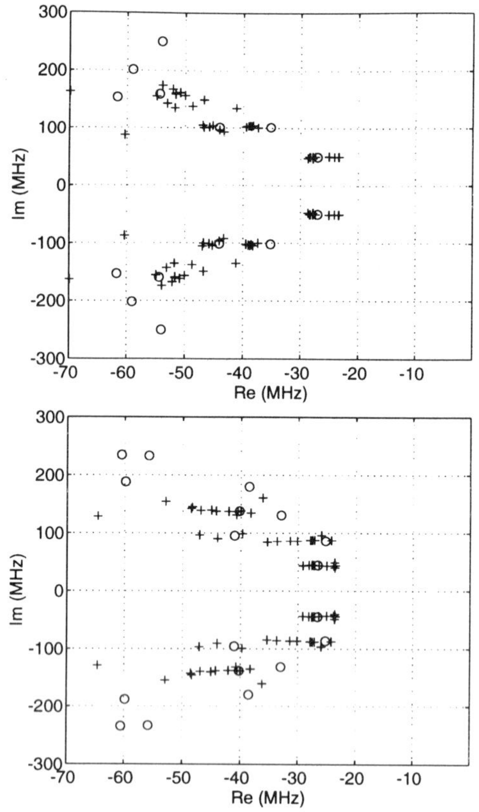

Figure 6 Estimated Cone (Upper) and Cylinder (Lower) Pole Clusters, IV Method, With True Poles

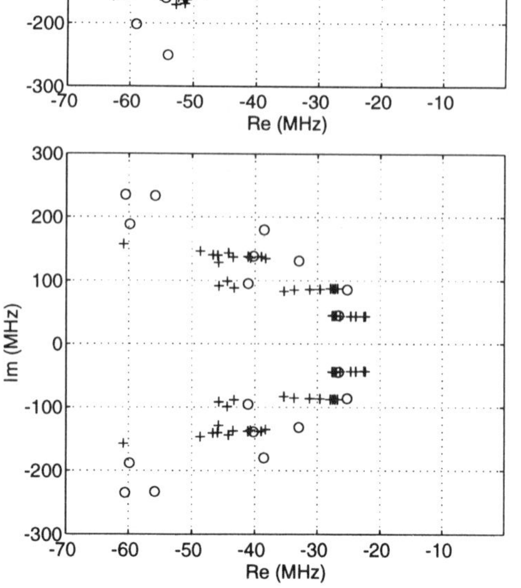

Figure 7 Estimated Cone (Upper) and Cylinder (Lower) Pole Clusters, ARX method, With True Poles

The estimation methods created several poles which are "deep in the left hand plane" and are highly damped, so only the first six estimated complex pole pairs are plotted. That is, the six pole pairs which are "closest" to the imaginary axis; these physically represent poles which are most dominant and thus could be most easily estimated from spectral measurements, for example. In general, the estimated poles lie in narrow bands along the imaginary axis, but are somewhat dispersed along the real axis.

Now, the estimated parameters will be used to provide a classification of the object into one of two classes, "Cone Class" or "Cylinder Class." More concisely, this is a hypothesis testing problem with two hypotheses. Let H_i represent the ith class;

$$H_0 = \text{"Cone"}$$

and

$$H_1 = \text{"Cylinder"}.$$

The features to be estimated are categorized as follows: For each of the two target classes, seven data sets are collected representing the target returns from aspect angles (0 to 90 degrees in 15 degree increments). Each data set consists of target returns from two orthogonal polarizations. Thus, there is a total of 84 parameter vectors available for analysis.

A two-layer backpropagation (Werbos,(16)) neural net was applied to the classification problem, using MATLAB 4.2 and the Neural Network Toolbox. A target sum-square error of 10^{-4} was desired, and the neural network was trained after about 730 epochs, or training cycles. A Monte Carlo simulation was conducted in which 126 feature vectors, for both the IV and ARX estimates for each BOR class, were generated. The variance of the white noise input to the model was 10^{-6}, which led to a nominal correlated SNR of about 25 dB. A two-layer neural network was trained with three feature vectors from each class per estimation method. The hyperbolic tangent sigmoid function was used for the hidden layer, and the Cone class was assigned a target output of -0.95, and the Cylinder class a target output of +0.95. The output layer was a pure linear function to permit a clear view of the operation of the neural network, and to permit determination of "shades of gray" in the classification.

There is a considerable "gray area" for both estimation methods; in such cases, one would normally place bounds about the two target values and exclude the points in the middle as satisfying neither hypothesis. A rejection band of $-0.8 < k < 0.8$ was used, where k is the ordinate of the plots. TABLE I and TABLE II are confusion matrices for the classification problem. There were two values falling within the rejection band form the IV Cone classification and 21 rejected values form the IV Cylinder classification. Also, 51 values were rejected from the ARX Cone classification and 0 from the ARX Cylinder classification. TABLE III and TABLE III summarize the probabilities of correct and incorrect classification for both estimation methods.

SUMMARY and CONCLUSIONS

We have shown that the SEM parameters of conducting bodies of revolution may be estimated using classical system identification methods, and those estimated parameters may then be used for target classification. The classification is achieved irrespective of target aspect. In addition, the probability of misclassification of targets for both methods is quite low. Although the ARX method provides good performance, the IV method was shown to provide good performance while also generally obtaining the best "fit" to the reflected component. The observation that the IV estimated time response closely fits the measured time response can also assist in the development of an automated parameter estimator. Finally, the backpropagation neural network was shown to provide adequate separation of the estimated parameters into the two target classes.

TABLE I-Confusion Matrix for IV Classification

(Numbers Exclude Rejected Values)		Estimated Class	
		Cone	Cylinder
True Class	Cone	100	24
	Cylinder	29	76

TABLE II-Confusion Matrix for ARX Classification

(Numbers Exclude Rejected Values)		Estimated Class	
		Cone	Cylinder
True Class	Cone	51	24
	Cylinder	0	126

TABLE III-Probabilities of Classification, IV

		Estimated Class	
		Cone	Cylinder
True Class	Cone	0.80	0.19
	Cylinder	0.23	0.60

TABLE IV-Probabilities of Classification, ARX

		Estimated Class	
		Cone	Cylinder
True Class	Cone	0.40	0.19
	Cylinder	0.0	1.00

CONTINUING EFFORTS

We are currently extending the concepts developed in this paper, and additional concepts, to the classification of non-metallic land mines.

ACKNOWLEGMENTS

The authors wish to express their appreciation to Dr. Scott Vechinski for his support in computing the data used in Figure 2 above, and his general advice on SEM.

REFERENCES

(1) 1994, "Unexploded Ordnance Advanced Technology Demonstration Program at Jefferson Proving Ground (Phase I)", Report No. SFIM-AEC-ET-CR-94120, (AD A294935)

(2) 1995, "Evaluation of Individual Demonstrator Performance at the Unexploded Ordnance Advanced Technology Demonstration Program at Jefferson Proving Ground (Phase I)", Report No. SFIM-AEC-ET-CR-95033, (AD A295074)

(3) Baum, C.E., 1971,"EMP Note Series, EMP Interaction Note 88", Air Force Weapons Lab., Kirtland AFB, Albuquerque, NM, (AD A0 66905)

(4) Smith, C.P., and Goggans, P.M., 1993, IEEE Ant. Prop. Magazine, Vol. 35, No. 2, pp 27-38

(5) Brooks, J. W., 1994,"Estimation of Singularity Expansion Method Parameters and Supervised Learning Applied to Target Classification", The University of Alabama in Huntsville AL

(6) Vechinski, S.R., 1989, "Natural Resonances of Bodies of Revolution", Auburn University, Auburn, Alabama.

(7) Baum, C.E., Rothwell, E.J., and Nyquist, D.P., 1991, Proc. IEEE, vol. 79, pp 1481-1491

(8) Kennaugh, E.M., 1981, IEEE Trans. Antennas Prop., vol AP-29, pp. 327-331

(9) Rothwell,E.J., Chen, K.M., Nyquist, D.P., and W. Sun, 1987, IEEE Trans. Antennas Prop., vol AP-35, pp. 426-434

(10) Chen, K.M., Rothwell, E.J., Nyquist, D.P., Behermeyer, R., Q. Li, Tsai, C.Y., and Norman, A., 1995, Proc. IEEE 1995 Int'l Radar Conf.

(11) Dowling, E.M., DeGroat, R.D., and Linebarger, D.A., 1994, IEEE Trans. Antennas Prop., vol 42, No. 5, pp. 590-599

(12) Van Blaricum, M.L., 1991, "A View of the Early-time Component in Impulsive Scattering", *Ultra-Wideband Radar: Proceedings of the Los Alamos Symposium*, CRC Press, Boca Raton, Florida, USA

(13) Morgan, M.A., and Larison, P.D., 1991, "Natural Resonance Extraction From Ultra-wideband Scattering Signatures," *Ultra-Wideband Radar: Proceedings of the Los Alamos Symposium*, CRC Press, Boca Raton, Florida, USA

(14) Ljung, L., 1987, *System Identification: Theory for the User*, Prentice-Hall, New Jersey, USA

(15) Fukunaga,K.,1990, *Statistical Pattern Recognition*, 2nd Ed., Academic Press, Boston, USA

(16) Werbos, P., *1990,* Proc. IEEE, vol. 78, No. 10, pp. 1550-1560

(17) Wahlberg, B., 1990, IEEE Trans. Acoustics, Speech and Sig. Proc., vol. 38, no. 7

(18) Stoica, P., Soderstrom, T., and Friedlander, B., 1985, IEEE Trans. Auto. Control. Vol. AC-30, No. 11

Anti Personal Mines, a world-wide problem: From political conscience towards humanitarian, research and industrial action

Jan Cornelis (Prof. at VUB), Hichem Sahli (Visiting Prof. at VUB),
Marc Acheroy and Yvan Baudoin (Profs. at KMS/ERM)
Contact: Prof.. J. Cornelis, VUB (ETRO-BIIP), Pleinlaan 2, B-1050 Brussel; Fax: 02-6292883; Tel. 02-6292930

This document points out the urgent need for solutions of the civilian demining problem and suggests action plans for enhancing the demining efficiency.

1) The nature of the mine problem

1.1. THE NUMBER OF MINES

- There are currently an estimated 50.000.000 to 300.000.000 mines lying hidden throughout the world, each of them a potential killer [1]. These will continue, blindly, to claim victims long after the war is over (children form a large fraction of them: e.g. in Afghanistan 23 %). In peaceful Angola, ten million mines are retaining 1.5 million people in the cities, preventing their return to the agricultural areas [5]. Afghanistan counts 8.000.000 antipersonnel mines, about 90.000 have been neutralised from 1990 till now, resulting in only 50 km^2 of cleared land. Former Yugoslavia is polluted by approximately 1 to 3 million landmines. In Cambodia, the estimated number of amputee's is 36.000; this means one amputee for every 236 citizens [1]. More statistics are given in [2].

1.2. THE DEMINING METHODS [3]

- The utilised demining strategies are mainly determined by the required Clearance Efficiency (CE) which is completely different for military (in war times a higher casualty risk is accepted) and civilian (99.96 % CE is imposed by the UN) applications.

- In war time it is often sufficient to punch a path through the mine field (mine breaching), which is done by "brute force" mechanical devices (ploughs, rollers, ...) mounted on tanks. Military research budgets have allowed to develop modern equipment.

- The high CE required for humanitarian actions can only be achieved by hand clearance. A demining campaign is a complex multistage procedure:

(1) mine field marking - a tedious procedure because, from a field of about 1 km^2, often only 100 m^2 is finally delineated as a real mine field,
(2) detection of tripwires,
(3) mine detection including gradual cutting of vegetation upto ground level and mine marking,
(4) mine prodding including exact mine localisation,
(5) destruction.

For stages (2) to (5), a team of two persons proceeds in corridors of 1 m large in steps of 0.5 m. One platoon of 30 trained people is able to demine approximately 1000 m^2 per day. The limited research funding has not allowed to develop adequate equipments, and hence the detector and the mechanical instrument technology date from world war 2.

1.3 THE DEMINING COST

- In 1993, the international community has spent about 70 million dollars for removing 100.000 mines. During the same period, 2.5 million new mines were disseminated. Clearing mines by hand is extremely expensive: upto $1.000 per unit, while the average cost of a mine is below $ 10 [4].

1.4 THE TECHNICAL EVOLUTION

- Mine technology is developing fast - much faster than demining techniques.
While mines were originally detectable because of their metal content, new technologies and materials have made it possible to produce practically undetectable varieties that may last for decades. The evolution in activation mechanisms (e.g. the incorporation of programmable devices), makes the behaviour of certain mines extremely vicious and unforeseeable (mines = "smart sensory devices").

1.5 THE POLITICAL ACTIVITY

- It is internationally acknowledged that - in the context of the growing public awareness of the mine problem - Belgium plays a pioneering role in the elaboration of treaty's against mine production, trade and usage. Belgium is also well known for its experience in demining procedures (e.g. the actions in the Gulf war, during the Yugoslavian conflict, in Cambodia and Somalia). A concrete action on the humanitarian, industrial and research level could further support this firm position.

- With the current document, we want to make decision makers aware of the fact that in our opinion, existing technologies (present in industry an research labs), could be combined to speed-up civilian demining procedures. Provided that the political authorities give the right incentives for an applied scientific research and development action, the efficiency of humanitarian aid programmes can be increased and industrial spin-offs can be generated. Currently, many research funding channels are closed for this kind of activity, due to various reasons. In our opinion the only way to deal with the problem of demining is convince governments to act simultaneously at various levels: humanitarian aid programs, industrial incentives, applied (goal oriented) research funding.

2) Humanitarian and Environmental aspects

- In countries, like Cambodia, Afghanistan, Kurdistan, Angola, El Salvador, large parts of the fertile land remain inaccessible until mines are removed, causing famine. In Afghanistan, for example, the UN's estimate that 15 years will be needed to clear the priority zones.

- The physical and psychological damage caused to children prevents them from finding a productive place in (agrarian) society and hence also ruins the future for several tens of years.

- Since mines are a major problem in poor countries, recovering from a conflict, the cost of the medical treatment of mine injuries is proportionally very high (if the monthly average income is $15, a

$125 prosthesis can easily be too expensive). The training of local medical and paramedical personnel (e.g. prosthetic technicians) is not a real solution, since many of the newly formed specialists "directly emigrate in search of higher paying jobs" [1]. In some parts of the world, the number of mine victims admitted to the hospitals is so high compared to the total number of patients suffering from other diseases, that obviously solving the mine problem will have a higher impact on the health improvement of the population than any other health care measure.

- In our opinion, facing a limited budget for humanitarian aid, the most efficient investment is tackling the mine problem at its roots, i.e.

(1) interdiction of fabrication,

(2) mine detection and neutralisation,

rather than funding measures for reducing the consequences. This thesis is supported by the numbers given to us by Comandant Lambrecht of the Belgian army - DOVO, "Dienst voor opruiming en vernietiging van oorlogstuigen" (see appendix 1).

- In our opinion, demining is not only a humanitarian problem but also an environmental pollution problem. In times, where many nations are taking actions to clean up the environment, no meaningful resources are allocated to remove the deadly pollution represented by tens of millions of mines.

3) Research aspects

- From [3], it is clear that some research topics are essential, some of them are highly desirable and others are wishful thinking. The essential objectives include "The ability to detect non-metallic and metallic mines with a single equipment in various environments". Since individual sensors are already exploited at the limit of their performance, this objective can be achieved either by inventing new sensor principles or building a multi-sensor probe [6]. The latter solution is thought to give meaningful results in short time by combining the industrial knowledge on the sensors with the university know how on signal processing and data fusion. In our opinion, improving the mine detection (reducing the number of false detections, enhancing the sensitivity especially for non metallic mines, enhancing the independence on the environment; speeding up the mine field delimitation) is the most urgent problem which can be solved by combining existing technologies.

- The phasing of the research in time should be guided by the goal to achieve improvements gradually:

(i) enhancement of the detection (multisensor probe),

(ii) speeding up the minefield perimeter determination (remote operator control),

(iii) mapping of minefields and mines (robotic platform).

4) Industrial aspects

- From discussions at the Brussels demining workshop (see appendix 2) it was clear that many industrial partners are aware of the huge potential market for demining equipment. The reluctant attitude to go into that market is mainly caused by the fear that organisations carrying out demining campaigns are mostly NGO's and the countries which are affected by mine problems are often very poor. It was also clear that several companies have various isolated pieces of enabling technology for demining.

- Existing companies working in the demining field either use expensive high technology equipment (mine breaching equipment) or sell single sensor equipments, e.g. the metal detector, which is the mostly used instrument but inevitably leads to a slow demining process.

5) Proposed actions

- Taking into account the size of the mine problem (§1.1), its medical, social and environmental consequences (§2), we think it is mandatory to launch a combined research-industrial-humanitarian action, in parallel with the already existing humanitarian demining efforts.

- We think it is more efficient to tackle the mine problem at its roots (i.e. neutralising the mines which are already present), rather than to invest in solving their effects. Since the existing demining techniques for civilian applications are too slow (§1.2), too costly (§1.3) and since their clearance efficiency is insufficient, new technology should be developed.

- We are trying to set-up a network involving research institutes, universities, military departments, industry, and in the near future NGO's. The primary goal of the network is to establish what equipment capabilities are needed (user specifications) and which performances can be obtained by integration of background technology existing in industry and universities. The network will foster exchange of information among EC experts from diverse backgrounds.

- We think that a political decision towards the support of solving the mine problem could incite industries to go into an obviously large potential market (100.000.000 mines; current cost/mine 1000 $), which is insufficiently penetrated nowadays because of its uncertainty (§ 4).

- We also think that a research project involving all the actors in demining could lead to useful instruments, improving the efficiency.

- Currently, we found a national funding channel thanks the ministry of National Defence (national) but no any international (European, UNO), which can support this kind of research and development activity. We therefore ask that a specific European channel for this kind of project would be created.

6) Consulted References

[1] Landmines: Time for action (March 1995 - Comité International Croix Rouge - ICRC)

[2] United Nations department of Humanitarian Affairs Land Mine Database, (5-7-1995)

[3] Survey of Mine Clearance Technology - The United Nations University and The United nations Department of Humanitarian Affairs - J.A. Craib

[4] Report on the United Nation International Meeting on Mine Clearance (J.D. Nicoud Geneva, 5-7-1995)

[5] De Standaard, 9-10-1995

[6] Proceedings Workshop on Anti-personnel Mine Detection and Removal, Lausanne, June 1995

Appendix 1

Brussels demining workshop: summary report of the discussions

Summary of the results of the Brussels Demining Workshop (Sept. 4-5-1995) at RMA

1. Participants

Mr. Jan Vermeiren	OIP
Mr. Axel Wagner	DLR Munchen
Mr. Jean Yves Catrus	THOMSON CSF / RII
Prof. Josef Bigun	EPFL
Prof. J.D.Nicoud	EPFL
Dr. Hichem Sahli	BIIP
Prof. Jan Cornelis	BIIP
Prof. Marc Acheroy	BIIP
Mr. Jean-Philippe Quin	DASSAULT electronique
Mr. H. Eggers	Daimler-Benz
Mr. P. Druyts	BIIP
Prof. Y. Baudoin	BIIP
Mr. U. Strunz	APS
Dr. F. Charpiller	CRIN/INRIA
Mr. E. Rose	MATRA
Prof. A. Preumont	BIIP
Mr. P. Alexandre	BIIP
Comandant Lambrecht - DOVO	Belgian army

* BIIP = {VUB ETRO-IRIS, RMA-SIC, ULB-SLN}

2. Summary of the results of the workshop

It was decided to support three complementary project proposals :

3.1 project (I)

Sensor survey and improvement of mine detection by multiple sensor data fusion : EC-programme SMT "health and safety" chapter (exposure and protection of workers to physical, chemical and biological agents).

3.1.1 Potential Partners:

Daimler-Benz, Dassault , Thomson DOI, DLR, OIP
EPFL, BIIP, Ministry of National Defence (Belgium)

3.1.2. Goals:

- a reliable, light weight and robust sensor system might be the most urgent need
- using multiple technologies for mine detection and answering the question which ones should be used for performant detection (the next generation of anti personnel mines will be completely undetectable by metal detectors)
- enhancing mine detection by multimodal information processing (including dynamic characteristics of the scanning)
- contribution to computer assisted remote mine detection
- reduction of the amount of false alarms in mine detection
- reduction of the environmental influence on mine detection capabilities

- definition of efficiency criteria in collaboration with demining specialists
- elaboration of a database for mine types including the signatures needed for (multiple) sensor mine-detection

3.1.3. Non limitative list of sensors supported by one of the partners in the project: (to be corrected and completed by the partners)

metal detector (active, ground based) BIIP
tactile sensors (active, ground based)
ground probing radar (active, ground based)
stimulated IR (active, ground based) OIP
X-Ray sensors (active, ground based) OIP
SAR (active, airborne) DLR
radiometric sensor 3...10mm (passive, ground based & airborne) DLR
odour (gas) detector (passive, ground based)
cameras (colour video & IR) (passive, ground based & airborne) BIIP
......

3.2. Project (II)

Tele-robotics solutions for mine field delimitation and marking of mine positions using existing hardware platforms - EC programme REACTIVE LTR.

3.2.1. Partners:

MATRA, DLR
EPFL, BIIP, CRIN

3.2.2 Prime contractor

BIIP

3.2.3. Industrial Steering Committee

OIP (Mr. Jan Vermeiren)
DLR Munchen (Mr. Axel Wagner)
THOMSON CSF / DOI (Mr. Jean Yves Catrus)
DASSAULT (Mr. J.P Quin)
Daimler-Benz (Mr. H. Eggers)
MATRA (Mr. E. Rose)
....

3.2.4. Goals

- Definition of realistic scenario's for mine field delimitation
- Definition of realistic scenario's for mine detection/ marking

3.2.5. Tasks

task 1: requirements of the application (scenario's, time limitations, ..)

task 2: definition of a minimal platform for mine detection based on existing sensors, existing robotics platform and existing remote operator environment.

task 3: technical modules :

(path planning, interactive map building, eigen motion determination of robotics platform, communication with the platform, user-interface, co-operation between multiple robots, demonstrator building, mine recognition/detection, very local navigation for obstacle avoidance)

3.3 Project (III)

Development of a (or more) mobile robot platform(s) carrying appropriate sensors for de-mining in real world situations.

Prototypes of BIIIP will be adapted

Potential partners:

BIIIP
EPFL (Switzerland)
FGAN (Germany)

NEUTRON ACTIVATION AND ANALYSIS

P. Bach[(1)], P. Le Tourneur [(1)]; B. Poumarède [(1)]; M. Brette [(2)]

[(1)] SODERN, France
[(2)] DGA/DSTI/STSAT, France

INTRODUCTION

Detection of abandoned land mines is a difficult problem. The only true mine detection systems in the inventory are based on visible and infra-red observation, on metal detection , on microwave detection, and on nuclear techniques. Each of these systems is able to detect a mine from one of its characteristics, i.e. presence of antennas or trigger wires, or presence of a metallic detonator, or anomalies in microwaves backscattering properties of a soil with a mine. The interest of nuclear techniques is related to their capability to detect in depth, and their capability to recognize sensitive elements of the explosives in the mines, such as nitrogen. Nevertheless, only few nuclear methods are of interest as soon as the false alarm rate is expected to be low : it is required to detect all the explosive constituents of the mines, to analyse roughly their relative concentrations, and to compare them with reference data from a data base. Taking these requirements into account, in addition with more operational considerations, we have selected and tested the more efficient way to detect the sensitive elements of the explosives hidden in the soil. Fast and thermal neutron interrogation analysis (FNA + TNA) has been qualified in our laboratories as a powerful close distance method for identifying the presence of an explosive, from all of its sensitive elements : carbon, nitrogen, oxygen, hydrogen, ... A ground mobile mine detection system would be designed, with a multisensor detection equipment including cameras, microwave sensors, metallic sensors, and FNA/TNA sensor.

SURVEY OF NUCLEAR METHODS

Nuclear methods are based on an excitation of elements (from the soil, and from the objects in the soil including explosives) by gamma rays or neutrons, and on the detection of emitted gamma rays or neutrons. Excitation by X rays is not of interest because of the too low penetration of interrogating X rays in the soil. Excitation by 13.5 MeV gamma rays makes possible the photoactivation of nitrogen of a mine. R.B. Miller (1) and his team have been working on a system using this principle, but the capability of the method in terms of false alarm rate and in terms of maximum depth is limited by two phenomena : explosive is detected from nitrogen only, and emerging gamma rays comes from positron annihilation at 511 keV, a low energy line which is not specific to nitrogen. Excitation by neutrons is more attractive, because all elements are excited together : using the appropriate detection means, and after a careful optimization of the signal/noise ratio, it is possible to detect carbon, hydrogen, oxygen, nitrogen (C, H, O, N) and other elements. Neutron detection from elastic scatter gives information on hydrogen contents, coming from water, snow, or explosives, and on the other elements using special neutron source and detectors, as described by H. Gomberg et al. (2). Spectrometry of gamma rays detection from different physical interactions shows all of the useful elements including silicon and calcium from rocks and sand.

Table 1 show the capability of the different neutron activation analysis methods as applied from mines detection, based on spectrometry of gamma rays between 1 and 12 MeV. For three different neutron energy categories, then for three kinds of neutron sources, prompt modes and delayed modes for detection are described. Delayed modes are often non interesting because the time for detection of elements is too long, but they are used to evaluate a background. This table shows that, when using only thermal neutron analysis (TNA), detection of oxygen from soil and from mines is impossible, and detection of carbon from wood pieces or from mines is hardly or not detected : see column 3 and 4. When using fast neutron analysis (FNA), all elements but hydrogen are detectable, and when waiting a few hundred microseconds for neutron slowing down, then hydrogen becomes detectable by TNA. In addition, neutron collimation is effective when using fast neutron, while thermal neutrons directions are randomly oriented.

Most significative works performed in explosives detection using neutron activation analysis are those of T.Gozani et al. (3, 4) in the TNA technology, those of D.R. Brown (5) and T. Gozani et al. (6) in the nanosecond pulsed FNA, and those of C. Rhodes (7) in the associated particle imaging with FNA. The pulsed fast neutron analysis (5, 6) needs a large size accelerator with a deuterium gas target as a neutron source, which is not well adapted from the operational aspects. The associated particle imaging technology (7) requires a special deuterium/tritium sealed neutron tube, which is not yet commercially available with the suitable lifetime and in a small size.

In order to take advantage of FNA, a study was made here using 14 MeV neutrons emitted from SODERN small size sealed neutron tubes. Neutrons are emitted all at 14 MeV, and in any directions with the same intensity (see P. Bach (8) for more details). As a consequence, all the common nuclear reactions of interest are excited. These neutron tubes may be used in pulsed mode, and of course may be turned off after use (this is not possible with radioisotopic sources). As a consequence, detection systems may be designed with following characteristics :

- no neutron emission when not in use
- pulsed emission allows to measure separately :
  - FNA gamma rays during the pulse, with a high efficiency
  - TNA gamma rays after the pulse
  - activation gamma rays (FNAA + TNAA) after a pulse train.
  - average emission of 10^8 n/s from a small size neutron emission module.

THE FNA/TNA MINES DETECTION SYSTEM

Principle of operation is based on the measurement of gamma rays induced by neutron interaction with irradiated nuclei from the soil and from a possible explosive of a mine or of an unexploded shell.

A typical composition of a soil and a mine is given here in terms of density (unit being 10^{22} atoms/cm^3).

Element	Soil density	Explosive density
Carbon	0.12	2.95
Hydrogen	1.5	6.4
Oxygen	6.9	2.56
Nitrogen	- (note 1)	1.28
Silicon	1.9	- (note 2)
Calcium	0.038	-
Aluminium	0.056	-

(1) possible fertiliser or chemical
(2) possible silicon from glass or steels

In order to characterise the explosive in a short time, it is necessary to detect the contributions of each element in a relatively small field of view, in order to evaluate roughly the density of the scene under examination for each element. The gamma emission from each of these elements is in the range of 1 to 12 MeV, either in FNA or in TNA modes. As a consequence, the maximum measurement depth is sufficient, while limited by the mean free path of neutron and gamma-rays, and by the noises coming from equipment and from the soil around the mine. A good screening of signals by a careful collimation and by time selection allows to detect the elements of interest in a relatively simple way.

The FNA/TNA detection system is composed by the following sub-assemblies :

- an irradiation/detection head (see fig. 1) including the neutron emission module with its small tube, the set of detectors, the shielding between tube and detectors, the collimators for neutrons and for gamma rays, and the minimum shielding for operators safety when not used in a tele-operation mode. This head is placed in front of the vehicle.
- an electronics including neutron tube power supplies, measurement electronics for detectors, and computers. These elements are located inside the vehicle in a local operation mode and connected to a transmission equipment in a tele-operated mode.
- the cables between these elements.

Experimentation of a mock-up for such a system was made, using a neutron tube emitting 10^8 n/s of 14 MeV and a BGO detector. This tube was used in continuous mode and in pulsed mode. Pulsed mode is more interesting : during neutron pulses (FNA), inelastic scattering of neutrons induces characteristic gamma rays of carbon, oxygen, nitrogen, silicon, aluminium, ... and after neutron pulses (TNA), neutron capture induces characteristic gamma rays of hydrogen, nitrogen, calcium, ... By splitting detection time windows in two classes, gamma ray pulses detected during FNA periods were directed to a multichannel analyser section A (FNA window), and those detected during TNA periods were directed to the multichannel analyser section B (TNA window). A careful comparison of obtained spectra in each window, with respect to preceeding spectra in memory and to stored spectra from a data base, allows to detect the N/O, C/O, H/O ratios and to compare them with stored values. An example of inelastic scattering spectra (FNA window) is shown on figure 2. Neutron pulses duration was 3 µs, repetition rate being 1 kHz. Contributions of oxygen at 6.13 MeV, carbon at 4.14 MeV and silicon from the soil at 1.78 MeV are easy to detect after a small computation time. The observed maximum depth is 20 to 30 cm, depending on hardware materials and on allocated time for detection.

PROSPECTIVE SYSTEM

The equipment results of a trade off between the level of emitted neutrons (more neutrons, more gamma rays), the maximum speed of the electronic gamma ray spectrometer (more neutrons, more pile up in electronics), the number and the sensitivity of detectors for 1 to 12 MeV gamma rays (the highest the best), the low sensitivity of materials for scattered neutrons (low neutron activation required), the shieldings and the radioprotection (more neutrons, more shielding and more weight). One of the key issue is to deal with a reasonable neutron output and to get the hardware lighter, by the development of an adequate measurement technology and a highly capable software.

The radioprotection is made with high density material, the thickness of which depending on the required maximum dose at a given distance from the source. For example, using a 10^8 n/s tube such as the small size and low cost SODITRON tube, and using a 20 cm shield thickness, the dose at 80 cm from the skin would be about 25 µsv/h and at 4 m about 2.5 µsv/h (public level), and this only when the tube in on. When not in operation, the equipment is easily transportable, while the residual activation of the head is very low (due to an irradiation limited to the useful time for detection only).

The weight of the detection head depends on the specifications. The main contribution is due to shieldings and radioprotection. For a tele-operated vehicle, this weight is in the 300 - 400 kg range. For a manned vehicle, this weight may be twice as much.

In case of an accidental explosion, the tube may be broken, but the tritium gas inside the tube is in a very low quantity (3.3 Ci). The consequences on environment are negligible, compared with the dispersion of a radioisotopic neutron source such as 50 µg of ^{252}Cf, emitting continuously 10^8 n/s.

It is now possible to design different kind of systems, taking into account the followings :

- the FNA/TNA system is a close distance detector, for confirmation of an alarm,
- the system may be implemented on a vehicle using a single detection head with a more or less large number of detectors, or a multiple detection head, but the weight of the head increases with its size,
- the time for detection is shorter if more neutrons are available : this iseasy to achieve using electrical sources, while it is impossible using ^{252}Cf or other isotopic source,
- the FNA/TNA system is safe, even in case of an accidental destruction of the tube,
- all the basic components of such a system are available.

We are now considering optimisation of data handling and computation algorithms. A fieldable demonstrator could be used in a multisensor concept and installed on a vehicle. The alarm would be given by observation in visible and infrared range, and by metal detectors and ground penetrating radars. The confirmation of the alarm would be given by the FNA/TNA system.

ACKNOWLEDGEMENT

This work was performed with the support of DGA/DSTI/STSAT (France).

REFERENCES

1. Miller R.B., Clifford J.R., Mc Cullough W.F., Habiger K., 1987
2. Gomberg H.J., Charatic G., Brundage J. ; 1982, Contraband and cargo inspection technology international symposium, Washington, USA
3. Gozani T., Seher C., Morgado R., 1989 ; Proc. 3rd Symp. on analysis and detection of explosives, Mannheim, Germany
4. Gozani T. ; 1993 ; NIM-B79.p. 601
5. Brown D.R., Loveman R., Bandahan J., Schulze M; 1992 ;Contraband and cargo inspection technology international symposium, Washington, USA
6. Gozani T., Brown D.R., Verbinski V.V. ; 1996 ; Port technology international, p. 189-194
7. Rhodes E. Dickerman C.E., De Volpi A. ; 1992; Neutron radiography (4), J.P. Barton editor, Gordon and Breach Science Publishers, p. 827-835
8. Bach P., 1992, 3rd symp. on applications of nuclear techniques, Mykonos, Greece (Gerald Printing Service, Kentucky, USA) p. 397-412

TABLE 1 -Some NAA measurement methods for explosives detection

Neutron energy	**14 MeV**	**1 to 8 MeV**	**<1keV and thermal neutrons**
Neutron sources examples	D-T tubes	D-D tubes (2,5 MeV) D-D accelerators (2,5 to 8 MeV) 252Cf sources (2,4 MeV) Am/Be sources	All kind of neutron sources + moderator
Detection modes: **1- Prompt modes** --->physical interaction	**FNA (Fast Neutron Analysis):** Elastic or inelastic scattering <13 MeV Nuclear reaction with threshold < 13 MeV Neutron capture after slowing down	**FNA (Fast Neutron Analysis):** Elastic or inelastic scattering <some MeV Nuclear reaction with threshold < some MeV Neutron capture after slowing down	**TNA (Thermal Neutron Analysis):** Thermal neutron capture
2- Delayed modes --->physical interaction	**FNAA (Fast Neutron Activation Analysis)** High, medium, and low energy activation	**FNAA (Fast Neutron Activation Analysis)** Medium and low energy activation	**TNAA (Thermal Neutron Activation Analysis)** Low energy activation
Detected explosives elements C, H, O, N, +.....	**C, H, O, N, Cl, P,Al ...**	**H, N, Cl, P,Al...** **C: hardly or not detected** **O: not detected**	**H, N,Cl...** **C: not detected** **O: not detected**
Neutron collimation **—>directed beam capability**	**yes**	**yes**	**no**

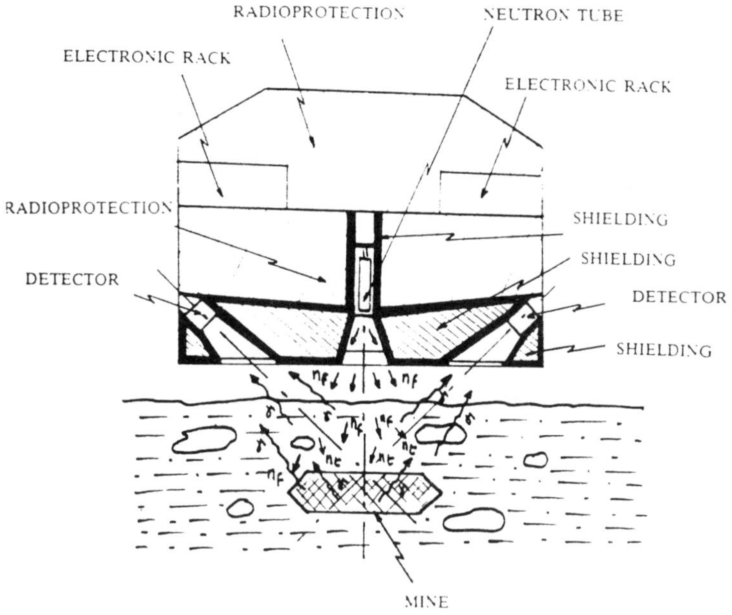

Fig.1:FNA/TNA mine detector: schematics of the irradiation/detection head

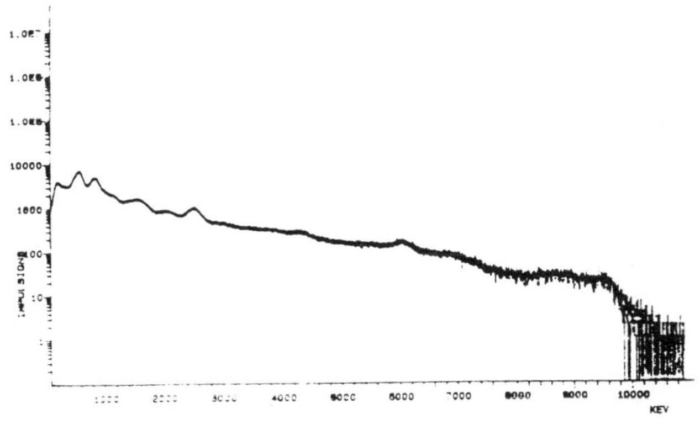

Fig.2: Example of an inelastic scattering spectra, from a mine simulant in the soil

MINE DETECTION BY NUCLEAR QUADRUPOLE RESONANCE

M D Rowe and J A S Smith

King's College London, UK

1. INTRODUCTION

Nuclear quadrupole resonance (NQR) methods of mine detection rely on the observation of radiofrequency (RF) signals from the ^{14}N nuclei present in the explosive (1). The frequencies of these signals lie between 0.5 and 6 MHz; they are characteristic of a given explosive and provide not only a positive identification but also an estimate of quantity or depth. Unlike the better known technique of nuclear magnetic resonance (NMR), no static magnetic field is needed, so that portable probes can be used. Furthermore, signals are only seen in solids or solid-like materials, and because of the highly compound-specific nature of NQR frequencies, there is little, if any interference from other nitrogen-containing materials which may be present in the mine casing or the surrounding terrain. These advantages have already been successfully exploited in the design of a practical detector of RDX-based explosives in airport luggage (2).

2. PRINCIPLES OF NQR

NQR is a technique in RF spectroscopy in which the applied radiation drives transitions between the quadrupole energy levels of an atomic nucleus. These levels arise because the charge distribution of many nuclei, of which ^{14}N is one, is not spherically symmetrical, but behaves as if it consisted of two electric dipoles back-to-back; in the electric field gradient generated by surrounding charges (figure 1), one orientation viz. (a),

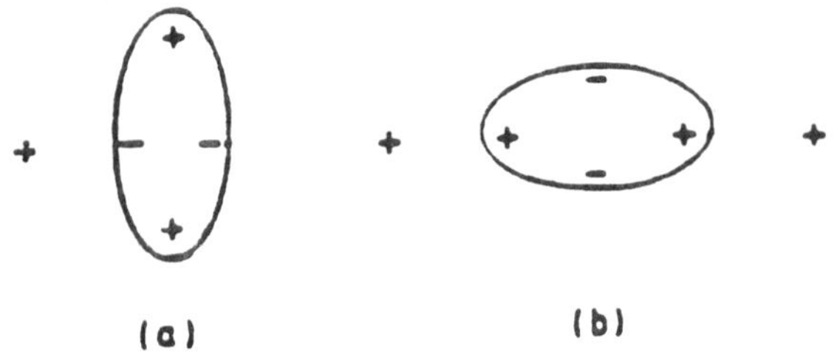

Figure 1 Electric quadrupole in an electric field gradient.

has a lower energy than the other, (b). Nuclei of the latter kind will therefore tend to move towards the orientation of lower potential energy. Because they also have spin, they behave as gyromagnets and respond by precessing about the axis of the electric field gradient at the quadrupole precession frequency, whose magnitude depends on the product of the nuclear electric quadrupole moment, Q, and the electric field gradient, q. Now the latter is a tensor quantity, and is usually discussed in terms of its three principal components, q_{xx}, q_{yy}, q_{zz} along three Cartesian axes. Since their sum is equal to zero, it is customary to define two independent quantities, q_{zz} or q, the maximum principal component (in magnitude), and the asymmetry parameter η, which is defined as the difference between the other two components divided by q. The product (e^2qQ/h) (e is the charge on the electron, h is Planck's constant) is known as the quadrupole coupling constant and has the units of frequency.

The NQR frequencies observed experimentally depend on both the quadrupole coupling constant and the asymmetry parameter. The appropriate energy level diagram and allowed transitions for a spin-1 nucleus such as ^{14}N are shown in figure 2; there are, in the general case, three levels and three allowed transitions,

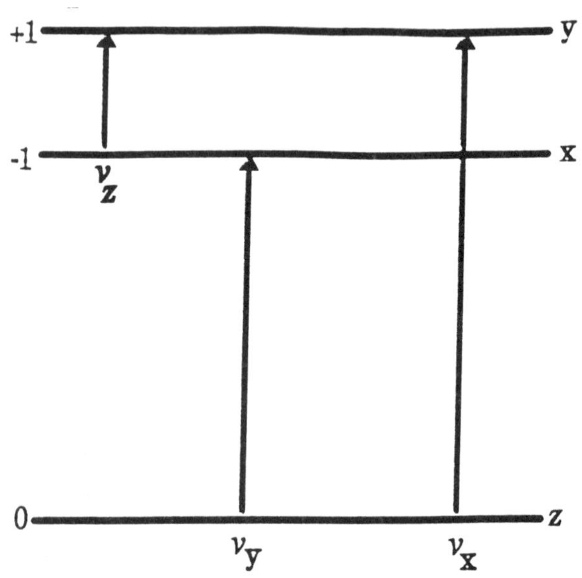

Figure 2 Quadrupole levels and allowed transitions for ^{14}N.

one frequency (ν_x), being the sum of the other two ($\nu_y + \nu_z$). These frequencies are related to the ^{14}N quadrupole coupling constant and asymmetry parameter by the equation

$$\nu_x = 3/4\ \left(\frac{e^2qQ}{h}\right)\ (1 + \eta/3)$$

$$\nu_y = 3/4\ \left(\frac{e^2qQ}{h}\right)\ (1 - \eta/3)$$

$$\nu_z = 1/2\ \left(\frac{e^2qQ}{h}\right)\ \eta \qquad (1)$$

Note that no stationary magnetic field is required and signals are only observed in solids. They have been detected in all three principal high explosives, RDX (I),

PETN (II) and TNT (III) but at considerably different

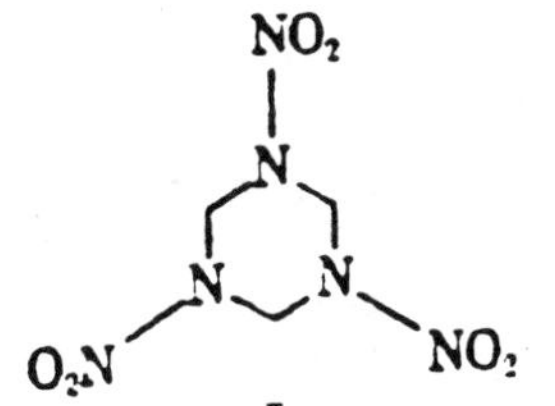

I

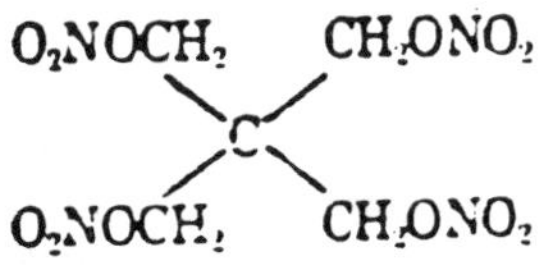

II

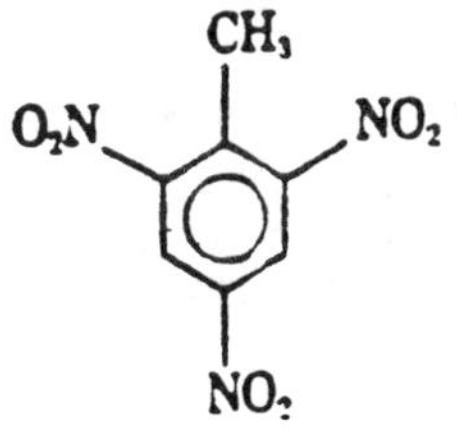

III

frequencies, because of the chemically different nature of the nitrogen atoms in these three compounds; for example, in RDX the ring nitrogen nuclei generate three sets of three frequencies near 5.2 (ν_x), 3.4 (ν_y) and 1.8 (ν_z) MHz (3), whereas in TNT (III), the nitro group frequencies are found between 860 and 900 (ν_x) and 730 and 800 (ν_y) kHz in both monoclinic and orthorhombic forms (4).

3. EXPERIMENTAL METHODS

Most NQR experiments at the present time are conducted by means of pulsed RF techniques, in which the ^{14}N nuclei are subject to bursts of RF radiation at or near to their NQR frequency, and the resulting signals monitored in the quiescent periods between pulses. These signals may be divided into two types: free induction decays (FID) and echoes. The first is the decaying signal observed immediately following the pulse, as shown in figure 3 for RDX; it is generated by

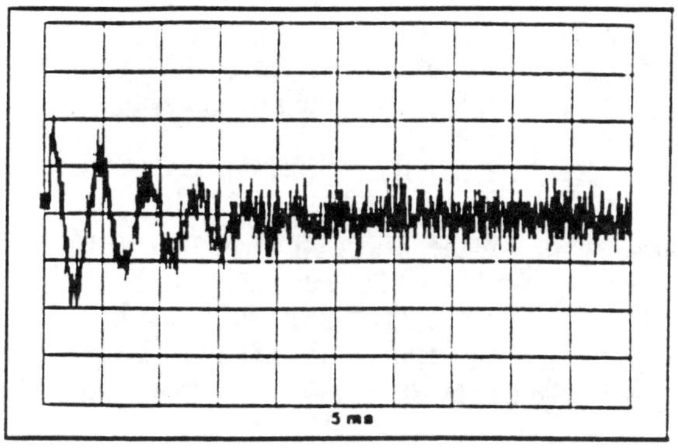

Figure 3 ^{14}N FID in RDX at 5.190 MHz

the interactions of the oscillating RF field of amplitude B_1 with the magnetic moment of the ^{14}N nucleus. In conventional spectrometers, this B_1 field is provided by surrounding the sample with a small induction coil, of (say) 0.1 dm^3 in volume. The pulse may be imagined to tip the nuclear quadrupole moments away from their equilibrium orientation in the electric field gradient of their surroundings by a so-called 'flip angle' α; they then precess at one of their allowed frequencies, inducing an RF voltage in the same RF coil as was used to excite them. This voltage is observed to decay with a time constant T_2^* the FID or spin-phase memory decay time, which in RDX is about 1.4 ms for the 5.190 MHz transition, as figure 3 shows. The initial intensity of the FID depends on the magnitude of the flip angle α produced by the pulse and this in turn depends on B_1 and the time t_w for which the pulse is applied, the relationship for ^{14}N being

$$\alpha = 2\gamma B_1 t_w \qquad (2)$$

in which γ is the gyromagnetic ratio of the ^{14}N nucleus, equal to 1.932×10^7 rad s^{-1} T^{-1}. In a polycrystalline sample, however, the electric field gradient at a given nitrogen site can assume all possible orientations with respect to the B_1 field and as a consequence, the maximum FID intensity is observed not at a value of α equal to 90°, as in pulsed NMR, but at 119° (0.661π). At larger flip angles, the FID intensity declines and is zero at 257° (1.43π). In the following text, it will be convenient to refer to a 119° pulse as "$90°_{eff}$", defined as that pulse length which generates the maximum FID intensity. The time taken for the spin system to return to equilibrium after the conclusion of the pulse is governed by the spin-lattice relaxation time T_1, about 10 ms in RDX at room temperature. RDX is somewhat of an exception in having single-exponential relaxation behaviour at ambient temperatures; typically, for the three level system of a spin-1 nucleus (figure 2), bi-exponential decays are expected. One important experimental consequence of the value of T_1 is that following the pulse, a time τ of about $5T_1$ is needed to produce a fully relaxed system. Note that in most nitrogen compounds, both frequencies and relaxation times are temperature dependent; in RDX, for example, the transition frequency of 5.190 MHz has a temperature coefficient of -0.52 kHz K^{-1}.

The modulation observed on the FID in figure 3 is due to the use of a phase-sensitive detector in the receiver system and represents the difference between the applied RF of 5.191 MHz and the actual resonance frequency of 5.190 MHz at 295K. Fourier transformation of this time domain spectrum produces an absorption signal at 5.190 MHz with frequency width of about 200 Hz, due almost entirely to inhomogeneous broadening, generated, for example, by defects or impurities in the material which disturb the electric field gradient in neighbouring molecules. The 'true' line width is close to 40 Hz, and this can be demonstrated by the existence of echoes, which are produced when two or more RF pulses are used. For example, the pulse sequence consisting of two $90°_{eff}$ pulses, time τ apart, the phase of the first being shifted by 90° with respect to the second, generates an echo at a time τ after the second, provided that $T_2^* < \tau < T_1$. Figure 4 shows an example for RDX at

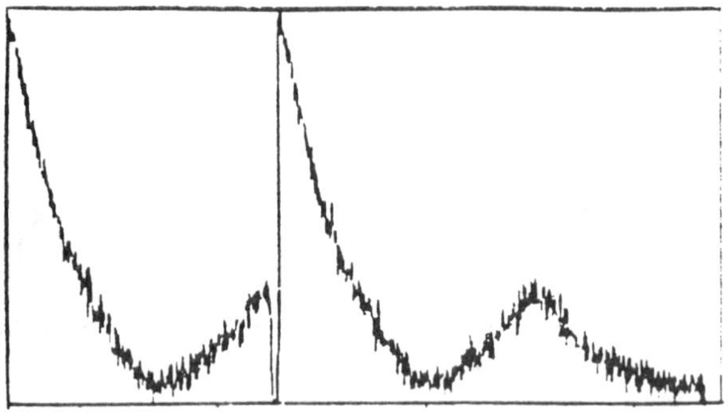

Figure 4 Echo signal in RDX (τ = 7.5 ms).

5.190 MHz, for which the pulse separation τ was 7.5 ms. As τ is lengthened, the echo intensity decays with a time constant T_2 known as the spin-spin relaxation time; in RDX, this is about 8 ms, corresponding to a "true" line width of 40 Hz. The effect of the second pulse may be imagined to be a refocussing of the dephased signals produced after the first FID, so removing the effects of the inhomogeneity broadening. The same argument would predict that the refocussing effect could be repeated by yet another pulse provided always that $\tau < T_1$. In fact, long echo trains can often be generated by multiple pulse sequences; one of the earliest to be discovered (5) was SORC (strong off-resonance comb) in which the RF is slightly off-resonance. One typical sequence can be represented by

$$\alpha_{0°} \text{-}(\text{-}\ \tau\ \text{-}\ \alpha_{90°}\ \text{-}\ \tau\ \text{-})\text{-}_n \qquad (3)$$

in which the subscripts '0°' and '90°' represent the RF phase and $\tau < T_2^*$; the echoes are sampled in the intervals 2τ between the second and further pulses. This sequence is a specific example of what is generally referred to as pulsed spin locking (PSL or SLS); the flip angle α is usually taken to be $90°_{eff}$, but other values may be used provided the first is kept at $90°_{eff}$ (6). PSL sequences are appropriate for compounds in which $T_1 >> T_2$.

Another group of rapid acquisition pulse sequences, referred to generally as steady-state free precession (SSFP) can also produce long echo trains, a specific example (7) being the phase-alternated multiple-pulse sequence (PAPS) shown below;

$$(\alpha_{0°}\ \text{-}\ \tau\ \text{-}\ \alpha_{180°}\ \text{-}\ \tau\ \text{-})_n \qquad (4)$$

which is appropriate for compounds for which $T_1 \sim T_2$. Other multiple pulse sequences have been described in the literature (8,9); their common effect is to lengthen considerably the time for which the NQR signal can be sustained. PSL sequences allow a definition of a new pulse echo decay time T_{2e}, where T_{2e} is usually longer than T_2; unlike T_1,T_2 however T_{2e} depends critically on the pulse parameters such as α and τ. The important advantage of multiple pulse sequences is that the many echoes which they generate can be summed in a very short time to produce an enhancement in the signal-to-noise ratio (SNR) which can be as large as $n^{1/2}$, where n is the number of equal intensity signals accumulated. Alternatively, the long echo train can be treated as a slowly decaying signal, governed by a time constant T_{2e}, Fourier transformation of which can give a much sharper signal of enhanced SNR.

The experiments to be reported in this paper were conducted on two variable frequency pulsed RF spectrometers operating between 0.5 and 40 MHz, one manufactured by SMIS of Guildford, U.K. and the other by Tecmag of Houston, U.S.A. The RF power amplifiers used were either a Kalmus of 1 kW or a Rodmell "Explorer" NMR of 3 kW peak power. RF probes were designed and built in the laboratory; for pulse operation, the receiver was protected from the high voltage pulses generated by the RF power amplifier by means of a duplexer placed between the antenna and receiver, and blocking diode pairs were used following the amplifier to act as a low impedance path during the pulse, and a high impedance isolator at other times (10). Following the pre-amplifier (Miteq, AU 1054-1103) and receiver, the signals passed to a phase-sensitive detector with in-phase and quadrature outputs. It is usual to accumulate a number of FIDS or echoes and Fourier transform the resulting time domain signal to obtain an absorption spectrum in the frequency domain. The success of the detection process is assessed by the SNR of the resulting signal, defined here as the signal height divided by the r.m.s. noise voltage.

4. NQR IN MINE DETECTION

Earlier work on the detection of NQR signals from remote objects (11, 12) has established some of the RF techniques which can be used; for example, various kinds of RF probe or antenna have been investigated, including planar spirals wound with copper ribbon (11) and receiver coils consisting of wire-wrapped ferrite rods (11, 12). Fast data capture and processing procedures have also been devised (13). Nevertheless, a number of problems still remain when NQR methods are applied in real situations; the remainder of this article will discuss three of these, together with possible solutions, for the particular case of the explosive RDX (I).

Firstly, the SNR in NQR experiments depends, amongst other factors, on $\nu_Q^{3/2}$, ν_Q being the NQR frequency; ^{14}N signals occur at relatively low radiofrequencies, some less than 1 MHz, leading to a poor SNR, often less than unity in a single-shot experiment. Secondly, variations in the distance of the mine from the probe produce considerable differences in both the B_1 field at the sample and the RF signal which it generates at the antenna. Thirdly, the presence of strong spurious signals or interference, generated for example by piezo-electric responses from quartz in sand or soil, may completely mask the weak NQR response. Correspondingly, it should be emphasised that the method also has its advantages; one is its high selectivity; it is most unlikely to find identical ^{14}N frequencies from other materials in the terrain, and the presence of moisture, vegetation, or even sea water will have little effect on the signal strength at these frequencies, nor will the nature of the terrain be of great importance.

Turning to the first of these problems, we note that the SNR in NQR experiments on polycrystalline samples (at room temperature) is given by the approximate expression

$$SNR = 2.12 \times 10^{-42} \, \gamma \zeta \, V_c N_o \nu_Q^{3/2} \qquad (5)$$

in which ζ is the fraction of the coil volume V_c occupied by the sample and N_o the number density of ^{14}N nuclei. The Q-factor of the coil has been taken to be 100, the preamplifier noise figure to be 1.6 and the receiver band width to be 10 kHz. In the laboratory, it would be practical to use a RF solenoid typically of volume 0.1 dm^3 and a filling factor of 0.8. For the 5.190 MHz transition of RDX, one then calculates a SNR of about 19 in a single shot experiment, which is an acceptable value in laboratory experiments. However, this value diminishes considerably when planar spiral coils are used; the B_1 field now lacks spatial homogeneity, although we may still refer to an optimum field-of-view which from analogy with nuclear magnetic resonance imaging (MRI) experiments we take to be the volume of a hemisphere of diameter equal to that of the RF antenna used. With a transmitting antenna of diameter 0.2 m (12) V_c becomes 0.0168 m^3 and the reduction in SNR by this factor alone is about 13. This loss in SNR can be countered by fast accumulation of a number of responses, either FIDS or echoes (14); figure 5 shows the maximum absorption signal obtained under

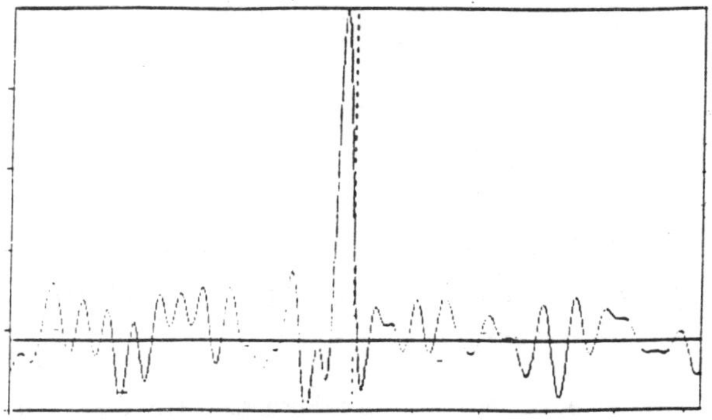

Figure 5 ^{14}N signal from an RDX sample at a distance of 14 cm from the antenna (the spectral width is 15.6 kHz).

these conditions from a sample of RDX in an amount typical of that found in many anti-personnel mines at a mean distance of 14 cm from the plane of the antenna. The SNR is restored to 12, which is more than adequate for detection of the mine.

Turning to the second group of problems associated with the spatial inhomogeneity of the B_1 field, we note that the magnetic field generated along the axis of a single turn of radius r at a distance d from the plane is given by the equation

$$B_1 \propto \frac{r^2}{(r^2 + d^2)^{3/2}} \qquad (6)$$

By the reciprocity theorem, the RF signal induced in this single turn by a sample distance d along the axis will be proportional to the same geometric factor. These equations have been used to show (12) that the distance at which a sample of volume V_s can be detected with a given minimum SNR depends roughly on $V_s^{1/3}$, so if the depth is doubled, roughly eight times the volume is needed to produce the same signal. This trend is observable in the trial experimental results so far published. Figure 5 shows the expected signal from a typical anti personnel mine at a distance of 14 cm from the RF probe. A comparable performance has been obtained by Hirschfeld and Klainer from RDX at 3.41 MHz from a 50 g sample at a distance of 10 cm from the probe (11). Grechishkin (14) has reported the detection of signals from RDX in anti tank mines, with considerably greater quantities of the explosive, at distances of up to 35cm from the probe. To obtain the best sensitivity for the detection of signals from remote samples, it is important to generate a flip angle α as close as possible to $90°_{eff}$ at that depth, and for a given RF antenna, this will depend on the RF power used. The likely depth at which mines are to be found is therefore an important factor to be taken into account in setting the pulse operating parameters.

The third instrumental problem to be discussed is that of RF interference from the sample, in this case both the explosive and the terrain in which it is concealed. The quartz present in sand and most soils will generate strong piezo-electric responses, as will other substances sometimes present, such as urea used as a fertiliser. In the laboratory such responses can often be removed by grinding the sample or immersing it in a silicone oil, but these methods are not applicable in non-invasive applications; another method is to use an electrostatic or Faraday shield between the probe and the terrain, made for example from 120 μm thick copper foil (15). In our experiments we have tested a slotted and earthed aluminium foil screen in front of the RF antenna, which almost completely eliminates the piezo-electric response from dry soil at a distance of 13 cm from the probe, without loss in the Q-factor, but with some reduction in the B_1 field. However, we have found that an equally effective method which does not affect the RF field is to use phase cycling. The basic principle of this technique is that the phase of the spurious response is largely determined by that of the preceding pulse, whereas that of the NQR signal depends at least on that of the two preceding pulses. Various versions are known in NMR, e.g. a composite pulse sequence with phase cycling has been used to reduce the probe ringing following a pulse (16). In NQR, the PAPS sequence defined in equation (4),

$$(\alpha_{0^\circ} - \tau^{A} - \alpha_{180^\circ} - \tau^{B} -)_n \qquad (4)$$

where A and B denote the captured signal in the quiescent period between pulses, has been used to reduce acoustic ringing from the sample (17). The method is to combine (4) with the non-phase-cycled version NPAPS

$$(\alpha_{0^\circ} - \tau^{C} - \alpha_{0^\circ} - \tau^{D} -)_n \qquad (7)$$

when the signal combination (A-B-C-D) provides a considerable reduction of the spurious signal without affecting the NQR response. We have shown that the same cycle will also produce a considerable reduction in piezo-electric responses. In laboratory experiments with a solenoidal RF coil full of dry sand, figure 6 shows the spectrometer response following a sequence of single

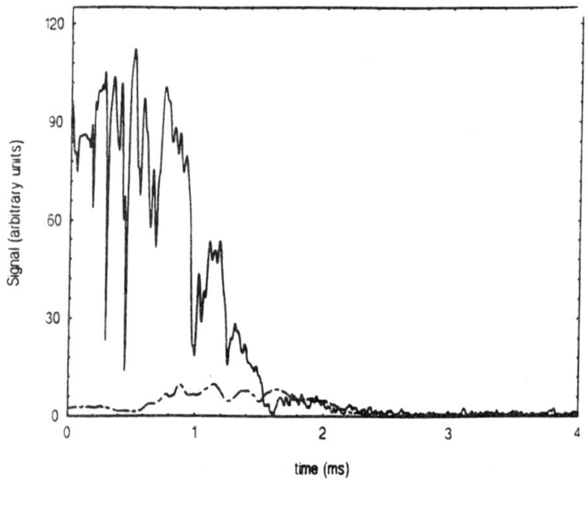

Figure 6 RF response from dry sand

pulses of the same phase (continuous line) compared with that generated by a PAPS/NPAPS phase cycled sequence with the same value of τ (dashed line). The phase cycling produces approximately 20 dB of attenuation of the spurious signal, sufficient to remove most piezo-electric responses in real situations, without any degeneration in the B_1 field or NQR signal strength. Any remaining interference could be removed by the use of a Faraday shield. Other problems remain but these preliminary experiments show the feasibility of using ^{14}N NQR to detect RDX-based anti-personnel mines in real situations.

ACKNOWLEDGEMENT

We thank the Ministry of Defence and the Defence Research Agency, Chertsey, U.K., for support of this project and Dr. Neil Peirson for figure 6.

REFERENCES

1. Smith, J.A.S., 1976, *Chem. Soc. Reviews 15*, 225-260.

2. Smith, J.A.S., 1995, *ECOS-95, IEE Conference Publication no. 408*, 288-292.

3. Landers, A.G., Brill, T.B. and Marino, R.A., 1981, *J. Phys. Chem. 85*, 2618-2623; Karpowicz, R.C., and Brill, T.B., *J. Phys. Chem. 87*, 2109-2112.

4. Marino, R.A. and Connors, R.F., 1983, *J. Mol. Struct. 111*, 323-327.

5. Kim, S.S., Jayakody, J.R.P. and Marino, R.A., 1992, *Z.f.Naturforsch. 47a*, 415-420.

6. Ainbinder, N.E., Volgina, G.A., Osipenko, A.N., Furman, G.B. and Shaposhnikov, I.G., 1983, *J. Mol. Struct. 111*, 65-70.

7. Osokin, D.Ya., Ermakov, V.L., Kurbanov, R.H. and Shagalov, V. A., 1992, *Z.f.Naturforsch. 47a*, 439-445.

8. Reddy, N. and Narasimhan, P.T., 1988, *J. Mol. Struct. 192*, 345-354.

9. Ermakov, V.L. and Osokin. D.Ya, 1984, *Mol. Phys. 53*, 1335-1353.

10. Griffin, D.D., Kleinberg, R.L. and Fukuhara, M., 1993, *Meas. Sci. Technol. 4*, 968-975.

11. Hirschfeld, T. and Klainer, S.M., 1980, *J. Mol. Struct. 58*, 63-76.

12. Grechishkin, V.S., and Sinyavskii, N.Ya, 1993, *Physics-Uspekhi 36*, 980-1003.

13. Grechishkin, V. S., 1994, *Appl. Phys. A58*, 63-65.

14. Grechishkin, V.S., 1992, *Appl. Phys. A55*, 505-507.

15. Gibson, A.A.V. and Raab, R.E., 1972, *J. Chem. Phys. 57*, 4688-4693.

16. Zhang, S., Wu, X. and Mehring, M., 1990, *Chem. Phys. Lett. 173*, 481-482.

17. Buess, M.L., Garroway, A.N. and Yesinowski, J.P., U.S.Patent No. 5,365,171 (November 15, 1994).

DETECTION OF SUB-SURFACE OBJECTS BY ELECTRICAL IMPEDANCE TOMOGRAPHY

C. J. Kotre

Regional Medical Physics Department, Newcastle General Hospital.

Abstract

A technique called INSEIT (Imaging Near the Surface by Electrical Impedance Tomography) has recently been developed for potential applications in the field of medical imaging. The technique allows changes in electrical impedance to be imaged in sub-surface planes below an array of surface electrodes. Although the adaptation of this approach to the task of detecting buried land mines involves some obvious practical difficulties with electrode contact, the method might find application in situations where large electrical impedance contrasts are present. The image reconstruction mathematics used for INSEIT and for general cross-sectional electrical impedance tomography has been shown to be suitable for industrial imaging using capacitive and inductive sensors, so alternative sensing methods based on the same image reconstruction strategy are a possibility. The viability of INSEIT in the practical situation of unknown sub-surface features and an imperfectly flat surface, and the overall performance of the approach in locating buried objects is examined using software simulations and experimental data sets.

Introduction

The objective of electrical impedance tomography (EIT) is to reconstruct images which represent the cross-sectional electrical impedance distribution within a three-dimensional object from measurements taken on its boundary surface. The main application of the technique is seen as medical diagnostic imaging. In medical EIT, 16 electrodes are commonly employed, placed on the skin surface around the cross-sectional plane to be imaged. A four-electrode measurement approach is used with a constant amplitude alternating current applied between each adjacent pair of electrodes in turn, and high-impedance voltage measurements made between all other adjacent electrode pairs. Due to the non-linearity of the reconstruction problem, images of absolute impedance are difficult to produce and dynamic images of impedance changes are normally used. This approach is followed in the present work.

The main limitation of EIT is its inherently poor spatial resolution. Although more information about the impedance distribution within an object can be retrieved by the use of larger numbers of surface electrodes, the additional information tends to relate to regions of the object close to the electrodes rather than at depth. This limitation is to some extent circumvented in an adaptation of EIT to the problem of sub-surface imaging called INSEIT (Imaging Near the Surface by Electrical Impedance Tomography) (1). The use of this geometry allows a chosen plane to be approximately uniformly sensed by a surface grid of electrodes leading to uniform resolution in the image and the possibility of exploiting the relatively high resolution information available at shallow depths. The technique has some similarities to methods used in geophysics (2) and previous attempts at the detection of used ordinance (3), but employs a different measurement and image reconstruction strategy.

Sub-surface imaging

The measurement geometry for INSEIT is shown in figure 1. A constant amplitude alternating current is injected via electrode pairs, m, and voltage measurements are made between electrode pairs, n. This measurement sequence is repeated for each linear electrode array, l, and the whole measurement cycle repeated with an orthogonal row orientation. To avoid problems with sensitivity null lines, the electrodes at the ends of each row are included as a pair both for current injection and voltage measurements (1). The images shown here were reconstructed using a sensitivity weighted filtered backprojection algorithm originally developed for EIT (4,5,6).

Simulated reconstructions of sub-surface structures

Some examples of the sort of imaging performance to be expected from INSEIT are given in figure 2. A 16 x 16 square array of surface electrodes with an equal spacing, g, was assumed in this simulation, with all electrodes available for measurement. Figure 2(a) shows a 64 x 64 pixel binary image representing a sub-surface planar network of pipes or vessels, or walls at depth on an archaeological site. This test distribution was used to produce simulated noiseless measurement sets at nominal depths of g, 2g and 4g which were then reconstructed using sensitivity weighted backprojection, followed by filtering with a filter kernel based on the calculated point spread function at the nominal depth. The reconstructed images are shown in figures 2(b), 2(c) and 2(d). The expected severe image degradation with depth is clearly demonstrated, with both the resolution and detected signal amplitude decaying rapidly.

Experimental reconstructions of buried objects

As an initial feasibility study of INSEIT applied to the detection of buried objects, the experimental task of sensing objects buried in wet sand was investigated. A test tank of dimensions 45 X 30cm was filled to a depth of 10cm with sharp sand (incorporating a range of particle sizes) which was dampened with tap water. An 8 X 8 electrode array was assembled with 1 cm diameter disk electrodes mounted on a flexible rubber sheet which was held lightly in contact with the sand. The electrode spacing was 3cm giving an overall electrode array size of 21 X 21cm. (Although dimensions related to the experiment are given in cm, it should be noted that the results obtained should hold for the case of all dimensions scaled in proportion to the electrode spacing.) Impedance measurements were made at a frequency of 50kHz and a constant amplitude current of 5mA p-p using equipment designed for medical EIT (7) in conjunction with an electrode multiplexer.

Figure 3 shows images obtained from sets of measurements on the sand with a buried object at a depth of approximately 1cm, referenced in each case to a measurement of just the sand. The object in figures 3(a) and 3(b) was a perspex rod 2cm in diameter and 19cm in length running at 45° to the image, 3(a) from top right to bottom left, and 3(b) from top left to bottom right. Figures 3(c) and 3(d) show an identically sized aluminium rod in the same orientations. Somewhat surprisingly, the aluminium rod is better imaged (giving a contrast double that of the perspex rod) indicating that the mean resistivity of the damp sand must be quite high even with respect to completely insulating objects.

Figure 4(a) shows three 1.5cm thick insulating disks, of diameter 5.5cm (top right), 4cm (centre) and 2.5 cm (bottom left) buried at a depth of approximately 1.5cm. Figure 4(b) shows the same arrangement this time for conductive objects. For this experimental run, salt was added to the water mixed with the sand and the ratio of water to sand was increased to lower the background resistivity, but the conductive objects are again imaged more strongly than the resistive ones.

Figure 5 shows resistive and conductive objects 4cm in diameter buried at depths of 5(a) 1cm, 5(b) 2cm, and 5(c) 3cm. The conductive object is imaged clearly in all cases, but the resistive one becomes masked by electrode contact artifacts as the depth is increased. The data set used in figure 5(c) was particularly contaminated with contact artifacts, causing the reconstruction program to reject a large number of measurements.

Discussion

When reviewing the results given above, a number of restrictions on the present experiment should be considered:

1) The data acquisition system was designed for EIT, not INSEIT, and is not set up to match the required dynamic range of data. Some measurements required for INSEIT are not made with this equipment, so that a number of data points have to be interpolated in each reconstruction.

2) The images obtained are effectively temporal subtractions between a measured 'blank' set of data and a data set where the various objects are present. This approach doubles the likelihood of electrode contact artifacts.

3) Measurements were obtained at a single frequency.

A prototype mine detection system based on INSEIT would not suffer from these restrictions. As well as purpose designed electronics, such a system could use a multiple frequency approach with higher current amplitudes, and a scanning linear electrode array possibly based on water jets to simultaneously wet down the area being scanned and provide a mechanically soft means of making electrical contact.

Conclusions

Despite the limitations of the apparatus used in this feasibility study, it appears that sub-surface electrical impedance imaging may be viable for the detection of buried objects such as land mines in cases where a strong impedance contrast between the object and its surroundings exists.

References

1. Kotre C J, 1996, IEE Proc.-Sci. Meas. Technol., 143, 41-46

2. Li Y and Oldenburg D W, 1992 Geophys. J. Int., 109 343-362

3. Wexler A, 1988, Clin. Phys. Physiol. Meas., 9 suppl. A 29-33

4. Kotre C J, 1989, Clin. Phys. Physiol. Meas., 10 275-281

5. Kotre C J, 1989, Clin. Phys. Physiol. Meas., 10 370-371

6. Kotre C J, 1994, Physiol. Meas., 8 A125-A136

7. Brown B H and Seagar A D, 1987, Clin. Phys. Physiol. Meas., suppl. A 91-97

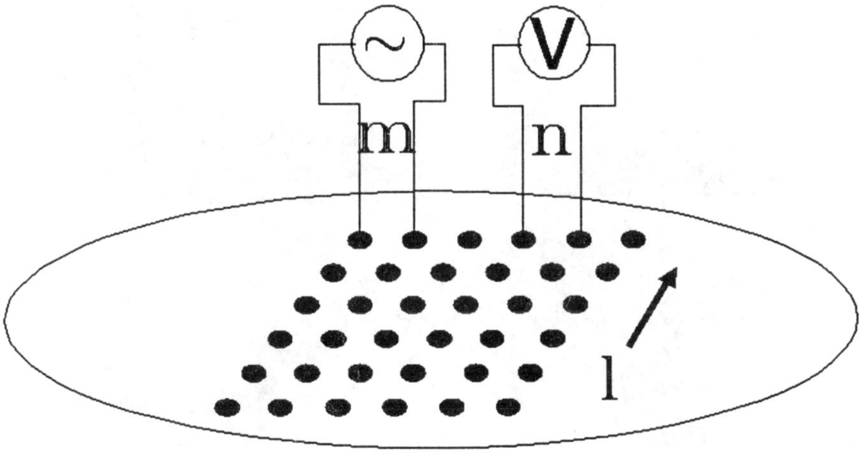

Figure 1

Measurement geometry for INSEIT. A constant amplitude alternating current is injected via electrode pairs, m, and voltage measurements are made between electrode pairs, n. This sequence is repeated for each linear electrode array, l, and the whole measurement cycle repeated with an orthogonal row orientation.

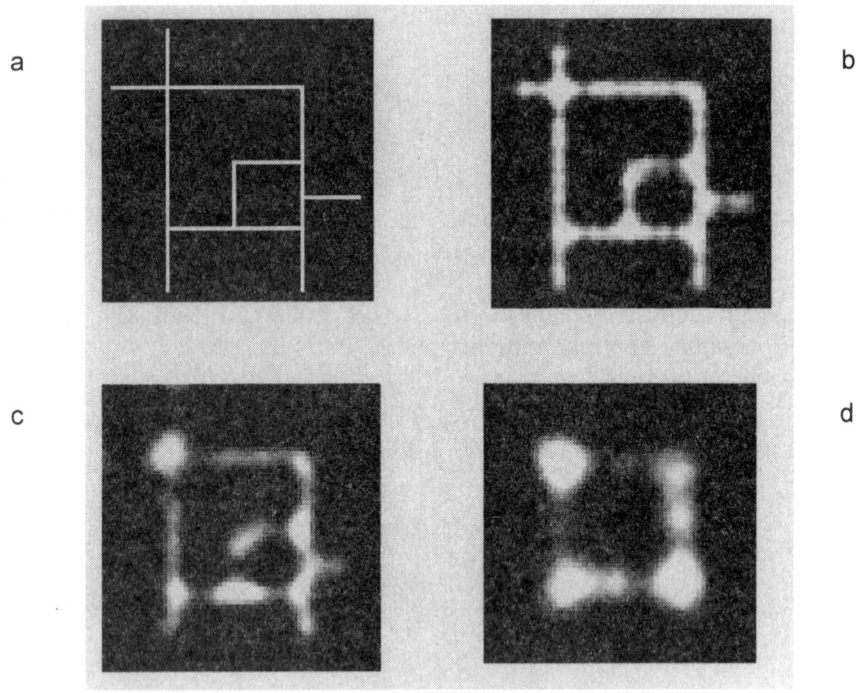

Figure 2
Software simulations of INSEIT using a 16 x 16 electrode array. The test object (a) is reconstructed at depths of g (b), 2g (c) and 4g (d), where g is the electrode spacing.

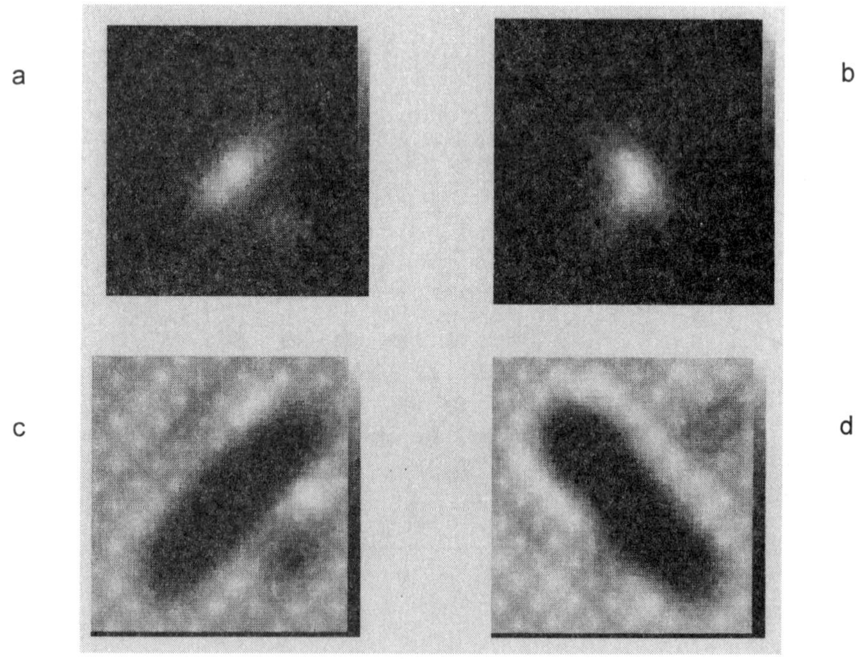

Figure 3
Images of a perpex rod (a)(b) and an aluminium rod (c)(d) buried in wet sand. The contrast of the conductive object is double that of the resistive one.

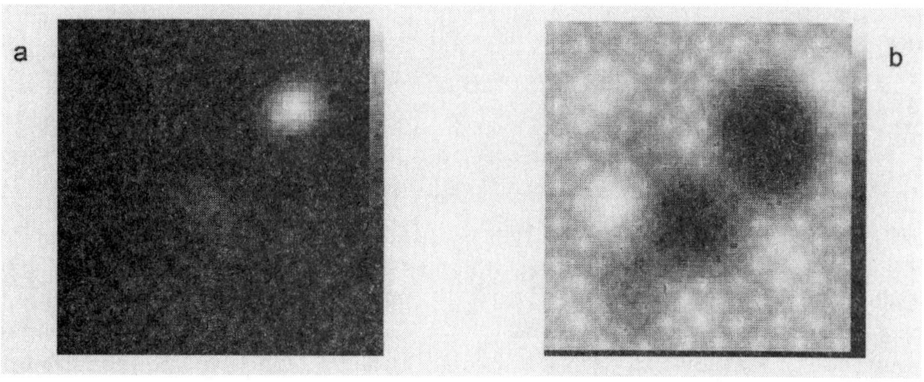

Figure 4
Images of (a) resistive and (b) conductive disks of diameter 5.5cm (top right), 4cm (centre) and 2.5c (bottom left) buried in wet sand at a depth of approximately 1.5cm.

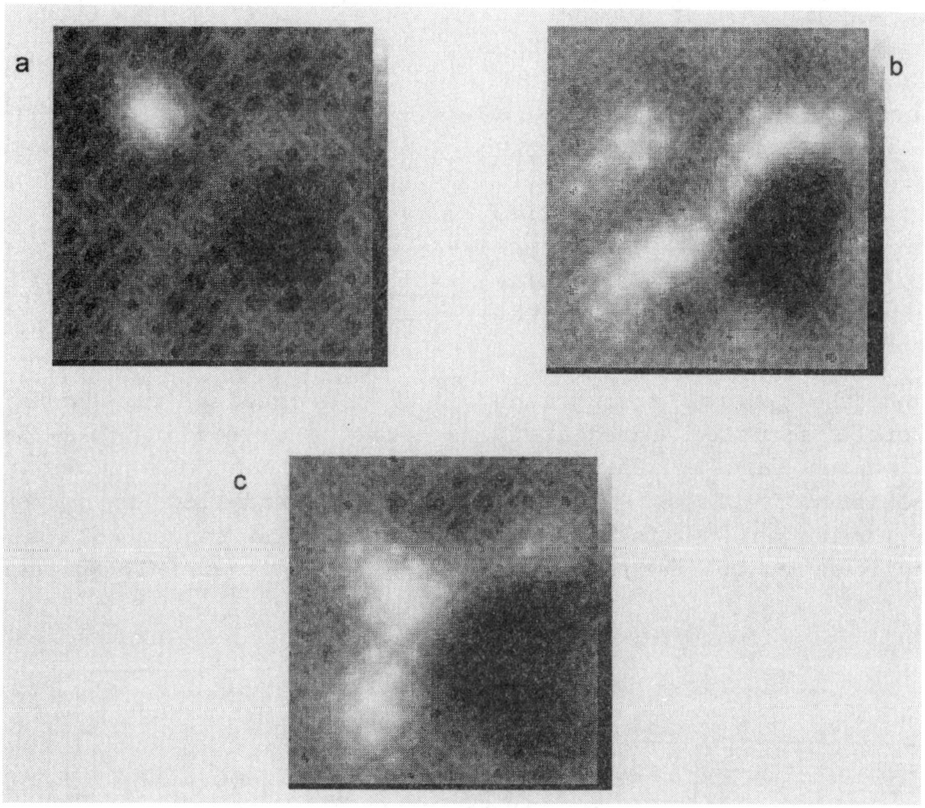

Figure 5
Resistive and conductive disks 4cm in diameter buried at depths of (a) 1cm, (b) 2cm and (c) 3cm. The conductive disk is detected in all cases but the resistive disk becomes masked by electrode contact artifacts as the depth is increased.

THE DETECTION OF BURIED PLASTIC LAND-MINES BY ELECTRO-MAGNETIC INDUCTION?

K. G. Spouse

A prototype detector has been constructed that is intended to be capable of locating a non-conducting object (e.g. a plastic land-mine) shallowly buried in earth.

The detector consists of an inductive probe the coil of which is mounted, with its axis vertically, in a manouverable search-probe. This, plus the associated circuitry and a battery power supply comprises the complete instrument.

The inductive probe and capacitor form a tuned circuit which is self-excited at resonance; if placed in close proximity to the earth, the circuit Q (= ωL/R) is reduced because of the energy loss incurred as a result of eddy currents induced in the comparatively conductive soil. When the coil is moved above a buried non-conducting object the losses are reduced and the increase in Q results in an appropriate increase in the output signal.

Fig 1. shows the general arrangement with the probe situated immediately above the buried target; the physically non-existent resistor Rs represents the equivalent resistance of the soil surrounding the target.

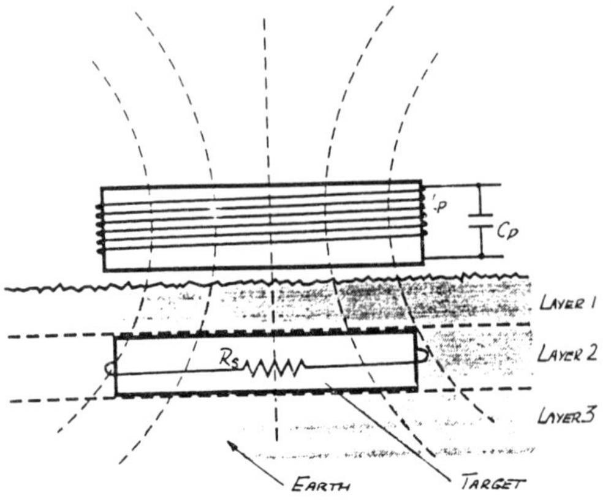

Figure 1 : Inductive transducer -

The transducer must obviously be capable of detecting this "buried hole in the ground" which is electrically analagous to a single-turn secondary winding resistively loaded and loosely inductively coupled to the resonant primary circuit.

Ignoring, for the moment, the effects of layers 1 & 3 the system may be modelled electrically as shown.

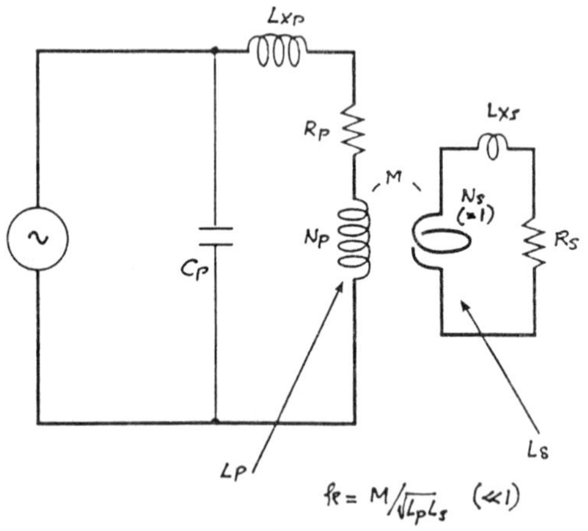

Figure 2 : modelled electrically -

When excited at constant current i.e. from a source of high impedance at the resonant frequency the circuit may be modelled as in Fig 3. where Rs' is the reflected loss resistance and X_L' is the total leakage reactance.

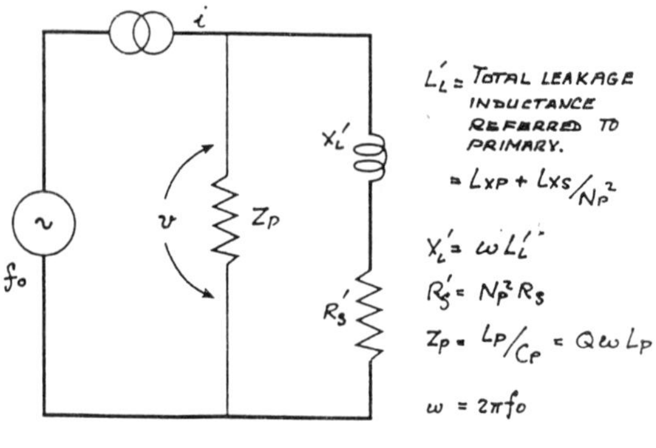

Figure 3 : and at resonance.

It may be seen (Fig 3.) that the sensitivity of such a passive system is limited by the intrinsic Q of the resonant circuit which is unlikely to exceed 80, given the adverse length/diameter ratio of the search coil. The effective Q can be increased if the circuit is included within a regenerative circuit oscillating in the marginal mode i.e. with the loop gain barely exceeding unity at the resonant frequency.

This is achieved, in the present instance, by connecting the tuned circuit as the frequency-determining element in a tuned-collector oscillator comprised of a cascode amplifier with automatic gain control as shown in the schematic diagram Fig. 4.

Oscillation is maintained at constant amplitude, by automatic level control i.e. by comparing the peak voltage excursion with a D.C. reference voltage and automatically varying the amplifier gain to reduce any error.

The (negative) dynamic output impedance at the collector of TR2 is determined by the loop gain and controlled by the voltage of the base of TR4.

The feed-back winding NF is only lightly loaded by the input impedance at the base of TR1, RT is not capacitively by-passed; the linearity of the amplifier is improved as a consequence.

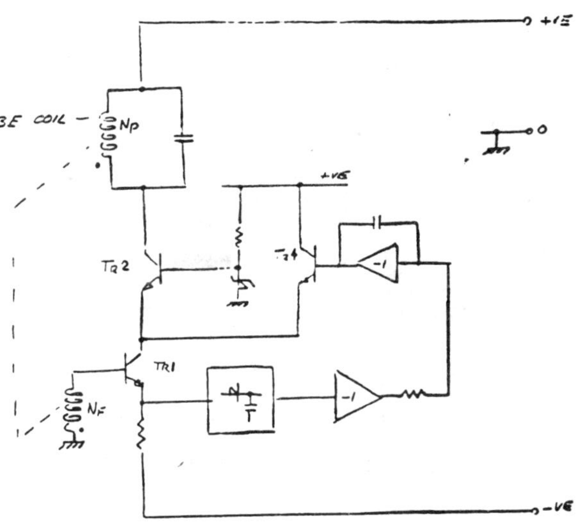

Figure 4 : Gain-controlled oscillator

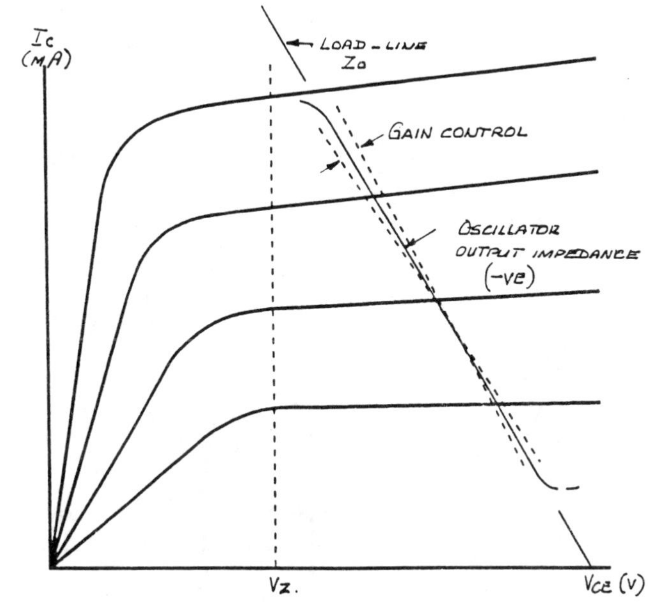

Figure 5 : Oscillator output

The detector's sensitivity has been assessed experimentally with plastic targets of various sizes immersed in saline solutions of known concentration. Typical results obtained using a 1% solution wt./vol. NaCl in water at 18°C (conductively 13 $mScm^{-1}$) and a non-conductive target 15cms. in diameter and 4cms. deep are shown in Figures 6 and 7.

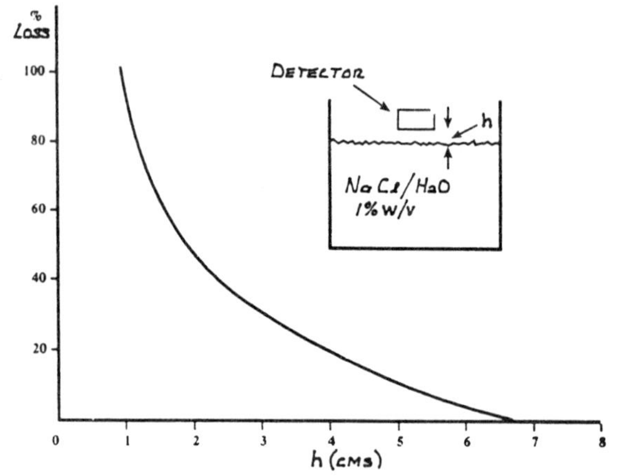

Figure 6 : Ground-clearance

Figure 6 shows the gradually reducing loss (i.e. increasing Q of the resonant circuit) as the detector's height above "ground" is increased.

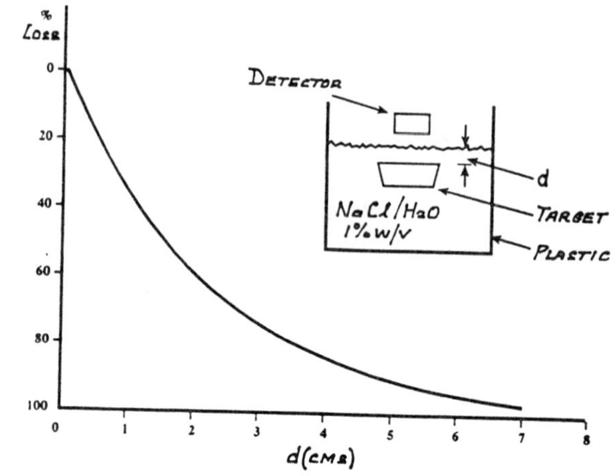

Figure 7 : Target depth

Figure 7 shows the gradual reduction in loss as the distance separating the target and the detector is reduced.

The results obtained with the primitive detector - circuit diagram Figure 8 indicate that, perhaps contrary to a first impression, electromagnetic induction may be able to assist in locating totally non-conductive buried land mines.

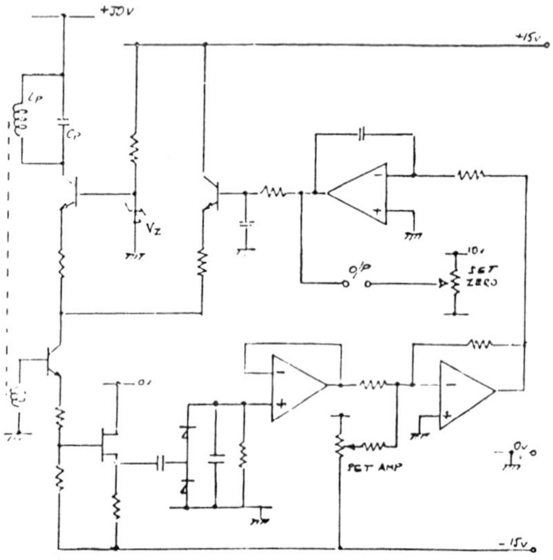

Figure 8 : Circuit diagram

References

1.Terman F.E. 1950 "Radio Engineers' Handbook", McGraw-Hill, pages 370. 439, 575, 624, 708.

2.Kwok M.C. and Pepper M.G. 1991 Med. & Biol. Eng. & Comp. 136-148 "Non-invasive detection of ventricular wall motion by electromagnetic coupling."

3.Burns R. 1995 "Radar Development to 1945" 438-439

TOWED, ACTIVE ELECTROMAGNETIC SENSOR ARRAY FOR ON-LINE TERRAIN MAPPING OF METALLIC SUBSURFACE OBJECTS

W. Rohrbeck, J.W. Leonhardt[1]
J. Laukemper, H. Neff[2]
G. Lehmann, H.-D. Ließ[3]

[1] IUT Institut für Umwelttechnologien GmbH, Germany
[2] TZN Forschungs- und Entwicklungszentrum Unterlüß GmbH, Germany
[3] Universität der Bundeswehr München, Germany

INTRODUCTION

The safe and effective detection of unexploded ordenance, including mines is a precondition for clean-up and renaturalization of large areas exposed to military training action over long periods, active combat or other military action. The situation in Germany is characterized by an unknown number of objects, which mainly have been produced during world war II and even I.

The pulse remance [1] principle is used for the detection array. A dc-current is fed to the transmitter coil via an electronic switch (fig. 1). By this current a magnetic field is built up around the transmitter coil. When the current has reached the steady state, the switch is opened, and the magnetic field decays. By the decaying magnetic field eddy currents are induced in metallic bodies inside the field. These eddy currents generate secondary magnetic fields decaying slower than the original magnetic field. The secondary fields are detected by a number of receiving coils. Their time dependend output voltages contain information on the coordinates, the volume, the resistivity and - to some extend - the shape of the metallic objects.

In order to reach a good sensitivity of such a device it is necessary to have a high field strength of the original magnetic field and a very rapid decay of the field after switching off to get a good separation in time between the signals of the original field and the secondary fields. High field strength means a large number of turns and a large aerea of the transmitter coil. This on the other hand leads to a large inductivity. Switching off high currents in large inductivities is difficult because this leads to high voltages at the electronic switch. These problems could be solved by special electronic means [2].

A comprehensive theory of the principle was worked out which will be published elsewhere. In the framework of MAXWELL's equations it can be shown that for a rectangular transmitting coil and rectangular receiving coils as well, which we used in our device, the time dependend signals in the receiving coils can exactly be calculated as functions of the geometry of the coils, their turn number, their relative position, the positions of conducting spherical metallic bodies with respect to the positions of the coils, the volumes of these bodies, the conductivity of the material of these bodies and the current flowing in the transmitting coil. Furthermore it can be shown that an exact solution can be found not only for sperical bodies but for elliptical bodies too. Then the orientation of the axes of the elliptical bodies with respect to the orientation of the coils comes into effect.

RESULTS

Fig. 2 shows the dependence of the peak value of the signal of a receiving coil on the distance R for a steel sphere with a diameter of 10 cm. The calculated curve is compared with experimental results (circles). The time dependence of the signal, S (t), is given in fig. 3 for different objects shapes: steel ring, shell in 3 positions and also a sphere. The conclusion is that the decay time T of the signal reflects some information about the shape of the object. Also it is shown that the sphere is a good approach to describe shells in the horizontal positions.

An other result is the correlation between the local half width of the received signal with the depth of the object. In accordance with the theory this correlation is shown in Fig. 4. Experimental results are circles. The agreement between theory and experiment is better than ± 10 %.

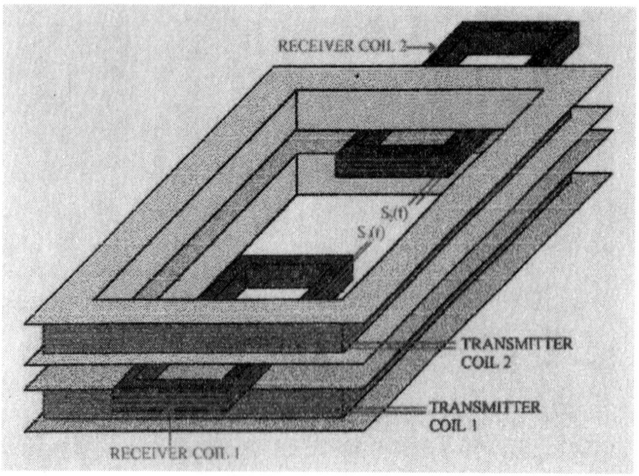

Fig. 1: Scheme of the used remanence systeme: transmitter coil und 2 receiver coils

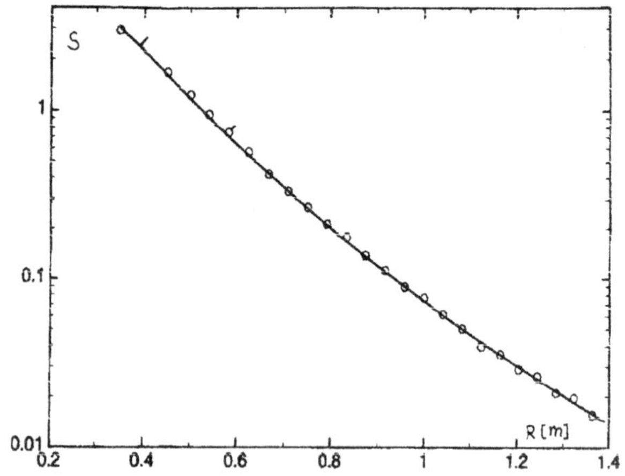

Fig. 2: Dependence of the received signal of a 10 cm steel sphere on the distance to the system - calculated curve and experimental results - cirvles

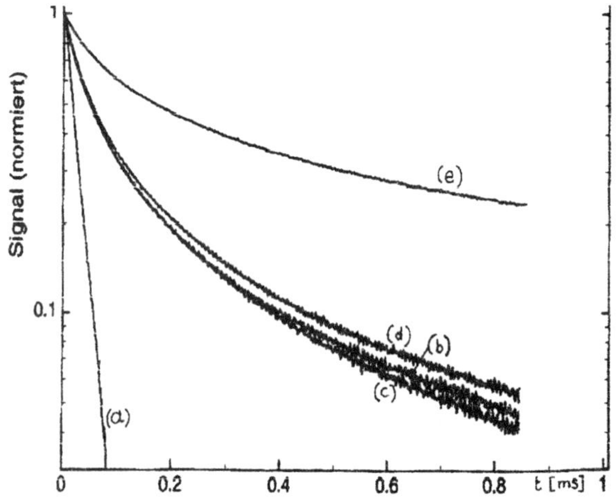

Fig. 3: Dependence of the signals of objects with different shapes on the fime: a) steel ring, b) steel sphere (10 cm), c) shell in horizontal position along the motion, d) shell in horizontal position - perpendicular to the motion, direction, e) shell in a vertical position

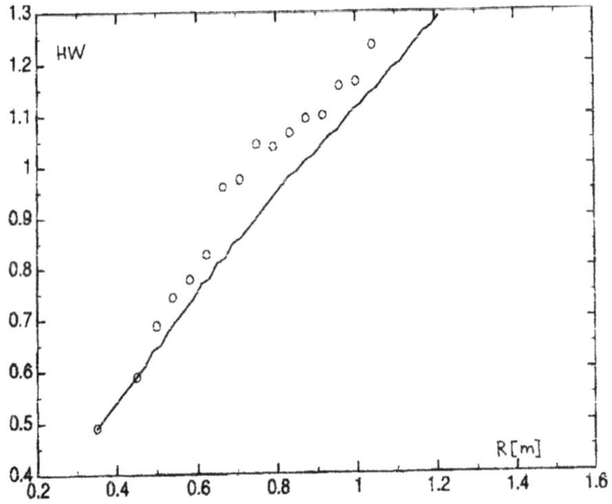

Fig. 4: The half width of the signal in dependence on the depth of the objekct - calculated curve and experimental results (circles)

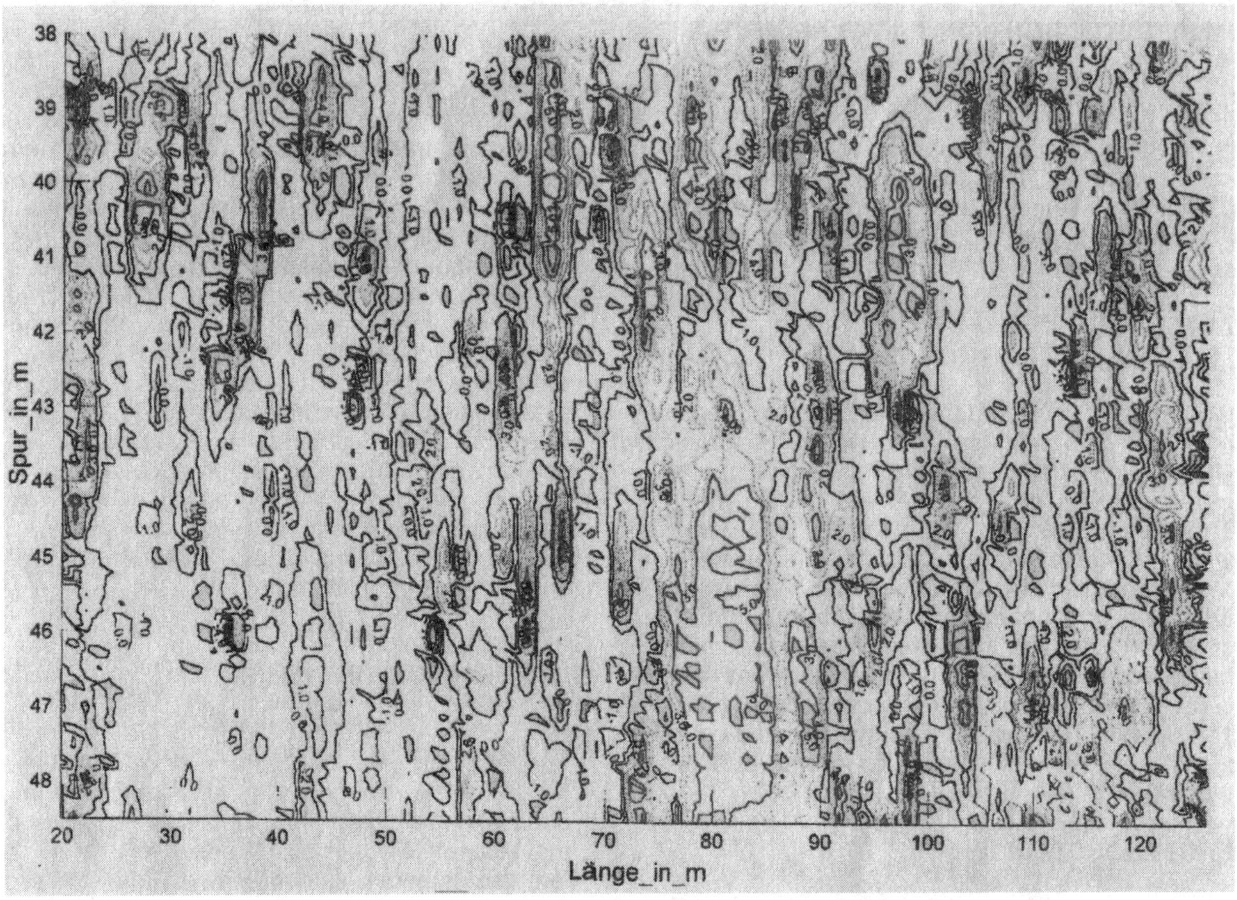

Fig. 5: The map of a part (10x50) of a former military training graound produced by the decribed system. Identified areas with UXO's are visible.

The productivity of the system is relatively high. It was possible to map 500 m^2 per hour under realistic circumstances. Fig. 5 shows an example for UXO detection on a previously used military training ground.

CONCLUSIONS

. The system should be able to perform in on line-mapping with a productivity of 5000 m^2/h in future using an array of 2.5 m in width and 1 m in length and 8 receiver coils which is available.

. The fully explored electromagnetic field theory will published in a forthcoming paper. Experimental results are in excellent agreement with theoretical data.

The work is supported by

- BMBF grants No. 1450 9401
- BMWi grants No. 439/94

[1] DD-AP 202 475
[2] DE 44 17 931

MINE DETECTION USING A POLARIMETRIC IR SENSOR

B. A. Barbour, S. Kordella, M. J. Dorsett,
B. L. Kerstiens
Nichols Research Corporation and U.S. Army
Space & Strategic Defense Command

ABSTRACT

Operation of current hand-held magnetic sensors for mine detection is a labor intensive process which involves a great deal of time and operator skill to detect near-surface and buried mines. The mine detection problem is made worse when plastic mines are sought, which are not easily detected by the magnetic sensors. The utilization of a polarimetric IR camera allows the detection of man-made objects, such as mines, by detecting the regularly polarized man-made objects which contrast with background clutter. This paper describes the system and shows results of tests in controlled mine-detection experiments.

I. INTRODUCTION

Following recent conflicts, there are now over 50 million land mines that have been abandoned in over 60 countries around the world but which are still active. While these mines remain uncleared, they will continue to kill some 10,000 civilians every year and injure and disable many more.
The technical problems in detecting these devices are formidable and no satisfactory system yet exists. The anti-armor land mines are often buried while the anti-personnel mines are relatively small and are naturally concealed by grass. Many abandoned mines contain few metallic components which evade detection by the inductive detectors. The area containing the mines is often littered with used ordnance which constitutes a formidable discrimination problem.

The detection of surface and near-surface buried mines from a stand-off sensor is a difficult problem. Several sensing approaches have been developed over the last several years which seek to exploit mine-specific features to increase the probability of detection and decrease the probability of false-alarm and false alarm rate in an automated mine detection system. Active and passive methods such as laser scanning and IR contrast of an area to detect mines are being pursued. For sensor detection a careful understand of the scattering and emission characteristics of mines and the surrounding environment is required to determine the optimal sensor suite to address the problem.

In this paper, we describe a unique technique which can be used either in an autonomous detection system, or in conjunction with other sensors to detect surface and/or buried mines from a distant platform. The sensing technique which we use exploits the polarimetric IR features of the scattered and emitted electromagnetic radiation from the mines and the surrounding ground clutter. The fact that mines are man-made and are of deterministic geometric regularities, versus the surrounding randomly oriented media results in a high probabilities of detection and low false alarms when utilizing IR polarization features. The IR sensor which is used to achieve these results is straightforwardly implemented and the signal processing techniques which are used are capable of resolving mines from standoff ranges in excess of 1 Km or more.

II. POLARIZATION THEORY

The majority of the polarization effects are dependent on the geometry of the surface under question, the sensor, and any light sources that may be reflected or scattered off of the surface into the camera's field of view. Figure 1. shows the geometry of reflection and emission from a surface. The surface normal n and the look angle to the sensor define the plane of incidence. The polarization components are either perpendicular (s- component) or parallel (p- component) to the plane of incidence and are defined by the direction of oscillation of the electric vector. The angle of incidence is given by the angle θ and is equal to the reflected angle for spectrally reflected light. The angle of incidence is also used to describe the angle between the surface normal and the sensor look angle and is also given by θ.

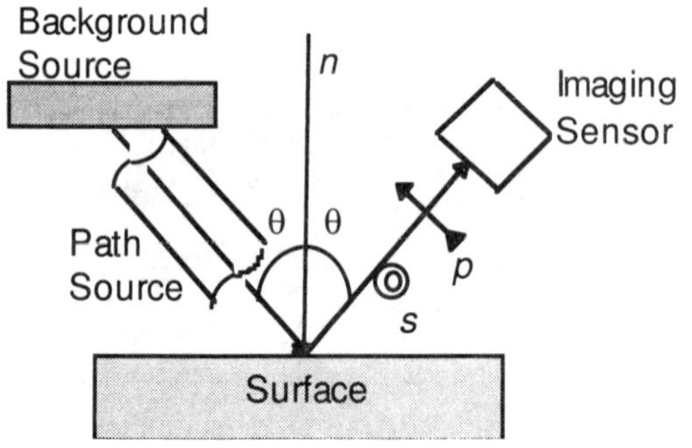

Figure 1. Geometry of Polarization Phenomena

Polarimetry is the measurement of the state of polarization of optical radiation, or the measurement of the change of state of polarization by propagation through an optical material. Polarimetry measures the complete

"Detection of abandoned land mines", 7-9 October 1996, Conference Publication No. 431 © IEE 1996

state of polarization including any unpolarized component. Alternatively, polarimetry measures the polarizing characteristics of a sample as well as any change in the degree of polarization resulting from propagation through the sample.

The polarization state of a light beam is described by the Stokes vector, a four element real intensity vector:

$$\vec{S} = \begin{pmatrix} S_0 \\ S_1 \\ S_2 \\ S_3 \end{pmatrix} = S_0 \begin{pmatrix} 1 \\ s_1 \\ s_2 \\ s_3 \end{pmatrix} \quad (1)$$

where the lower case letters represent elements normalized by the first element of the vector S_0. The first element of the Stokes vector S_0, gives the intensity of the beam and is the only element that is directly measurable by experiment. The other three elements of the Stokes vector describe the polarization state of the light and give the "preference" for horizontal, +45°, and right circular polarized light. Formally, the S_1 element represents the difference in intensities for horizontally and vertically polarized light, the S_2 element is the difference in intensities for light polarized along the ±45° axes, and S_3 is the difference of right and left circularly polarized light. For the normalized Stokes vector, the elements range from +1 to -1. The S_1 element takes on a value of +1 for completely horizontally polarized light and is -1 for completely vertically polarized light. Similarly, the S_2 element is +1 and -1 for +45° and -45° polarized light, and S_3 = +1 represents right circular light and S_3 = -1 is left circular. The degree of polarization is found from:

$$\mathrm{DOP} = \frac{\sqrt{s_1^2 + s_2^2 + s_3^2}}{s_0}$$

$$\mathrm{DOLP} = \frac{\sqrt{s_1^2 + s_2^2}}{s_0} \quad (2,3,4)$$

$$\mathrm{DOCP} = \frac{|s_3|}{s_0}$$

where DOP is the degree of polarization including linear and circular polarization, DOLP is the degree of linear polarization, and DOCP is the degree of circular polarization. DOP = 1 represents totally polarized light, DOLP = 1 represents totally linear polarized light, and DOCP = 1 is totally circularly polarized light. Fresnel's equations describe the magnitude of the electric field reflected, emitted, or transmitted by a smooth interface in terms of the incident magnitude. Fresnel's equations are given for the s- and p- polarization states for the interface defined by the plane of incidence.

III. POLARIMETRIC INFRA-RED SENSING

The breadboard IR Polarimeter is currently a device used to test the proof of concept for using IR polarimetry as a discrimination tool for the detection of mines. The IR Polarimeter uses an infrared camera with a 256x256 pixel focal plane array. The IR Polarimeter has been developed by Nichols Research Corporation (NRC) as an Internal Research and Development project. NRC holds the patent on the IR Polarimeter and has many additional claims still pending.

The NRC IR Polarimeter is a passive device detecting the thermal emission from its viewing area. Passive IR Polarimetry allows detection of targets at large angles of incidence. Active polarimetry has a major limitation in that it can only accurately detect the true polarization aspects which are returned in the specular reflection. The diffuse reflection becomes unpolarized quickly and becomes a noise source for thermal emission.

IV. POLARIMETRIC IR FEATURES OF MINES AND PHENOMENOLOGY

The U.S. Army Space and Strategic Defense Command supplied NRC with a variety of inert plastic antipersonnel mines for NRC to demonstrate the utility and effectiveness of the IR Polarimeter for the detection of plastic mines. Figures 2. and 3. are samples of IR Polarimetry data to demonstrate the phenomenology of simple shapes. Figure 2. has four images all of which were collected at the exact same time. The photograph of the scene shows a metal cube and a blackbody reference. The blackbody is a standard cavity blackbody which is randomly polarized. The other three images were all collected with the same frames of data and same sensor. The Infrared scene is the Stoke's vector S_0 term which is the thermal image. The DOLP scene is derived from the complete Stoke's vector. The Polarization Orientation Image is also derived from the same Stoke's elements. All the data presented in this paper will be displayed and processed in the same manner and no additional image processing has occurred on these raw images. Figure 3. is data on a metal cylinder. From these simple shapes a prediction can be made as to what to expect from an antipersonnel mine. The mines have cylindrical bodies with flat tops and edges at various angles. Four plastic antipersonnel mines were used to collect data in a controlled laboratory setting to show the polarization characteristics of the mines and to show the shape and contour enhancements that the polarization adds to the image. An Italian VS-50, US M-14, USSR BRMA-2, and CZEK PMA-3 SRB plastic antipersonnel mines are shown in Figure 4.

V. EXPERIMENTS WITH IR POLARIMETRY

Field tests with inert plastic antipersonnel mines have been performed. Figures 5. and 6. is test data from several of the tests. An Italian VS-50, US M-14, USSR BRMA-2, USSR POMZ-2 and CZEK PMA-3 SRB plastic antipersonnel mines are shown in Figure 5. All the test data has been taken at night with no external lighting. The passive IR polarimeter was used to collect the data and to display the results. The photograph on the left in Figure 5. is a daylight photograph of the sensor look angle and relative field of view. The mines are extremely difficult to visually detect from the high grass. The second photograph is the plastic antipersonnel mine used and relative positioning. Figures 7. and 8. demonstrate the ability of IR polarimetry to effectively distinguish manmade objects from heavy background clutter. Figure 7. also demonstrates that IR polarimetry can effectively discriminate manmade targets with no thermal contrast. The metal pipe was allow to come to thermal equilibrium with the surrounding background and cannot be detected in the thermal image. The DOLP image indicates high signal strength for the manmade metal object. The DOLP image is a normalized polarization image and is independent of thermal amplitude.

VI. CONCLUSIONS

Nichols Research Corporation (NRC) has developed and patented an innovative technique to detect and discriminate plastic antipersonnel mines with IR camera technology. Standard IR cameras can only detect the amplitude of the thermal emission. IR polarimetry will detect and analyze all orientation vectors, thus providing additional and complementary discrimination data to standard IR images. Several key benefits of IR Polarimetry is the high signal-to-noise or signal-to-clutter detection capability, the natural dual mode capability of thermograph and polarization in the same image frame, the natural surface mapping and object orientation detection capability and the ability to capture real-time data. IR Polarimetry is an innovative sensor technique which offers additional tools for target detection. NRC has collected limited field test data which clearly indicates that IR polarimetry provides a unique signature for plastic antipersonnel mines in the presence of high grass and heavy background clutter.

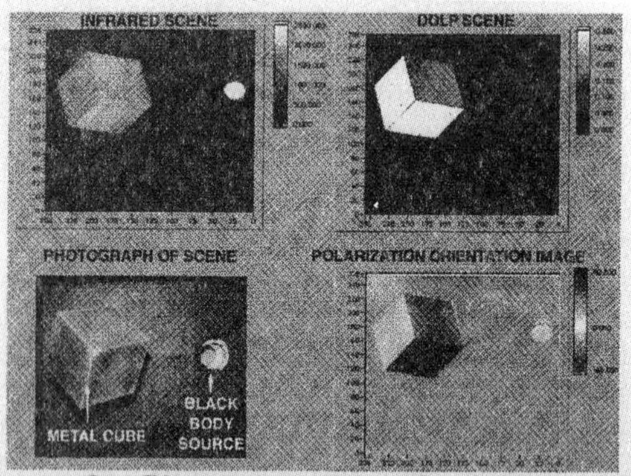

Figure 2. Phenomenology of a metal cube.

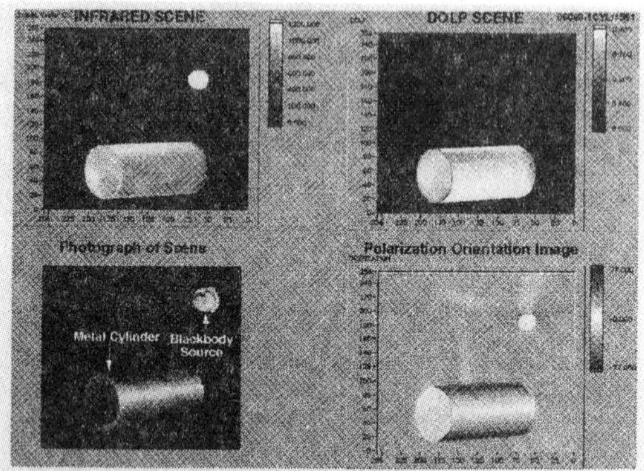

Figure 3. Phenomenology of a metal cylinder

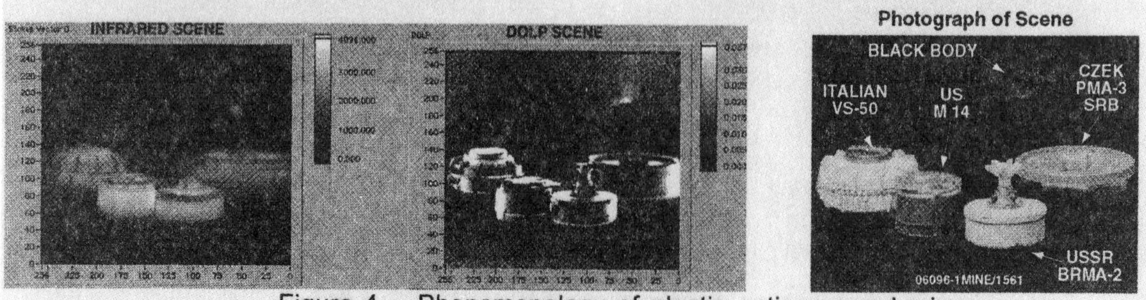

Figure 4. Phenomenology of plastic antipersonnel mines

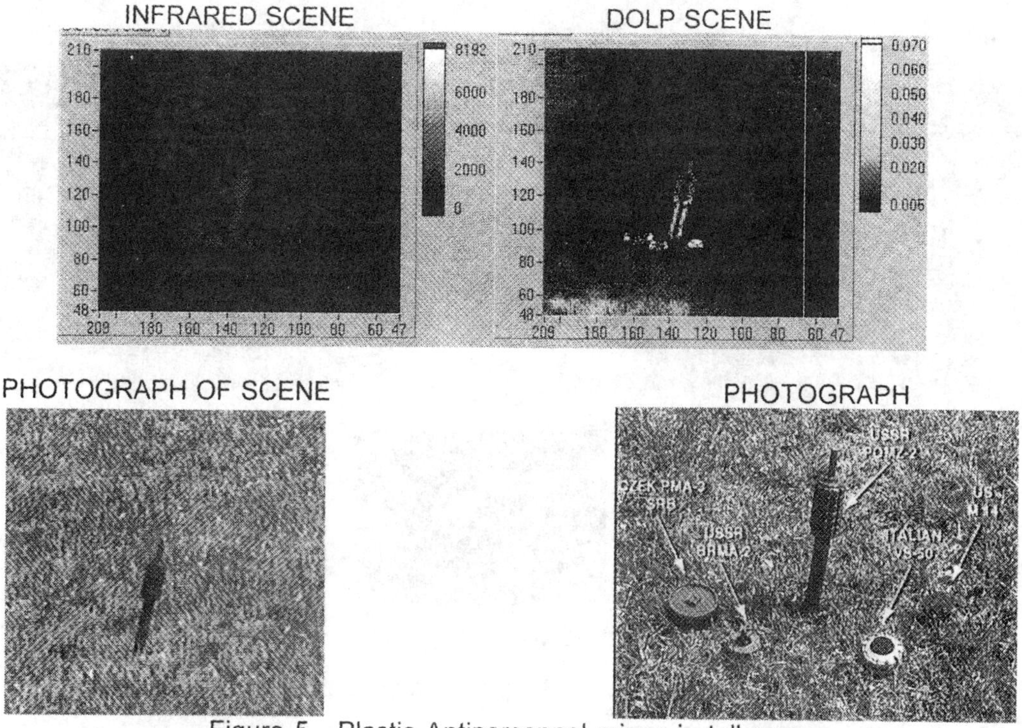

Figure 5. Plastic Antipersonnel mines in tall grass.

Figure 6. Plastic Antipersonnel mines close up.

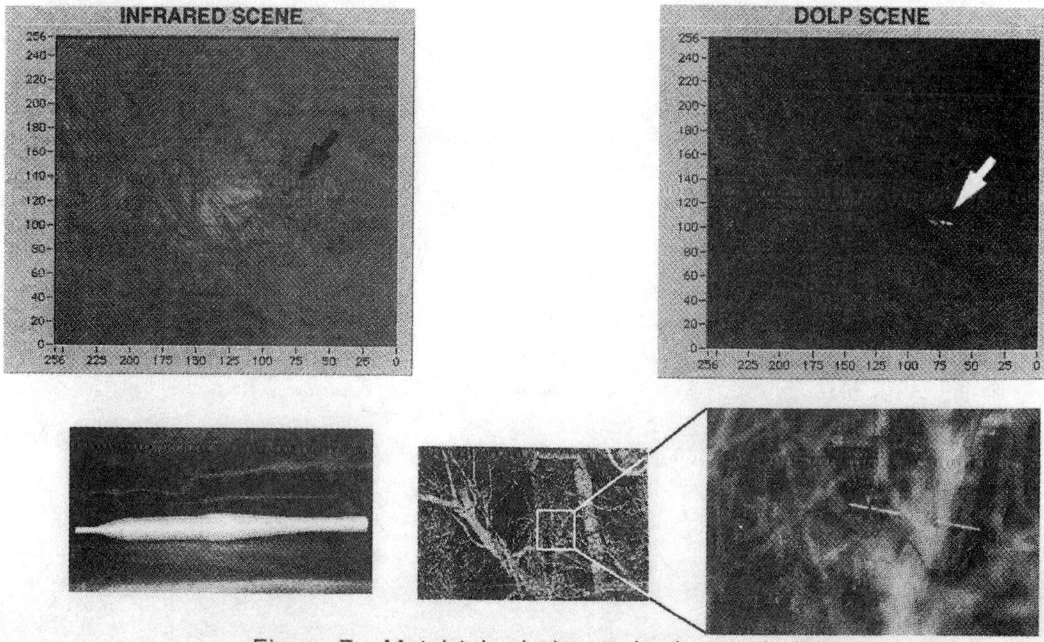

Figure 7. Metal tube in heavy background clutter.

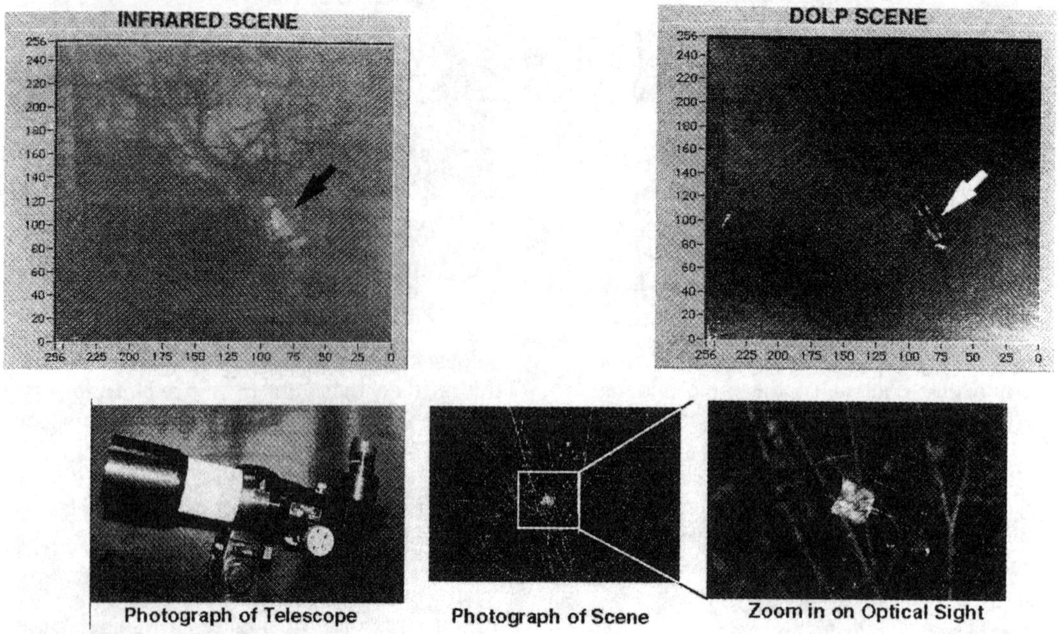

Figure 8. Optical Telescope in heavy background clutter

References

R. M. A. Azzam and N. M. Bashara, Ellipsometry and Polarized Light, North-Holland, Amsterdam (1977).

W. A. Shurcliff, Polarized Light, Harvard University Press, Cambridge, MA (1961).

E. Hecht and A. Zajac, Optics, Addison-Wesley, Reading, MA (1974).

IEE Web Page of “The Detection of Abandoned Land Mines” Program Need

MOISTURE AND LANDMINE DETECTION

L.J. Carter, G.H. Bryant, M. Le Fevre, and W.C. Wong

University of Auckland, New Zealand.

ABSTRACT

A landmine detection technique is proposed, based on differences in moisture content of the mine and the surrounding soil. Microwave radiation is used to heat the soil containing the mine. The moist soil is warmed far more than the mine. This produces temperature differences at the surface which can be detected by infra-red scanning. Laboratory tests show that buried non-metallic objects can be located with this technique.

INTRODUCTION

It has been estimated that there are about 110 million landmines in the ground worldwide, with about ten million in Afghanistan alone. Mines are laid during periods of conflict, but can remain dangerous for years afterwards. Their effects in terms of human suffering and damage to rural economies are well known. (1)

Many modern mines have very little metal in them, and are therefore very difficult to detect by conventional means. Manual probing, which is dangerous and extremely slow, is still the most common method of detection. (2)

There is a great need for a method of detecting landmines which have little or no metal in them. Such a method should be low-cost so that it can be available to the greatest possible number of clearance agencies.

POSSIBLE DETECTION TECHNIQUES

Various types of ground-penetrating radar (GPR) have been proposed for mine detection. A signal is reflected from the target mine due to a dielectric discontinuity between the mine and the soil around it. A significant problem with all GPR systems is that other discontinuities also give reflections, and in particular the air/ground interface reflects much of the incident energy. Very short pulses are needed to be able to resolve targets just below the air/ground interface, and the radar standoff may be critical. The choice of operating frequency is a compromise: high frequencies are needed to allow resolution of small mines, but signal attenuation in the ground is greater at the higher frequencies. GPR variants which have been used with various degrees of success include synthetic aperture radar, ultra-wideband radar, and recently micropower impulse radar(MIR). MIR combines low power with low cost and appears to be an attractive technology for this application.

Acoustic sounding has been proposed: difficulties with this include the problem of coupling energy into the ground, and the variability of sound propagation in the soil.

Other proposed methods include nuclear magnetic resonance, nuclear quadropole resonance, the use of radioactive sources, and x-rays. All of these are relatively high-technology approaches which for various reasons (including in some cases power consumption) do not lend themselves readily to application in a remote minefield.

Trace gas detection has in some cases been used for mine detection. The principle is to identify the location of the mine by the presence of traces of explosive vapour leaking from the mine. Vapour concentrations may be very small: perhaps as low as 0.1 ng/m^3. Dogs have been successfully trained to respond to such trace gases, but may not be able to sustain their performance for prolonged periods. Other technologies are available which could be used to identify explosive vapour, for example gas chromatography and photo-acoustic spectroscopy, but these are not yet developed to the point where they could easily be employed in a minefield.

Infrared detection methods, both active and passive, have been proposed. Active methods typically involve scanning with a suitable laser and looking for specular reflections from surface -laid mines; passive methods typically rely upon the sun's radiation to produce a characteristic ground-surface signature above the mine. False alarms are produced from background clutter, and there is perhaps some difficulty in obtaining good contrast between the mine and the surrounding soil, since all are similarly warmed slowly by the sun. Li *et al.* (3) have reported results obtained with active IR imaging, using infrared heating lamps. Craib (4) and Waschl (2) have reviewed proposed mine detection techniques.

A MOISTURE-BASED TECHNIQUE

Modern mines (and this would include the minimum-metal and zero-metal mines) are well sealed against moisture ingress, and in some cases are specified to be capable of being laid under water. Agricultural land typically has a fairly high moisture content. A mine buried in such land would therefore be likely to have a significantly lower moisture content than the soil around it. We propose this as the basis for a method of detection. If the ground is irradiated at 2.45 GHz, then (as in a microwave oven) the presence of moisture will produce a temperature rise. The mine, with little or no moisture in it, is likely to stay cool. The resulting heat flows are then likely to give, after a short delay, a characteristic temperature signature on the surface above the mine. This can be found, and with it the location of the mine, by infrared detection.

THEORETICAL CONSTRAINTS

This section presents the results of an initial analysis of the likely constraints on the proposed technique. We have assumed for this purpose a uniform soil, irradiated from above by 750 W of microwave energy at 2.45 GHz, supplied from a horn antenna immediately above the soil surface. A landmine is buried a few centimetres below the surface, and consists of a uniform plastic case with flat top and bottom, filled with TNT explosive. We consider the interaction of the microwave energy with the soil, and consequent thermal effects.

Reflection and attenuation of microwave energy

We assume that there will be some reflection of microwave energy at the following interfaces: air/soil, soil/plastic, and plastic/TNT. Reflected power P_r is given by

$$P_r = \rho^2 P_{in} \quad . . . (1)$$

where P_{in} is incident power, and ρ is reflection coefficient, given by

$$\rho = \frac{\sqrt{\frac{\varepsilon_1^*}{\varepsilon_2^*}} - 1}{\sqrt{\frac{\varepsilon_1^*}{\varepsilon_2^*}} + 1} \quad . . . (2)$$

where ε^* are complex permittivities of the media 1 and 2 at the interface.

$$\varepsilon^* = \varepsilon' - j\varepsilon'' \quad ... (3)$$

where ε' is relative dielectric constant, and ε'' is effective loss factor of the medium. ε^* is affected by the moisture content of the medium. The *loss tangent* is defined by

$$\tan\delta = \frac{\varepsilon''}{\varepsilon'}$$

and allows materials to be classified as dielectrics ($\tan\delta < 0.01$) or conductors ($\tan\delta > 100$). Typical values for soils are given in the Appendix.

Absorbed power P_a is given by

$$P_a = P_{in} - P_r$$

$$= (1 - \rho^2)P_{in} \quad ... (4)$$

We expect that the microwave energy will suffer exponential attenuation with distance travelled through the medium. We define a penetration depth D such that

$$D = \frac{1}{2\alpha} \quad ... (5)$$

where α is attenuation factor. α is related to permittivity by the equation

$$\alpha = \frac{\pi\varepsilon''}{\lambda_0\sqrt{\varepsilon'}} \quad ... (6)$$

where λ_0 is free-space wavelength. Table 1 shows expected values of reflected and absorbed power at the air/soil interface for various types of soil and moisture levels. Clay soils are defined here as those having particles less than 2 μm in diameter. Incident power is taken as 750 W, and moisture percentages are defined by weight.

TABLE 1 - Power reflected/absorbed at ground surface

SOIL, moisture	$\lvert\rho\rvert$	P_r W	P_a W
Sandy			
dry	0.23	39.7	710.3
2.18%	0.23	39.7	710.3
3.88%	0.35	91.9	658.1
16.8%	0.63	297.7	452.3
Loamy			
dry	0.22	36.3	713.7
2.2%	0.30	67.5	682.5
13.77%	0.63	297.7	452.3
Clay			
dry	0.20	30	720
20.09%	0.54	218.7	531.3

Note that the moisture content of the soil affects the reflection coefficient and hence the amount of power which it is possible to get into the ground. See for example Metaxas and Meredith (5).

Heating effect of microwave energy

By considering the power flow through a layer of soil of thickness Δx, it is possible to derive an expression for the temperature rise T in a layer of soil at depth x after irradiation for a time t. This is given in Le Fevre (6) by

$$T = \frac{P_{in}\, e^{-x/D}\, t}{AdhD} \quad ... (7)$$

where A is horn antenna aperture,
d is soil density,
h is specific heat of soil,
and D is penetration depth.

Note that both h and D are dependent on moisture content. Wet soils have higher specific heats, implying that more energy is needed to raise the temperature. Figure 1 shows expected temperature rise of sandy soil, 3.88% moisture content, with respect to times of exposure up to 60 seconds. The microwave power is set to 750 W and the antenna aperture is 15 x 15 cm.

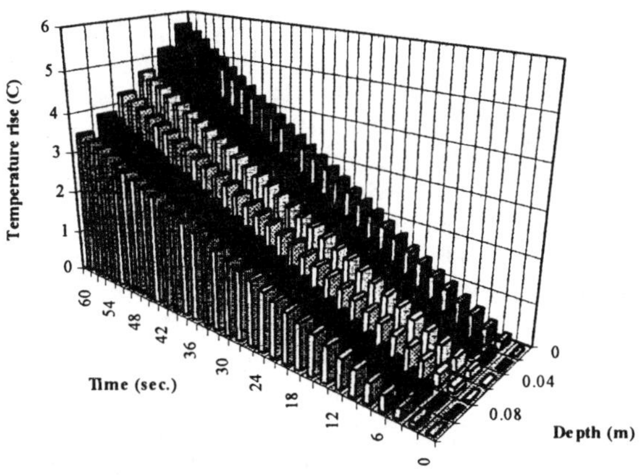

Figure1: Temperature rise of sandy soil

The temperature rise at any depth increases linearly with irradiation time, and decreases exponentially with soil depth. Table 2 shows expected temperature rise for different soil moisture levels at a depth of 10 cm, after one minute of heating.

TABLE 2 - Effect of soil moisture on temperature rise

Sandy soil moisture %	Temperature rise at 10cm depth, °C
dry	0.72
2.18	2.57
3.88	3.49
16.8	0.63

Note from Table 2 the inhibiting effect of a higher moisture level, due to increased reflection of energy. There will be an optimum moisture level which gives a maximum temperature rise.

Figure 2 shows the expected temperature rise at various depths for a uniform sandy soil with 3.88% moisture content. Microwave power is 750 W, antenna size is 15 x 15 cm, and irradiation time is one minute.

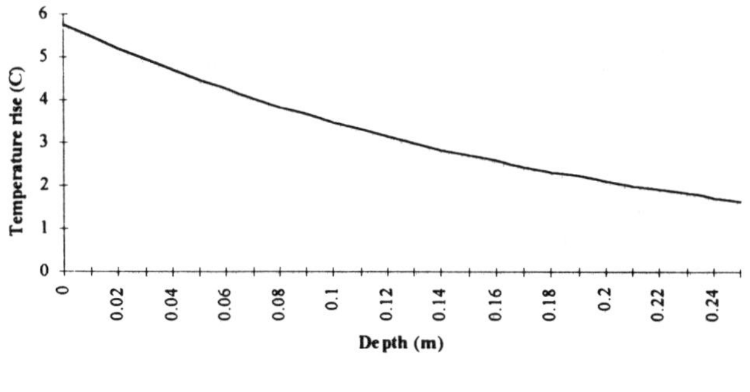

Figure 2: Temperature rise in sandy soil vs. depth

Figure 3 is similar to Figure 2 but this time allows for the presence of a landmine 5 cm below the surface. The mine is taken to be 10 cm in diameter and 5 cm in height. The plastic case is taken to be PVC, 5 mm thick, and the contents are taken to be TNT, of thickness 4 cm. Reflections from the soil/plastic and plastic/TNT interfaces are neglected, as are microwave attenuation and temperature rise inside the mine.

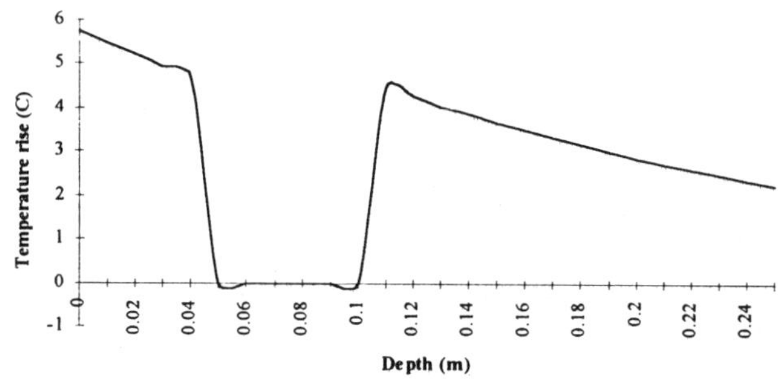

Figure 3: Temperature profile in soil with mine.

Temperature change at surface

With the creation as above of a cold mine surrounded by a relatively warm soil, we would expect there to be a consequent flow of heat in the soil. Assuming for simplicity a one-dimensional conduction process, this will be governed by an equation of the general form

$$\frac{\partial T}{\partial t} = K_T \frac{\partial^2 T}{\partial x^2} \quad \text{... (8)}$$

where T is temperature, t is time, x is distance, and K_T is thermal diffusivity of the soil. See for example Baver (7).

Soil may be thought of as solid particles surrounded by pores containing air and/or water. Heat transfer through soil is largely by conduction, with better conduction taking place through the solid particles. The presence of water rather than air in the pores has a major effect on heat conductivity since water conducts heat approximately 25 times better than air.

A finite difference equation was used to represent the one-dimensional heat flow in the irradiated soil:

$$\frac{[T_j^{n+1} - T_j^n]}{\Delta t} = \frac{[K_T(T_j^{n+1} - 2T_j^n + T_{j-1}^n)]}{\Delta x^2} \quad \ldots (9)$$

where j = depth interval, cm
n = time interval, s
K_T = thermal diffusivity, m^2s^{-1}
t = temperature, °C
x = depth, cm

Equation 9 was used to estimate the temperature difference at the surface after irradiation, this being the difference between surface temperature for no landmine, and surface temperature with a landmine. Figure 4 shows the result for sandy soil with 3.88% moisture, for a mine at depths of 1 and 4 cm. For the shallowly-buried mine, it can be seen that the 'signature' develops over a period of time to a maximum temperature difference, and then fades.

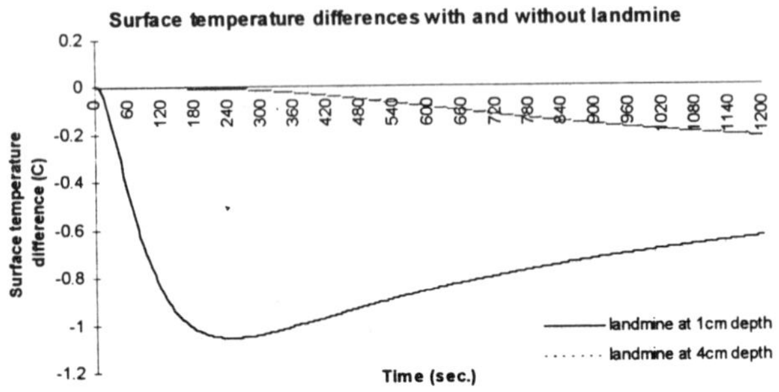

Figure 4: Surface temperature difference vs. time.

From the above it may be seen that moisture content plays an important part in all aspects of the development of the surface 'signature': in determining the amount of incident microwave energy reflected and attenuated; in the heating of the soil and (by its absence) the lower level of heating of the mine; and by affecting the heat conductivity of the soil, determining the rate at which the signature forms on the surface.

EXPERIMENTAL RESULTS

Some preliminary tests have been carried out in the laboratory to determine whether there is any basis for this hypothesis. A dummy mine was made, consisting of a 50 mm diameter plastic pot filled with petroleum jelly. This was buried in a container of moist sand, which was then irradiated at 2.45 GHz. A 750 W commercial magnetron was used as the source of the radiation. A hand-held infrared detectior ("Infratrace 1000") was used to determine the surface temperature a short time after irradiation.

With the 'mine' close to the surface, relatively large temperature differences (of the order of 8 degrees Celsius) were found between the surface temperature immediately above the mine and temperatures elsewhere on the surface. When the mine was buried at the bottom of the container under about 30 mm of sand, a discernible temperature difference of about 2°C was found. When the mine was buried in different places, the cold spot signature moved with it, increasing confidence that this was not the result of standing waves or mode effects. (8)

Tests were also carried out in very dry sand, and in sand saturated with water (with water standing on the surface). In both of these extreme conditions the characteristic signatures were unobtainable, suggesting that there is a 'window' of ground moisture levels for which the method is appropriate.

DISCUSSION

It is evident that the proposed moisture-based method of mine detection will work only for certain values of ground moisture. For humanitarian mine clearance, however, this is not necessarily an insurmountable problem, since clearance can be timed to coincide with the season when ground moisture levels are likely to be optimal. For extremely dry soils, it may be quite feasible to moisten the ground artificially before scanning.

It is likely that the technique would have difficulty in finding an older mine which was waterlogged. Such a mine, though, might well have sufficient metal content to enable its detection by more conventional means.
A disadvantage of this technique is its need to use significant quantities of power; perhaps of the order of 1 kW. This is much less than an x-ray system, but would still be difficult to provide in remote locations. A portable generator would probably be needed.

Any system which emits hundreds of watts of microwave energy must be regarded with caution from the perspective of operator safety. One possible solution to this problem is to imagine the sensor mounted on an autonomous robot, which could crawl over the minefield, marking the likely positions of mines.

By irradiating at 2.45 GHz and then scanning at infrared, this approach avoids the ground-reflection problems associated with GPR. The use of commercial, microwave-oven magnetrons is attractive since these are mass-produced and readily available.

CONCLUSIONS

A moisture-based technique for finding landmines has been proposed. The ground is irradiated with microwave energy, and characteristic 'cold-spot' signatures, found with infrared detection, show the positions of the mines. Laboratory tests using a dummy mine buried in uniform sandy soil show that the technique works under certain conditions of soil moisture. Further testing is required to determine the validity of the technique under progressively more realistic conditions, and this is being pursued.

REFERENCES

1. U.S. Department of State, 1994, Hidden Killers: The Global Landmine Crisis: Report to the U.S.Congress on the Problem with Uncleared Landmines and the United States Strategy for Demining and Landmine Control

2. Maley, W. *ed*, , 1994, Dealing With Mines: Strategies for Peacekeepers, Aid Agencies and the International Community, Australian Defence Studies Centre, Canberra

3. Li, P., Maad, A., Moshary, F., Arend, M.F., and Ahmed, S., 1995, Applied Optics,34, 5809-5816

4. Craib, J.A., 1994, Survey of Mine Clearance Technology, The United Nations University and The United Nations Department of Humanitarian Affairs

5. Metaxas, A.C., and Meredith, R.J., 1983, "Industrial Microwave Heating", Peter Peregrinus Ltd., London

6. Le Fevre, M., 1995, Landmine Detection, Report EE95/67, Department of Electrical and Electronic Engineering, University of Auckland, New Zealand

7. Baver, L.D., 1966, "Soil Physics", Wiley, New York

8. Schouten, H., 1996, Wellington *Evening Post* , New Zealand, February 17th 1996

9. Wong, W.C., 1995, Landmine Detection, Report EE95/66, Department of Electrical and Electronic Engineering, University of Auckland, New Zealand.

APPENDIX: SOIL PROPERTIES

Soil, Moisture %	Relative dielectric constant ε'	Loss tangent tan δ	Effective loss factor ε"	Penetration depth D m	Reflection coefficient |ρ|	Density kg m^{-3} x 10^3	Specific heat J kg^{-1}K^{-1} x 10^3	Thermal diffusivity m^{2}s^{-1} x 10^{-6}
Sandy								
dry	2.55	0.0062	0.01581	1.96	0.23	1.60	0.80	0.24
2.18	2.5	0.03	0.075	0.41	0.23	1.62	0.87	0.27
3.88	4.4	0.046	0.2024	0.20	0.35	1.64	0.93	0.29
16.8	20	0.13	2.6	0.03	0.63	1.77	1.29	0.45
Clay								
dry	2.27	0.015	0.03405	0.86	0.20	1.60	0.89	0.18
20.09	11.3	0.25	2.825	0.02	0.54	1.80	1.44	0.35

Values are taken from Wong (9).
Soil dielectric properties are measured at frequency 3 GHz and temperature 25°C.
Penetration depth and reflection coefficient are calculated at the air/soil interface.

OPTRONIC LINE SCANNING REMOTE SENSING FOR INITIAL DETECTION OF LAND MINES

Mr C H Hamon
SAGEM S.A. (formerly SAT) - France

INTRODUCTION

Optronic sensing is one of the main contributors to the real challenge of detecting abandoned land mines.
Contribution of line scanning will be discussed in this paper.
Detection is a very difficult problem and needs to be envisaged faced with the tasks or functions to be performed and constraints to be met.
Humanitarian demining has not the same constraints as military demining : time constraints are less demanding, both in terms of time needed to perform the tasks and period of the day or even of the year, if seasonal effects significantly enhance the detection performances; on the other hand, the probability of detection could be more stringent when military demining does not cover anti-personnel mines because the mission consists in clearing a path for armoured vehicles. Sometimes also political reasons may lead to reject certain categories of techniques proposed by the scientists, e.g. techniques using nuclear physics or involving mounting on certain types of platforms, ...
Before describing the optronic line scanning technique, we will discuss the mine threat to be detected, the environmental factors which may affect the detection and the various functions needed for mine disposal.

THE MINE THREAT

This threat is very multiple. In the case of optronic detection, we can consider two main forms, i.e. the surface and buried mines. The main point is thus the type of detection : direct detection for surface mines, indirect detection for buried mines.
Direct detection might be affected by environmental factors like snow or dense vegetation. It requires high geometrical resolution, especially to detect anti-personnel mines (APM) which could be small, for instance less than 50 mm in diameter.
Indirect detection requires less geometrical resolution. It detects not the mine itself, but a change, a disturbance in the mine environment explained by the hole where the mine could be buried.
We can also mention a kind of indirect detection of APMs when they are used for preventing or deterring soldiers from lifting anti-tank mines (ATMs) : in such cases, the detection of ATMs is easier, especially thanks to their usual pattern of lay-out. Hence, it could be of great interest for the subsequent detection of the associated APMs.
The presence or not of metal within the mines is less important for the considered type of sensing : the real impact is linked to the optical characteristics of the external parts, in terms of temperature, spectral reflectivity and emissivity. As a matter of fact, these parameters will affect the contrast detectable by the optronic sensors. Due to the evolution of the threat, this point could be of great interest thanks to that specific complementarity with other techniques like Surface Penetrating Radar or especially ElectroMagnetic Induction.
Other forms of threat are more difficult to deal with, for example trip wires and remote control of the mine ignition which is then away from the demining place, at least out of the zone of the focus of attention. These mines could be bounding fragmentation mines or directional fragmentation mines.
The ignition could then be triggered through the contact and entangling in a trip wire made of a very small diameter electrical wire, fiber optics or simple textile. Thus, detection is very difficult to achieve, especially against a vegetation environment.
Finally, special attention must be paid to unexploded ordnances, which are very usually found in some mined areas. Similarly improvised mines are a real threat, especially because they usually have a familiar shape like a can of drink.

ENVIRONMENTAL FACTORS

Usually several factors are identified as environmental factors, which have to be considered for the mine detection and identification :
- magnetic environment,
- urban environment,
- nature of the soil,
- roughness of this soil,
- presence of humidity,
- presence of water,
- presence of boulders,
- presence of stumps,
- presence of scraps,
- microclimate factors,
- surface and ambient temperatures,
- light conditions.

Like any other candidate technique, the optronic passive detection could be affected by environmental factors, especially those which hide the target (like dense vegetation, snow, puddles, fordable zones, paddy fields or sunlight conditions for ultra-violet to near infrared passive detection) or fade the contrasts (like icing or rain). For the microclimate factors like snow, ice or rain, the solution is just to wait for better weather conditions. Temporary drainage or diversion could be proposed to overcome fords or paddy fields. Sometimes, dense vegetation could be burnt off or mechanically eliminated when there is real interest to enter such zones.

THE FUNCTIONS TO BE PERFORMED

We can propose the following set of functions to be performed facing the possible presence of abandoned land mines :
- alarm or clue detection,
- mine detection,
- mine identification,
- explosive identification,
- mine location,
- mine marking,
- mine neutralisation and disposal.

Clearly, passive optronic techniques may perform :
- the detection of alarm or clue like the presence of a suspect hole in the soil, a pothole in the asphalt, a local disturbance of the contrast,
- the detection of surface mines,
- the identification of surface mines, when it is possible to assess the typical features (shape, accessories, ...), and even sometimes to read the written indications,
- the relative position of elements in the surroundings of a single mine.

Mine marking, neutralisation and disposal could be facilitated. Active optronic techniques are more appropriate for such functions.
The explosive identification could be necessary for defeating dummy or improvised mines.

ABOUT THE OPTRONIC DETECTION

General

The problem is so difficult that a wide benefit will arise from the fusion of several well selected spectral bandwidths for a better discrimination of the mine and its environment. This data fusion is a critical point and has to be widened to the main candidate techniques like Surface Penetrating Radar and Electromagnetic Induction. Even if detection is indirect, it is worth it, for instance, to detect a clue of a potential mine like some disturbance in the closest vegetation compared to the surrounding one. This could be achieved by different optronic spectral bandwidths (from ultraviolet up to far infrared) and/or using the polarisation effects of reflected and emitted thermal infrared.

Types of optronic sensors

Three families of optronic sensors could be presented :
- TV cameras and FLIR (Forward Looking InfraRed), also named thermal imagers with an image rate of 25 to 60 images/second and fields of view between a few tenths of degrees to fifty degrees,
- panoramic sensors (360° in azimuth), also named IRST (InfraRed Search and Track) with an image ra-te of 0,1 to 5 images/second and a vertical field of view between 5° and 20°,
- linear cameras or line scanners, especially the IRLS (InfraRed Line Scanner) with line rate of 50 to

1000 lines/second and a lateral field of view of 40° to about 180°; Side Looking derivative equipment could lead to a narrower lateral field of view.

Contribution of line scanning

The main difference between the image of a camera and a line scanner is like the difference between a video image and a photography : as a matter of fact, a line scanner scans laterally, i.e. transversally, only a line or a limited group of lines and needs the movement of the platform to perform scanning in the other longitudinal direction. Scanning is performed either opto-mechanically thanks to a rotating mirror or electronically by means of large linear arrays of photodetectors with integration, possibly time delay, and CCD read-out.
The advantage is to avoid the redundancy of scanning the scene. This point is fundamental in terms of data rate : the gain is very significant when considering the requirements of data link bandwidths and of storage capacity. As a consequence, data processing, data fusion and georeferencing are made easier, especially in terms of time and volume of needed hardware. Among the other interests of line scanning, this very specific and attractive aspect has to be particularly pointed out, compared to the very redundant sensing of the TV cameras.
The detection of mines is well adapted to the use of line scanners because the mines are fixed targets and the whole minefield has a certain architecture or certain characteristics (like density along some directions, statistical bidimensional density of scattered mines, ...) only reachable by the analysis of the complete minefield.
The principles of line scanning are always very elementary with a single direction of scanning, making the line scanner simple and easily ruggedised.
Another typical feature is the very wide lateral field of view, up to 180°, frequently 120°. Such a field provides a swath of 3.4 times the height above ground of the sensor.
This large coverage is only one interest of line scanners.
Another interest lies in the flexibility of integration, because it is adapted to any vehicle speeds, from near-zero to very high speeds (mountings on-board supersonic combat aircraft are in service). Associated to a tilt head, it could be used on fixed platforms like for chemical plant monitoring.
This type of sensors is traditionally mounted onboard any flying platform (rotary or fixed wings, high-speed or low-speed vehicles, military or civil aircraft, drones, UAVs, ...). The scanning process is usually performed vertically, but could be made tilted for stereoscopic sensing through a pair of sensors, typically with a 20° angle between the lines of sight.
The constraint of a moving platform with a quite constant speed could be overcome by mounting the line scanner on a tilt head : the large value of the lateral field of view is maintained while the instantaneous field of view is always relatively sharp compared to TV cameras.

Evaluation of CYCLOPE line scanners

The CRECERELLE French Army UAV system comprises a CYCLOPE sensing suite of 2 line scanners. A first one operates in the visible spectral bandwidth while the second in the Long Wave InfraRed (8 to 12 microns).
Following first trials, the French MoD has awarded to SAGEM S.A. a contract for the evaluation of the payload of CRECERELLE, the French battlefield surveillance UAV system. This payload consists of one forward looking visible TV camera and two line scanners : one operates in the visible spectral bandwidth, the other in the thermal infrared (Long Wave InfraRed : 8 to 12 microns).
Present results are very encouraging for future mine detection through line scanning.
The detection is clearly an indirect one against burried mines. Burrying causes some detectable disturbances in the soil.
The first trials have shown a detectability during all day for mines mechanically laid out with depths of hole between 10 and 35 centimeters. The detectability is possible for flight heights up to a few hundred meters.
Unlike other results described in the

literature, there is no perceptible contrast change during four hours of trials through the morning. The reason might be the nadir viewing the of the sensor.
Results and especially images will be presented during session 6 of the conference.
Sensor design optimising, under test enhanced mine image processing as well as algorithms for automatic mine detection are all very promising topics.

CONCLUSIONS

The use of line scanning is quite interesting because of the intrinsic qualities of this kind of sensors :
- simple architecture, robust design,
- very large lateral field of view, i.e. quick and wide area coverage,
- quite high geometrical resolution,
- reduced data rate for data link, data processing and storage,
- wide flexibility in choosing the spectral bandwidths,
- high flexibility for integration into flying or ground platform,
- wide ability for data fusion and georeferencing.

Following first trials, the French MoD has awarded to SAGEM S.A. a contract for the evaluation of the line-scanning payload of CRECERELLE, the French battlefield surveillance UAV system. This payload consists in particular of two line scanners : one operates in the visible spectral bandwidth, the other in the infrared.
Present results are very encouraging for future mine detection through line scanning : sensor design optimising, under test enhanced mine image processing as well as algorithms for automatic mine detection are all very promising topics.
Today, SAGEM heliborne line scanning systems already perform visual mine detection.

THE DETECTION OF MINES USING RF / MILLIMETRIC RADIOMETRY

R Frost, R Appleby, S Price, F Nivelle,
M Allin, G D H Hawkins.
TME Ltd. UK, DRA UK, TME SA France

1 SUMMARY

Radiometric images of anti-tank and anti-personnel mines have been taken at several detection ranges in a variety of conditions. This paper gives brief descriptions of the equipment used and the radiometric detection mechanism involved. Samples of the images and the results obtained are presented. It is qualitatively concluded that, with a radiometric sensor, surface mines are clearly visible and there is a significant degree of visibility for buried or covered mines.

2 INTRODUCTION

Radiometric imaging equipment, known as MITRE, has been developed by TME Ltd, under contract to DRA Malvern, for use as a demonstrator and data gatherer for a range of potential applications. The equipment, shown in Figure 1, can operate at 94 GHz or 140 GHz using eight parallel receiver channels. It has a maximum thermal sensitivity of 0.1°K with an angular resolution approaching 3 mrad at 94 GHz and can be focused from 7 metres to infinity.

Figure 1 -MITRE Imaging Radiometer

DRA Malvern and DRA Chertsey have now used MITRE in 94 GHz mine detection trials carried out at Sennybridge Range.

TME SA have used passive millimetre-wave technology since the beginning of the eighties for all weather detection of tanks. It has been shown that thermal contrasts of several dozen °K are observed between a terrain background and tanks or many other man-made objects. Such contrasts are still observed in bad weather conditions, smoke and camouflage.

In parallel with the MITRE trials work TME SA have used an experimental 94 GHz brassboard radiometer, known as MELCHIOR, in similar mine detection trials. MELCHIOR is shown in Figure 2 and its performance is summarised in Table 1.

Further trials work has also been carried out by TME Ltd in which data were gathered at 140 GHz and 94 GHz.

This paper gives an explanation of the radiometric detection process, presents results and describes the findings of the mine detection trials described above.

TABLE 1 -Characteristics of MELCHIOR

CHARACTERISTIC	VALUE / UNITS
Spectral Band	3.1-3.3 mm
Field of View	min : 15°x15° (75x75 pixels) max : 60°x 30° (300x150 pixels)
Focus	2m to ∞
Thermal sensitivity	0.5°K
Resolution	12 mrad
Image frequency	0.008 Hz 120s
Weight	60 kg

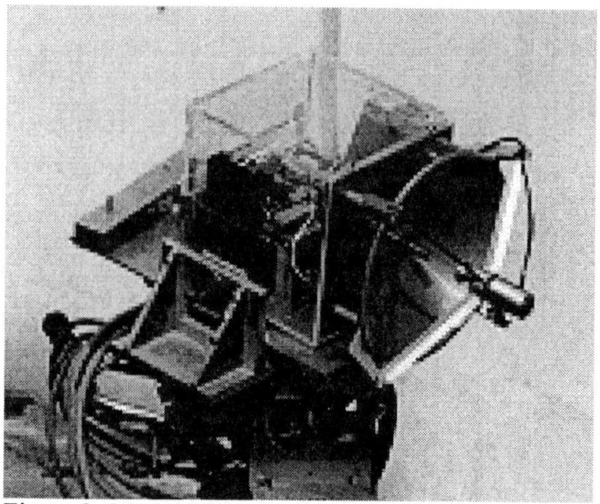

Figure 2 -MELCHIOR Radiometer

3 THE RADIOMETRIC DETECTION PROCESS

The radiometric temperature of an object is a measure of the level of radiation emanating from the object within the bandwidth of the radiometer viewing it. This radiation is a combination of emitted radiation, in proportion to the object's emissivity and absolute temperature, and reflected radiation in proportion to the object's reflectivity and the radiometric temperature of its surroundings.

In an outdoor scenario, the predominant source of reflected radiation is the sky. At millimetre wave and microwave frequencies in particular the sky is very cold and thus objects with high reflectivity, such as metal objects tend to appear much colder than their surroundings. Dielectric objects also generally contrast with their surroundings provided that there is some difference in dielectric constant. Since the radiometric temperature of the sky is largely unaffected by the time of day, and still remains significantly cold in the presence of fog, cloud and rain, radiometric imagers can operate at any time in a wide range of weather conditions.

Radiometric imagers are able to achieve a degree of penetration through dielectrics and, as the operating frequency is reduced the penetration increases.

For the mine detection application, radiometric imagers are able to detect small changes in the level of sky reflection from the ground caused by mines on, or close to, the surface. Because the sky essentially covers a hemisphere, detectable specular reflections occur for a wide range of mine orientations. Its ability to penetrate the ground allows a radiometric imager to complement Ground Penetrating Radars (GPR) which have difficulty detecting close-to-surface mines because of the strong reflection from the air-ground boundary which may even cause blinding.

The visibility of mines to optical and IR sensors is significantly reduced by paint camouflage and even a few blades of long grass. The radiometric imager's ability to penetrate foliage and paint allows it to detect mines under these conditions.

A theoretical model which describes a buried mine has been developed. This represents the ground and mine as cascaded lossy transmission lines. The characteristic impedances of these transmission lines are calculated from the complex dielectric constants of the ground and mine while the reflection coefficients at interfaces are calculated using the Fresnel equations thus allowing for angle of incidence and polarisation effects. This model shows some consistency with the presented results.

4 SURFACE MINE RESULTS

Images of a variety of surface laid, deactivated anti-tank and anti-personnel mines have been collected at Sennybridge Range by a team of staff from DRA Malvern and DRA Chertsey, using 94 GHz MITRE.

The MITRE imager, shown in figure 1, utilises object space scanning; the picture is constructed by scanning a series of contiguous 8-channel vertical swathes. The antenna is of the Cassegrain type with the primary reflector being a 1.2m square paraboloid giving a half-height beamwidth of 0.18° for each of the beams. The imager is programmable and allows a choice of thermal sensitivities ranging from 0.1 to 0.8K and a range of fields of view. The focus of the imager can be adjusted from 7m to infinity. The shortest frame time is about 1s for a 32 by 32 pixel image at 0.28K temperature sensitivity. A more typical image would be 256 by 256 pixels with a temperature sensitivity of 0.4K and would take 32s to collect.

Examples of the 94 GHz images collected at Sennybridge are presented in Figures 3 to 6. Figure 3 shows an image taken at a range of 7m of a plastic SB81 mine located on a smooth tarmac surface. The top edge of the mine reflects cold sky to the imager and thus appears as a cold object against the more emissive and hence warm tarmac (in these images white is cold, black is hot). Figure 4 shows the same mine situated in rough grass and partially concealed by tussocks of long grass. Figures 3 and 4 serve to demonstrate that mines can be detected passively at 94 GHz due to their temperature contrast with the background..

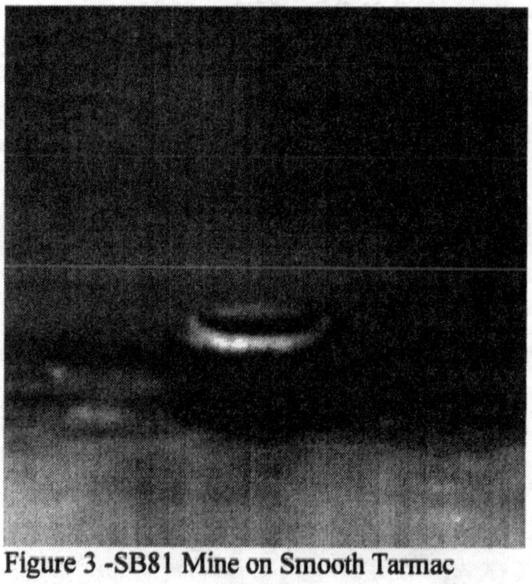

Figure 3 -SB81 Mine on Smooth Tarmac

An image of a selection of mines placed in the tyre tracks of a Landrover on a grassy bank some 75m from the MITRE imager is shown in Figure 5. The mines appear as cold objects against the surrounding grass.

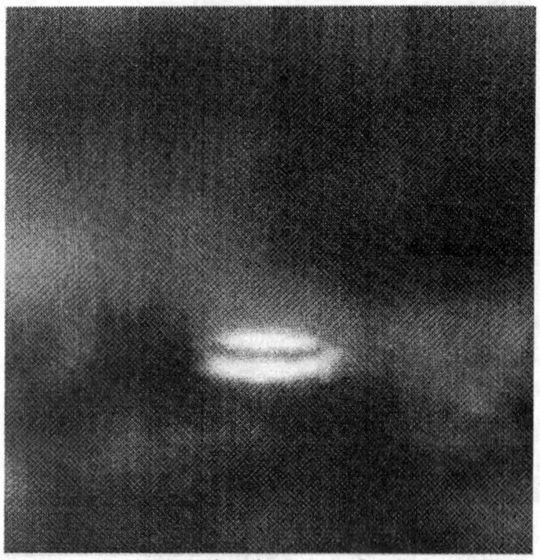

Figure 4 -SB81 Mine in Long Grass

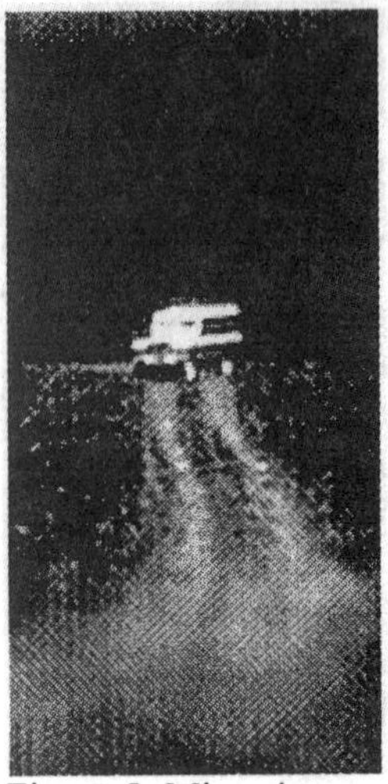

Figure 5 -Mines in
Tyre Tracks

Figure 6 demonstrates detection of three mines located in front of a Landrover on a shale forest track flanked on both sides by trees with the imager 70m from the mines. This image was taken in very wet conditions through heavy drizzle with poor sky illumination. The images given in Figures 5 and 6 demonstrate the ability to detect mines at an appreciable range with the MITRE equipment. They also suggest how the scene might appear to a vehicle-mounted passive millimetre-wave sensor

MELCHIOR is being used in France to gather data on both metal and plastic anti-tank and anti-personnel mines.

Figure 6 -Mines on Shaded Shale Track

These trials are divided into three stages. The first stage concerns close range detection at about 2 metres which is potentially applicable to hand held applications. The second stage covers mid range detection from 5 to 50 metres which potentially applies to vehicle mounted systems (this phase is currently in progress). The final stage will provide data at range of over 200 metres for stand-off detection systems.

In addition to MELCHIOR, three IR sensors are used for comparison. The line-of-sight of all these sensors are harmonised to allow future comparison and data fusion processing. The three IR sensors used are a MWIR line scan camera, a MWIR array detector and a LWIR uncooled camera.

In the close-range trials a metallic mirror was placed near the mine and observed with the mine. This mirror was oriented to directly reflect the sky into the first columns of the imager. These columns were used to calibrate the whole image and to eliminate the effects of fluctuations in sky conditions.

Figure 7 shows sample images for surface anti-tank mines and Figure 8 shows sample images of anti-personnel mines on sand and on soil.

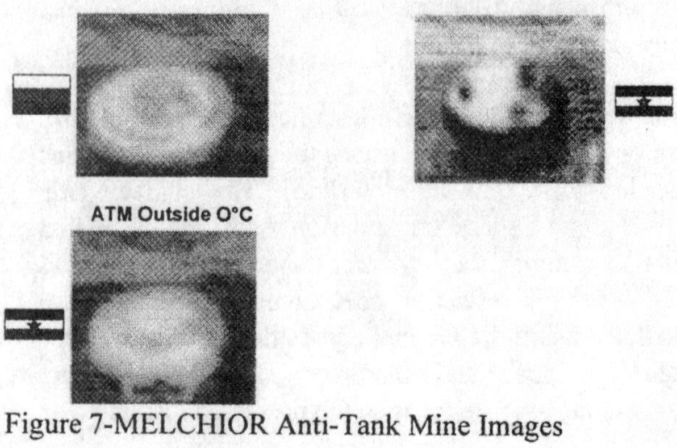

Figure 7-MELCHIOR Anti-Tank Mine Images

Mines	On the Sand	On the Soil
VS50		
NR409		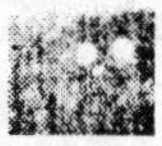
PMN		
PMN2	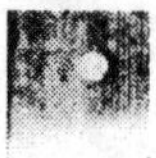	

Figure 8-MELCHIOR Anti-Personnel Mine Images

It was found that all anti-tank and anti-personnel mines were visible on both surfaces.

Tests are in currently in progress at mid range. Figure 9 shows two metal anti-tank mines laid on grass, 10 cm high, at a range of 35 m, with a depression angle of 13°. These preliminary results demonstrate the visibility of mines at ranges which are appropriate to a vehicle mounted system.

Figure 9 -Two Anti-Tank Mines at 35m Range

5 BURIED/COVERED MINE RESULTS

MELCHIOR was used to image covered mines. The trials were performed in winter with an external temperature of about 0°C. The mines were laid and covered 24 hours before any measurements were taken in order to ensure that the results were not influenced by any initial temperature differences between the mines and the ground.

A range of conditions were tested for anti-tank and anti-personnel mines and a sample image for an anti-tank mine, covered with about 10cm of wet leaves is shown in Figure 10.

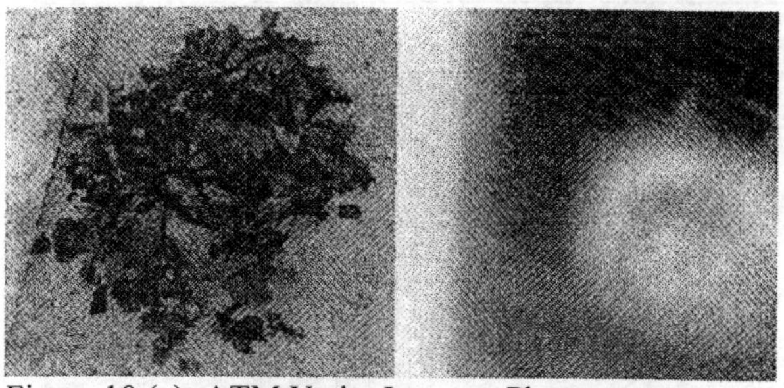

Figure 10 (a) -ATM Under Leaves - Photo
(b) -ATM Under Leaves - Image

It was concluded that, for all mines, detection seems possible. The plastic anti-personnel mine images were, however, of less well defined shape and appeared smaller, probably due to attenuation and bias produced by the foliage.

These effects were not noticeable on the larger anti-tank mines or with any objects which had emissivity coefficients around 0.1 (ie. metal objects).

Trials are continuing in order to determine the effects various weather conditions, particularly moisture content, and the effects of seasonal changes.

TME Ltd have independently carried out close range trials, using MITRE receivers, at 94 GHz and 140 GHz. The findings were consistent with those of TME SA and DRA in that all surface mines were visible. Mines covered with leaves were still visible but noticeably less distinct and mines buried in sand or soil could still be seen but, as anticipated, penetration was much better at lower frequency.

This work was aimed primarily at the detection of anti-personnel mines, which may be buried or covered with leaves, using hand held equipment.

Table 2 gives a qualitative summary of all the results, using the following mines, spaced at 300mm intervals :

1. Plastic anti-personnel mine
2. Metal anti-personnel mine
3. Plastic anti-personnel mine
4. Plastic / metal anti-personnel mine
5. Metal anti-tank mine

Figure 11 shows a typical 94 GHz image of mines buried in sand.

At both 94 GHz and 140 GHz, covered mines were visible but, in some cases difficult to distinguish from clutter. Further data and statistical analysis would be necessary in order to determine detection and false alarm probabilities.

It is particularly significant that covered plastic mines were often clearly seen.

TABLE 2 -Summary of Results		
Condition	Mines Visible (%) / Mine Types (nos.)	
	140 GHz	94 GHz
Tarmac - surface	100% 2 3 4 5	100% 1 2 3 4 5
Grass - flush	100% 1 2 3 4 5	100% 1 2 3 4 5
Long Grass	67% 3 5, not 4	-
Grass + 5-10mm leaves	80% 2 3 4 5, not 1	100% 1 2 3 4 5
Sand - flush	100% 1 3 4 5	100% 1 3 4 5
Sand 2-5mm cover	100% 1 3 4 5	100% 1 3 4 5

1 3 4 5

Figure 11 -94 GHz Image of Mines Beneath Sand

6 CONCLUSIONS

The applicability of passive RF / millimetre wave imagers to mine detection has been demonstrated. The detection mechanism has been described and a variety of trials measurements have produced promising results which include the detection of buried plastic mines.

The potential mine detection applications have been grouped into the categories of long medium and short range systems. Long range systems will require the best angular resolution and are likey to operate at millimetre wave frequencies in order to achieve high resolution with the smallest possible antenna. Data gathering trials for such applications are currently in progress.

Medium range systems will also require good angular resolution. The visibility of surface mines and mines obscured by foliage, in both good and unfavourable environmental conditions, has been demonstrated.

Short range systems are assumed to be hand held. Some ability for such systems to detect buried mines has been demonstrated. A passive millimetre wave sensor could, for example, be used in a dual mode system, with GPR, to improve performance against mines which are buried close to the surface in the region where the GPR cannot discriminate between surface reflections and mine reflections. It may also provide some additional capability for more deeply buried mines.

Optical Mine Reconnaissance at the National Defence Research Establishment. Multispectral Imaging and Classification: Thermodynamic Soil Modelling.

A-L Christiansen and S Ringberg
National Defence Research Establishment, Department of Sensor Technology, Linköping, Sweden

ABSTRACT

Optical mine reconnaissance is carried out with sensors in the spectral band 0.2-12 μm. The procedure is multispectral in that a number of different sensors, i.e. IR cameras and a multispectral camera, are combined or that a number of spectral bands are analyzed with a camera by filtering methods. The sensors can be used to discover both surface-laid and buried mines.

A thermal simulation model is used for theoretical studies of the physics behind the flow of heat between the mine and its surroundings. The purpose of the model is to show the influence of the different parameters. The studies are carried out in a laboratory and field measurements are also performed in relevant types of terrain. Some introductory measurements have been made using both the multispectral system and the thermal simulation model.

INTRODUCTION

The demand for new technologies in mine reconnaissance increases when there is not only a need to detect both metallic and non-metallic mines but also to separate metallic mines from harmless metallic fragments. During a long period, the National Defence Research Establishment, FOA, has been active in research on radar technology for mine detection. Recently, efforts have been made using optical techniques and spectroscopical methods. Fig. 1 shows FOA activities in the development within the mine detection area. More than one technique is needed to get the most reliable result.

Optical mine reconnaissance at FOA consists of two parts: a multispectral system and thermodynamic soil modelling. The multispectral system consists of two camera systems. One using six wavelength bands, the multispectral camera, and one using two, thermal scanners. This gives eight wavelength bands from UV to thermal IR, 0.2-14 μm. A mine placed on the surface and a mine buried in the ground can be found due to contrast variations in the collected images. In the latter case, mines will be difficult to detect due to smoothing and dithering of the thermal characteristics of the soil surface. The variations in contrast can arise due to secondary effects. In most cases, it will be necessary to register these effects which can include digging marks, loose earth, variations in the vegetation because of mine burial, and even disturbances in the ground's natural drainage. These traces will most probably be primarily detected with thermal cameras but also in the UV-VIS-NIR wavelength band.

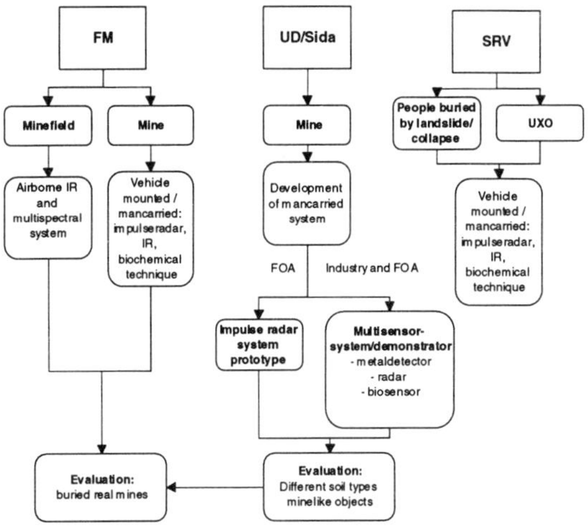

Fig. 1. FOA activities in the development within the mine detection area. FM = Swedish Armed Forces, SRV = Swedish Rescue Agency, UD/Sida = Swedish Ministry of Foreign Affairs/Swedish International Development Authority.

Spectral information from different types of mines and different local backgrounds are collected and analyzed using classical statistical classification algorithms and algorithms of properties of shapes. Previous experience, (1) and (2), has shown that the problem with the detection of objects in the ground using IR methods is very complex. Our intention is that laboratory activities will successively turn to more field-like conditions by measurements of realistic terrain conditions. One important part of our activities consists of improving the methods for classification after the discovery of potential mines by means of effective signal processing. In reconnaissance applications using common imaging equipment, it can be difficult to differentiate the desired object from the background. Multispectral techniques can then be used since it contains more information than images from common cameras, owing to better object detection and identification than a broad-band sensor. The investigation is concentrated on a development of a method for minefield detection based on the multispectral technique.

SYSTEM DESCRIPTION

Multispectral System

The *multispectral camera* shown below the two thermal scanners in fig. 2 has a sensitivity from ultraviolet to near infrared, 200-800 nm. The total spectral bandwidth is currently divided into six narrow spectral wavelength bands with central wavelengths/bandwidths of 350/80, 400/40, 500/20, 550/10, 680/10, and 760/40 nm. The filtering method makes it is easier to find and separate man-made objects from natural objects in the scene in question. Several algorithms are tested and analyzed to separate objects from the background. The rotation of the filter wheel is synchronized with the video frame rate. The camera is a second generation image amplifier and the photocathode is of UV type.
The *thermal scanners* in fig. 2 are radiometric imaging devices. The shortwave scanner has two InSb detectors and the longwave scanner has one CMT detector; both are cooled using the Stirling technique. The scanners have three separate built-in temperature sensors and two temperature references to compensate for the radiation emitted by the scanners themselves. An additional temperature sensor is located at each interchangeable lens to compensate for radiation from the lens unit itself due to e.g upheating effects from the surroundings e.g solar radiation. This is performed by a microprocessor controlled system, the references are scanned and the temperature sensors are measured. The scanning is non-interlaced and can operate at two different frame rates, 15 Hz and 30 Hz. It is also possible to use a line scanning mode.

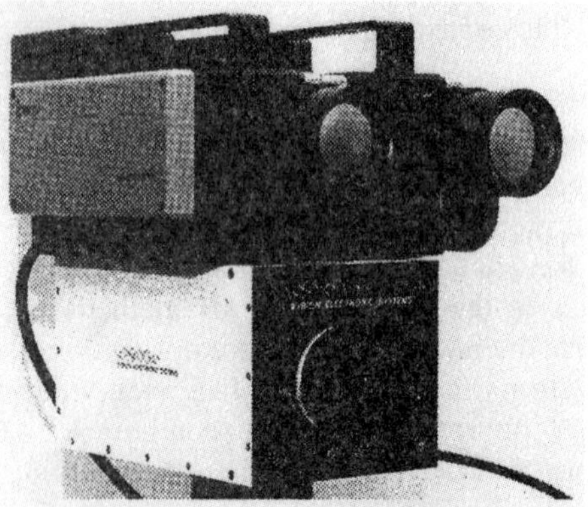

Fig. 2. The multispectral system: two thermal scanners, 3-5 µm and 8-12 µm, and a 6-band multispectral camera.

Laboratory sand landscape

FOA has a sandbox with a solar panel, and a digital temperature recorder to carry out controlled laboratory experiments. The sandbox is shown on the left picture in fig. 3 and consists of a box of sandwich construction of glassfibre-reinforced plastic, 1 m x 1 m x 0.4 m, isolated from the surroundings.
The light ramp is movable and consists of 64 light sources that give a maximum irradiance of approximately 1000 W/m^2 at the surface of the sand and has a spectral distribution which corresponds well with that of the sun's in the wavelength range 0.3-3 µm. The ramp can be split in two, making it possible to mount IR cameras between the two parts. The irradiance can be varied using a transformer.
Examples of how the mine and the temperature probes can be placed in the sandbox are shown in cross-section to the right in fig. 3. They have been arranged to minimize disturbance of the measurement. The digital temperature recorder and its accompanying logging program, PC-based, can be used for 16 temperature probes.

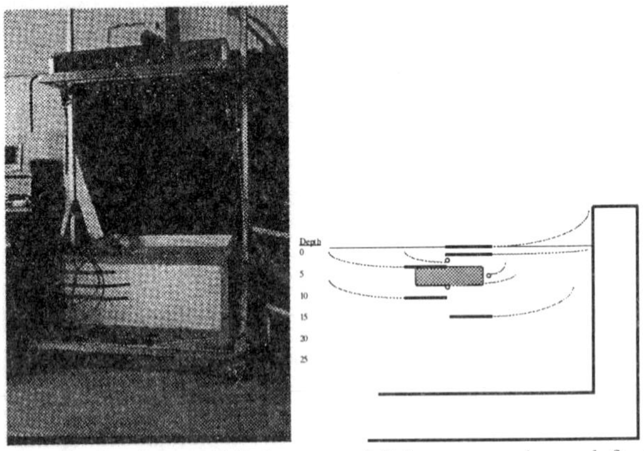

Fig. 3. Sandbox and light ramp, 0.3-3 µm, are shown left; the placement of the mine and the temperature probes in the sandbox are shown in cross-section to the right.

Experimental landscape

In the outdoors experimental landscape a number of sample areas with grass, asphalt, sand, earth, clay, and coarse sand mixed with stone will be constructed. These will have an area of 2 m x 2 m and a depth of 0.5 m and will be constructed in the natural ground. Furthermore, a number of different soils in wooden boxes 3 m x 5 m and 0.7 m deep will be constructed to be able to control the moisture content in these soils. The boxes will be protected by a removable cover. Mines and stones of different sizes will be buried in each soil at different depths. Surface registrations of temperature will be made at different times of the day and night.

Weather Station

A number of weather and radiation parameters will be measured using a weather station. Studies of how the variations in the collected parameters influence the classification of the results will be carried out, including how variations effect the thermal contrast between the object and the background.
The weather station is a stand-alone system with an automized logging facility, either battery or AC-powered. Measured weather and radiation parameters are visibility, wind speed, wind direction, air temperature, soil temperature, relative humidity, air pressure, rain, net

radiation, solar radiation (pyranometer), and sky radiation (pyrgeometer).
The pyranometer measures hemispherical irradiance in the wavelength band 0.3-2.8 μm. The pyrgeometer measures hemispherical thermal irradiance in the wavelength band 3-50 μm. The net radiometer measures net radiation in the wavelength band 0.3-60 μm.

METHOD DESCRIPTION

Image sequences from the multispectral system shall be registered during different times of the day/year, of both buried and surface-laid mines of different types, and in different types of soils: asphalt, earth, sand, grass, etc. Different types of weather and radiation parameters shall be registered with the weather station.
The data registered by the weather station will be used to correlate the influence of these parameters with the images of the different wavelength bands. Analysis of the number of wavelength bands necessary and the choice of the optimal center wavelengths and bandwidths used in the filter shall be performed. A number of classification and object detection algorithms shall be evaluated, e.g. maximum likelihood, neural network as well as how the classification is influenced by variation in the surroundings.
To be able to simulate temperature contrasts on the ground surface which are a result of buried mines, FOA purchased a software *simulation model* and continued development in cooperation with SAAB Missiles, now SAAB Dynamics. The model was acquired to be used in a theoretical study of the physics underlying thermal flow and temperature contrast between mines and their surroundings (1) and (3). It is to be used as an aid in optical mine reconnaissance, to compare theory with experiment. The results are relevant for the sensors which work in the thermal IR bands. The program and the underlying theory is described in (1), (3), and (4). The model takes thermal conductivity into consideration in the ground and how thermal exchange with the surroundings occurs.
Using the model, primarily the processes of the thermal cameras should be understood. Secondly the model is used to describe and predict what will happen and get a feeling for the meaning of different factors, e.g. thermal factors and the nature of the ground.
Fig. 4 shows the thermal interaction on which the model is based and includes the following factors: solar radiation, 0.3-3 μm, wind chill, evaporation, heat capacitivity, c_p, thermal conductivity, λ, for the ground and the mines; and radiation exchange with the sky, 3-60 μm. The radiation exchange consists of atmospheric radiation, εI_{atmos}, and ground radiation, $\varepsilon\sigma T^4$.

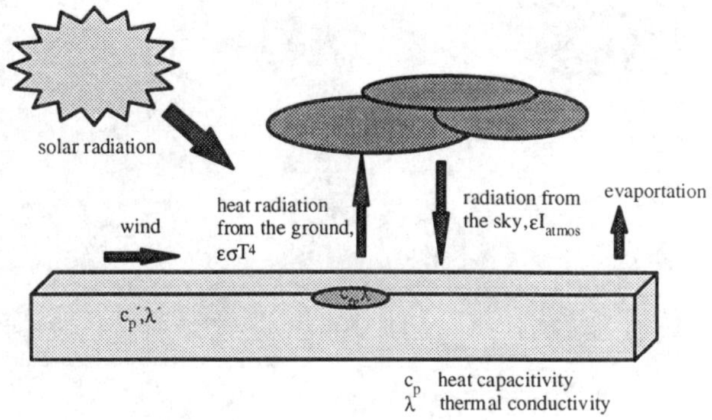

Fig. 4. Thermal soil model.

PRELIMINARY MEASUREMENTS

Multispectral system

Vegetation has a low reflectance in the wavelength band 0.35-0.7 μm due to absorption by chlorophyll. The reflectance in the spectral band 0.7-1.35 μm is higher as a result of the high scattering caused by the porous cell structure of the chlorophyll cells. The reflectance is approximately 50% in this wavelength band and accounts for the differences in contrast seen in the 3rd and 4th pictures in fig. 5 below.

Fig. 5. Images from a multispectral camera with the following filtration values: 450, 550, 650, 750, 850, and 950 nm.

Figs. 6 and 7 are from the thermal scanners and show some backgrounds and different types of mines in the longwave band. The graphs show apparent temperature vs. position along the line. The starting point of the line in the graphs is at gravel and continues to grass, anti-personnel mine, anti-tank mine, grass, and finally asphalt. The black body air temperature reference in the image is a double cone mounted on a tripod.
At daytime, the mines give positive peaks due to warming effects from the sun and at nighttime negative peaks due to reflection of the cold sky. The grass surface shows a relatively large variation in apparent temperature, primarily because of angles dependance of emissivity and reflectance. This would contribute to a masking effect of any buried mine.

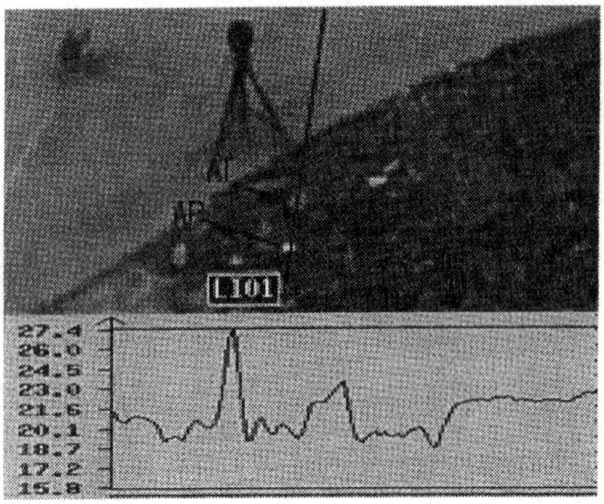

Fig. 6. Daytime image and corresponding graph, range 8-12 μ m, showing apperent temperature vs. position along the line. (AT=anti-tank and AP=anti-personnel mine)

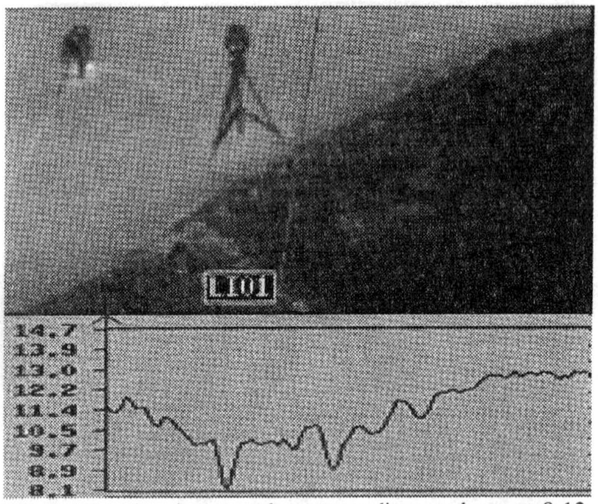

Fig 7. Nighttime image and corresponding graph, range 8-12 μm.

Fig. 8 shows grass and an anti-personnel mine (thick line). The values are from the mean values of the grey level of the contrast in the collected images inside a box placed over the object of interest (8-12 μm). The variations in the curves are largely sinus shaped, and the curves are modulated by factors such as clouds and rain showers. Rapid variations are caused by the sun going behind clouds, while slower variations are caused by rain showers.

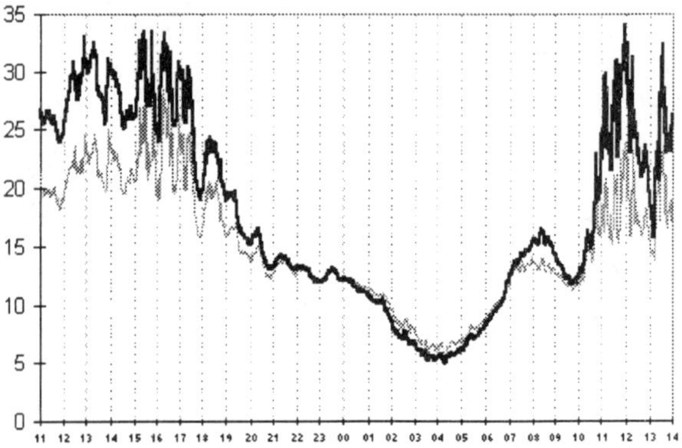

Fig. 8. Apparent temperature vs. time of day from the thermal scanners showing grass and an anti-personnel mine, thick line.

Thermal soil model

A simulated run with a generic aluminium mine. Simulations have been made with a generic aluminium mine of the same dimensions as an anti-tank mine. Fig. 9 shows a simulation and its measurements. The left figure shows the results from an aluminium mine buried between 7 and 12 cm. Examples of parameters are thermal conductivity, albedo, atmospheric radiation and emissivity. The coordinates in Fig. 9 are depth, time, and temperature.

The simulation begins at 16:30 and continues until about 18:30. Solar radiation is a step function. It is turned off after 17:00. It can be seen that the surface temperature slowly follows the step function of the solar simulation.

At the beginning of the simulation, the temperature is homogeneous in the vertical dimension of measurement. After a few minutes, it can be seen that the ground temperature is rising. After approximately 10-15 minutes, a plateau begins to form at approximately 7-12 cm where the simulated mine is placed. The mine in this case is made of aluminium, a good thermal conductor.

At the lower border between the mine and the sand another decreasing flank can be seen the heat is slowly spreading down in the ground. The surface layer reacts more quickly to variations in incoming radiation than those layers further down in the sand. The plateau, indicating the position of the mine, is still increasing in temperature.

Fig. 10 is a rotation of fig. 9 and shows clearly that when the temperature in the surface layer decreases, the heat is retained further down and the temperature of the mine continues to increase.

A measurement with a generic aluminium mine. To the right in Figs. 9 and 10 is an actual measurement shown with corresponding simulation on the left. The lines for constant depth correspond to the different positions of the temperature probes. The figure with the actual measurements is similar to that of the simulation. Two things seen in the previous simulation, the temperature slowly spreading downwards and the plateau between the edges of the mine, can also be perceived in the experiment. In Fig. 10, it can also be seen that when the temperature in the surface layer decreases after the sun has disappeared, heat remains further down and the mine plateau continues to increase in temperature.

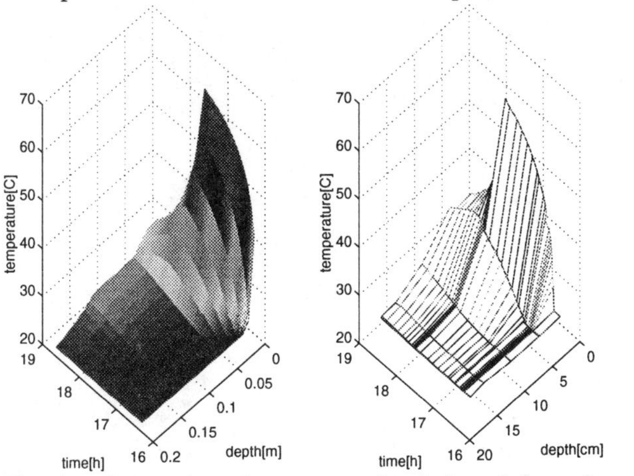

Fig. 9. Comparison between a simulation, left, and an actual measurement, right, of a generic aluminium mine buried in a sandbox at 7-12 cm's depth

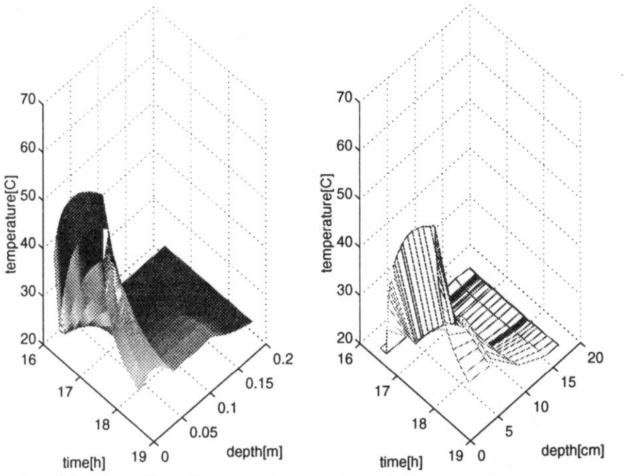

Fig. 10. As fig. 9, rotated, a simulation and actual measurement of a generic aluminium mine buried in a sandbox.

Classification and object detection

Fig. 11 describes the three steps in multispectral classification. First, a series of images is taken of a scene and entered into a computer. Each series consists of eight images in different wavelength bands. Each pixel therefore has eight digital values, one for each wavelength band. Training data was collected at an earlier stage and describes different known objects and the spectral distribution of natural phenomena. The training data can be seen as a database of "spectral fingerprints".

In the classification stage, each pixel position in the series is compared with each category of training data: sand, water, stone, etc. Each pixel position is assigned to a specific category and given that classification in the classification image. If the pixel position does not correlate with any of the categories, the position is assigned to an unclassified category. The classification image is an image where each pixel has a label, e.g. grass or mine.

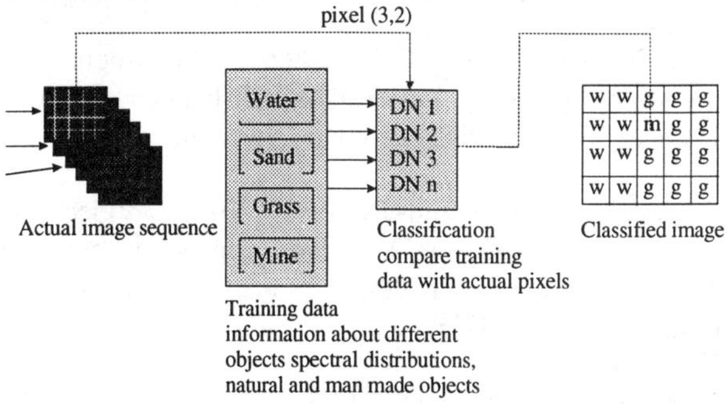

Fig. 11. Description of typical classification flow.
DN stands for digital number.

The first image in Fig. 12 shows the classified result of a multispectral set with the database classifications *grass* and *extracted anti-tank mines.* The second image shows the result where a simple object detection algorithm has been used to analyze the classified image and localize the position of the mines and extract them even more.

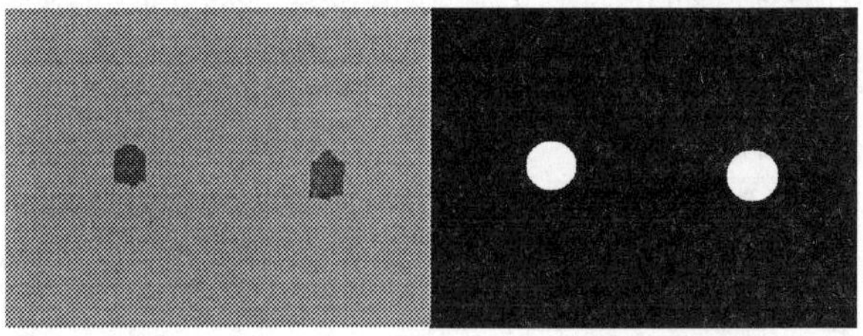

Fig. 12. Classified and object detection images.

CONCLUSIONS

Using measurements obtained from a multispectral system and peripheral equipment such as a weather station and soil models, we shall evaluate and attempt to determine where, when, and how it is possible to detect mine fields with optical methods and which classification and object detection methods are necessary to yield the desired result.

References

1. Peterson L et al., 1992, "IRmm-IT4 slutrapportering av simuleringar av yttemperaturen hos enklare markytor", SAAB Missiles, Sweden.

2. Petterson H, 1989, "Värmeutstrålning från terräng", National Defence Research Establishment, Sweden.

3. Peterson L, 1994, "Simuleringsmodell för minletning med IR-teknik, 2-dimensionell modell", SAAB Missiles, Sweden.

4. Axelsson S, 1980, "On the accuracy of thermal inertia mapping by infrared imagery", 14th Symposium on Remote Sensing of Environment, Costa Rica.

5. Petterson et al., 1992, "Mätningar mot markbakgrunder i de termiska IR-banden", National Defence Research Establishment, Sweden.

6. Witherspoon N and Holloway, 1992, "Video based multispectral detection and land mines, a technology applicable for use in law enforcement technologies", Orlando, Florida, USA.

7. Witherspoon N et al., 1995, "The coastal battlefield reconnaissance and analysis (COBRA) program for minefield detection", proceedings of the SPIE Detection Technologies for Mines and Minelike Objects, Orlando, Florida, USA.

8. Del Grande N et al., 1993, "Dual-band infrared capabilities for imaging buried object sites", proceedings SPIE, Orlando, Florida, USA.

9. Del Grande N, 1991, "Airborne detection of buried minefields", article in E&TR, December.

GROUND PENETRATING RADAR - A SENSOR FOR MINE DETECTION

Dr Richard J Chignell
EMRAD Limited, Godalming, Surrey

1. INTRODUCTION

The reliable detection of all types of mine in a wide range of terrains under highly variable conditions is one of the most demanding problems to challenge technology. The widespread use of both plastic anti-personnel and plastic anti-tank mines, which contain so little metal that they can not be detected by conventional means, is at the core of the problem. Such mines are comparatively inexpensive to lay (a few dollars) but, typically, at least a hundred times more expensive to eliminate by currently available manually intensive, low technology methods. The low cost of laying such mines is attractive to a poor third world country engaged in a war, but the costs of clearance with currently available techniques are not affordable, preventing the re-habilitation of debilitated communities.

The challenge to technology is thus two fold, firstly to realise a detection system that is capable of meeting the very high detection rates required to deal with these deadly targets and, secondly, to achieve a search rate that is sufficiently rapid to reduce the massive costs, that currently prevent clearance operations being funded. Intimately linked with these two requirements, is the need to achieve a very low false alarm rate. In some clearance operations, with conventional metal mine detectors up to one thousand fragments of metal have to be removed, to ensure that one mine is detected. Elimination of such false positives is a key factor in achieving the search rates required to achieve financial acceptability.

It is recognised that mine detection is only one part of the process of minefield removal, which includes disposal and many requirements for logistic support. However, detection is such a key issue in directing other activities, that it is the only topic covered in this paper.

Over fifty different technologies have been suggested or trialled for plastic mine detection. Of these, very few, and possibly only ground penetrating radar or ground probing radar, have demonstrated even under highly controlled trial conditions, the reliable detection of plastic mines. The first effective work was carried out by Prof. Leon Peters (Reference 1) of Ohio State University at the time of Vietnam. He demonstrated the detection of plastic anti-tank mines with simple radar and classic predictor-correlator detection techniques.

In the mid-1980's the present author led the Falkland Island research project into ground-probing radar for mine detection. The project was the first major demonstration of the use of digital signal processing to analyse ground-probing radar data. There are nine mine types laid in the Falkland Islands, six of which are plastic or "minimum metal" and three of which are metallic types. Of the plastic mines, three are anti-tank and three are anti-personnel. The project involved extensive measurement of the characteristics of all the mine types in a wide range of different terrain, followed by the formulation of detection procedures. The project was successfully concluded, when a trial minefield laid by 33 Engineer Regiment was searched and every mine found. This included a high percentage of plastic anti-personnel mines.

The Falkland Islands project conclusively demonstrated the feasibility of using ground-probing radar to find plastic and metallic mines but, at that stage, there was no requirement or commitment by the military to take the technology forward (Reference 2).

The Falklands project was successful because of the small number of mine types and the limited range of terrain involved. This allowed effort to be highly focused upon a small range of parameters which allowed real progress to be made. Mine clearance operators would like universal detectors that can be used anywhere in the world without special precautions. There is however a huge development gap between being able to reliably operate in a narrow range of terrain against a limited number of targets and the universal requirement.

To illustrate the scope of the work required, it is interesting to consider EMRAD's experience in developing a universal pipe and cable detector, able to reliably detect pipe and cable in any soil type or installation. The requirement is again to detect close to 100% of all targets, but this has only been realised after eight years of development with three generations of hardware and four generations of software. However the key to success was the analysis of representative radar data from over ten thousand locations, representing many of the ways a pipe or cable could appear on a radar.

The other key point to note, is that the Pipe Hawk programme, EMRAD's universal pipe and cable locator programme was only fundable by New York Gas Group, because realistic detection criteria were set at the outset. Prior to the work starting, the Gas Research Institute of Chicago had examined the performance achieved by conventional metal detectors operating against metal cable. The conclusion was, that a radar system that achieved a 90% detection rate against all types of pipe and cable in 90% of all conditions, would represent a step function improvement over previous practice. In reality the development goal has been to achieve a performance of effectively 100% detection, but the 90/90 criterion provided an achievable benchmark that allowed funding to be authorised.

In examining the requirements of a mine detection system, similar criterion will apply. It should be noted, that the same metal detection technology that contributed to the formulation of the 90/90 criterion is currently used for conventional metal detection, and the detection rate achieved is not 100%. Generally safe operations are realised by carefully controlled site procedures and the area may be searched more than once. In the past the setting of desirable but unrealistically high detection figures for mine detection has stopped the funding of much development. A more pragmatic approach allows progress to be made.

2. MODERN GROUND-PROBING RADAR SYSTEMS

Since the first effective demonstration of plastic mine detection, using research standard equipment in the Falkland Islands project, the development of ground-probing radar technology and systems has advanced significantly. The objective of this paper is to illustrate that development, and indicate how it now can be re-directed to the realisation of effective mine detection equipment.

In the original Falkland Island demonstration of plastic mine detection, it took approximately a week to search an area of less than 50 m^2. Today specifications are being generated for highway inspection systems that search this area in around one minute. Cracks in highways are long thin objects that are easier to detect than mines, but when crossed by a radar, have a radar cross-section comparable to that of a plastic anti-personnel mine.

The engineering of radar systems has advanced significantly and the paper will illustrate this with a discussion of three different systems built by EMRAD Limited, before going on to discuss the types of systems that can now be proposed for plastic mine detection. The three systems to be described are:-

(i) the man portable real-time Concept Demonstrator Mine Detection System, built at the time of the Gulf War and shown in figure 1. Development of this system stopped at the end of the war.

(ii) Pipe Hawk, EMRAD's universal pipe and cable detection system, designed as a street machine to be operated by utility company personnel and shown in figure 2.

(iii) A Multi-Channel radar capability mainly used for highway inspection, but which demonstrates an area search capability.

Each of these systems has contributed towards the many significant advances that have been made in the technology which include:-

(i) Electronic Components. All the major electronics modules used in ground-probing radar have been subjected to design review resulting in much enhanced sensitivity and stability, compared to earlier generations of equipment.

(ii) Processing Power. The availability of digital signal processors has made available processing power that was unimagined even a few years ago and opens up the scope for real-time data analysis.

(iii) Power Consumption. The development of modern components has dramatically reduced power consumption.

(iv) Packaging. All the hardware developments have allowed greater freedom for Design Engineers to most appropriately package equipment. Pipe Hawk is designed as a robust street machine, designed to withstand the environment in the back of a utility truck.

(v) Productivity. One of the major benefits of the development in system electronics is a major increase in productivity. Productivity is the key enabling step for many commercial systems.

(vi) Multi-channel Operation. The other development is in the use of multi-channel radars, illustrated in figure 3. The figure shows a deployment used in highway monitoring, in which four antennas are deployed across the width of a traffic lane. Sequentially each of the antennas is used in turn as a transmitter with all four antennas employed in reception.

(vii) Real-Time Automatic Analysis.

Traditionally ground-probing radar data has been analysed by expert interpretation, but as systems become more productive, now generating Gigabytes of data, expert interpretation becomes overwhelmed. EMRAD have always taken the view that the only long term route to effective systems is via automatic data analysis. The Concept Demonstrator Mine Detection System included this feature. All sixteen channels of data were analysed for every cycle of the radar system, and an audio tone generated. The radar cycled approximately four times a second, fully analysing the data from the 150 mm square footprint of the antenna every cycle. When an anomaly was found that could correspond to a mine, the pitch of the audio output changed.

(viii) Mapping Software.

Combining data from different radar scans allows volumes of interest to be examined automatically and maps or images generated. This is often an essential element of operating in areas where there are many different targets. Data from either multiple scans with a single channel radar such as the pipe locator or from multiple search heads is combined to produce information corresponding to the volume searched. The key to combining data from different areas, is first to focus the hyperbolic responses characteristic of radar targets back to a point source. This process known as migration requires that the characteristics of the ground are determined by the analysis to give the "best" focus. This also provides automatic depth calibration.

Once the response from, for example, a small pipe is focused to a point, the responses from a series of parallel scans can be combined to give the layer information, as shown in figure 4. The layers may then be searched for long thin objects corresponding to pipes or cables. This search is automatic, with the position of pipes being appended to the layer images.

The operator may scroll through the ground looking successively at different layers, or in comparatively simple situations use the sum map which combines all the data.

The performance of the imaging software illustrates some important principles with regard to automatic detection. Combining radar data allows effective signal to noise or signal to clutter ratios to be increased to a point where effective discrimination between target and no target decisions can be made in the detection domain. For example some New York roads are built of ungraded stone allowing many interstitial air voids to form that clutter the radar output. These sites present major detection problems, but by arbitarily increasing the number of scans made to thirty, on a test at forty seven such locations the pipe detector achieved a 100% detection rate with a 3% false alarm rate. The imaging allows an arbitrary number of radar scans to be combined, before the critical detection decision is made.

The principle is clear, if the detection performance of the system is marginal, increase the number of measurements made and combine them to increase the signal to noise or signal to clutter ratio, before the important detection is made. This has a major impact upon how mine detection systems should be configured.

(xi) User Interface.

The User Interface is a key element. In allowing systems to be used by operators with minimal training, the design of the User Interface must present radar data to the operator in a readily understood format, which encourages the highly disciplined operating procedures necessary for accurate detection.

To summarise, modern ground-probing radar systems make use of modern electronics components, including digital processors. The power consumption and other parameters of these systems allows then to be packaged in lightweight, robust configurations. Modern systems are highly productive with the possibility of raising the productivity further, through the use of multi-channel operation. A large volume of data is generated, which may be automatically analysed directly in real time, to give a variety of outputs including audio, or analysed in close to real time (few seconds), to give three dimensional images. Trials show that automatic detection allows plastic targets to be reliably detected. Further software is provided to generate a user friendly interface.

3. MINE DETECTION SYSTEMS

Although the basic feasibility of using ground-probing radar to detect plastic anti-personnel mines in a research context was demonstrated over a decade ago, it has taken the commercial pressure for productive systems, to advance the technology to a point where minefields may be searched at rates that are economically attractive, compared to low technology methods.

The question remains as to how the technology should be deployed and a range of options are likely to be required that reflect the scope of different problems faced by clearance operations. The requirements vary from the need to routinely search roads or tracks, through the need to search fields, to requirements in difficult terrain only accessible on foot. A number of different items of equipment may be necessary, based upon the individual needs of particular countries. It is suggested that these may include the type of systems that have previously been produced to detect only metallic mines and include:-

(i) a vehicle mounted plastic and metallic mine detection system. Essentially this would be mounted upon the front, or in some cases off the side of a vehicle, and allow more than the width of the vehicle to be searched in one pass. With the system operating from the front of the vehicle, the rate of search would be limited by the braking characteristics of the vehicle, to ensure that the road wheel of the vehicle stopped before reaching any detected mines. This is a severe constraint on forward speed, but one that has to be accepted in opening up the first paths through minefields in almost a minefield breaching role.

One of the advantages of operating off the side of a vehicle, is that higher speeds can be used to widen an already established "safe route" through a minefield. This allows areas of land to be progressively searched from the perimeter, with mines found in one pass, being disposed of by the time the vehicle completes a circuit of the area being searched.

This vehicle could have an all terrain capability, be autonomous and manned, but also have extensive blast protection for the crew. It will also be fitted with a marking system, both to define the area searched and any mines found.

(ii) a remotely operated vehicle controlled from a mother vehicle. The objective with this type of approach is to entirely remove the need for a vehicle crew, by operating the search vehicle remotely. Traditionally the thinking here has been along the lines used by the British in dealing with terrorist devices. A large van arrives on site and a "wheelbarrow" crawls out of the back to deal with the bomb.

The most important advantage of this approach is that no crew have to enter the hazardous area, the disadvantage of the "traditional" approach is that the vehicle is small, so the width of the track is small and the overall search rate is slow. However, it should be noted that the true requirement is for a minimal remote vehicle. No bodywork or crew compartment is required, so that a much larger vehicle could be considered, which need be only a skeletal chassis.

(iii) an alternative to the use of remotely controlled vehicles is to use a collection or swarm of small autonomous robotic vehicles, which would be designed to be so lightweight that they should not initiate the detonation of mines. However, if such an event occurred, the cost of the loss would be affordable and no personnel would be present.

(iv) a hand held Mine Detector. There will always be areas where a vehicle cannot reach and to deal with those areas, a hand held mine detector similar to the Gulf War Concept Demonstrator is desirable. A suitable design of such a system has now been established.

The problems of mine clearance are significant and the choice of operational methods in any particular country or region will need to be made, on the basis of detailed assessment of the problems and terrain to be encountered. Experience both in the Falkland Islands and in the development of Pipe Hawk strongly indicates, that the way forward is for a pilot scheme to focus clearly on one small area, so that the problems can be precisely defined and solutions sought. Humanitarian mine detection involves a new and large learning curve, it is essential that this is addressed in a step by step method.

The solutions that are likely to emerge should be significantly different from those that have been established by the military for hot war applications. Large armoured vehicles are not necessary and present logistic support problems that can not be justified in the humanitarian role. Returning land to agriculture is essentially an agricultural task and requires a specialised agricultural vehicle, similar to a modern combine harvester or crop spraying machine. For operation in third world countries the system must be easy to maintain. It may also be of skeletal nature that can be disassembled and easily freighted to different areas.

4. CONCLUSIONS & RECOMMENDATIONS

The problem of humanitarian mine clearance, particularly that posed by plastic anti-personnel mines, is most challenging and requires an innovative approach in almost all areas of activity, if the overall cost of operations is to be reduced to a level where they become affordable. The most demanding aspect is in the detection of these devices and many technologies have been suggested or trialled. However, ground-probing radar is one of the few technologies that has demonstrated, albeit in a research context, a 100% detection rate against these targets. This was in the Falkland Islands project over a decade ago.

In the interim, commercial developments of the technology have been aimed at increasing the productivity of systems to where costs savings may arise, compared to a labour intensive low technology approach. Ground probing radar technology has also developed the range of systems necessary from hand held devices to vehicle mounted equipment, required to deal with the various logistic problems to be encountered.

Many of the building blocks required for a solution of the problem of detecting abandoned plastic and other land mines are now in place. The problem needs to be specifically addressed by a multi-disciplinary team, involving detection, vehicles/robotics, mine and logistic expertise. Pilot schemes are required that provide an initial focus and allow the different activities to effectively interact.

The problem of mine detection is initially daunting, much development is still required. Some of the steps required have been successfully completed and now it can be seen that a continued step by step approach will yield an effective armoury of solutions to deal with this insidious scourge.

References

1. Chan L.C., Moffatt D.L. and Peters Jr L.
"A Charterization of Subsurface Radar Targets"
Proc IEEE, Vol 67, No 7, July 1979, pp 991-1000

2. Craib J.A. and Chignell R.J.
"Research into Electronic Detection of minimum metal mines laid in the Falkland Islands"
The Royal Engineers Journal, Vol 102 No 3, Dec 1988, pp 240-247.

Acknowledgement

The author has been fortunate in having many years experience leading and working with teams of dedicated professionals addressing demanding ground probing radar problems. All these interactions contribute to the formulation of the views expressed and are gratefully acknowledged.

Figure 1. Gulf War Concept Demonstrator
Real Time Mine Detection System

Figure 2. Pipe Hawk - Pipe Locating System for Utility Company Operation

Figure 3. Multi-Channel Operation

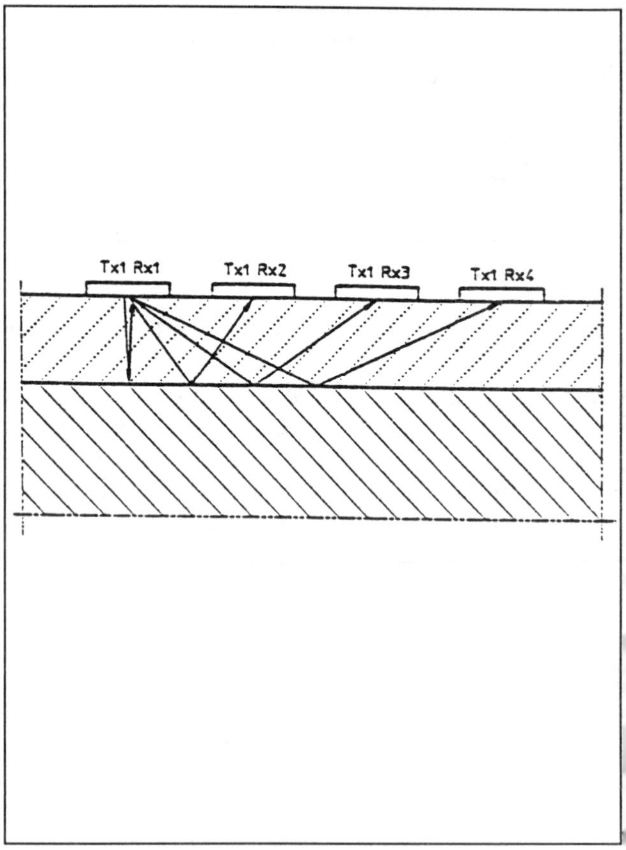

Figure 4 Layer Information Generation

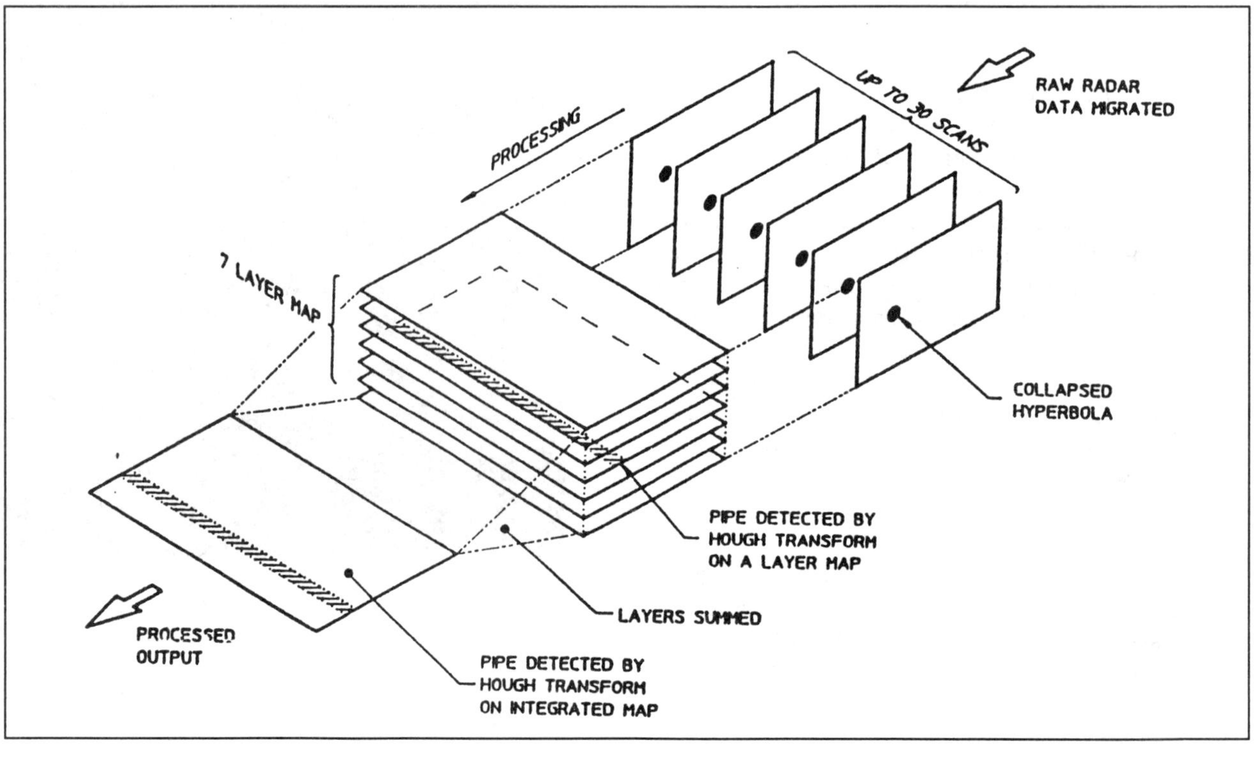

A STEPPED FREQUENCY CW POLARIMETRIC RADAR FOR MINE DETECTION

A Langman and M.R. Inggs
University of Cape Town, South Africa

Abstract

An important requirement of any mine detection radar is the ability to discriminate mines from other targets. By considering the polarization invariant properties of targets this can be achieved. In this paper the general theory of extracting these parameters from a stepped frequency continuous wave radar are derived.

Real measurements of various targets are also presented.

INTRODUCTION

One of the most important requirements of a mine detection radar system, is the ability to identify or determine the shape, size and material of buried targets. This is essential in order to separate the mine radar echo from the radar reflections from rocks, subsurface layers, ammunition shells and other radar clutter. In most GPR systems the only information available to perform this task is the magnitude of the target backscatter [5, 4, 18]. Some systems extract the phase of the radar returns providing the full vector nature of the received information and hence an extra dimension of information to meet the aforementioned requirements [8]. However this is only for illumination with a single polarization.

The next logical development is to extend the received information to include the full vector nature of the scattered electromagnetic wave, and hence allow for the use of polarization sensitive processing of the GPR returns. The idea of polarization measurements in GPR is not new [7, 1]. Previous impulse systems have implemented simple polarimetric measurements by physically rotating a linear polarized antenna [7, 1]. However this is cumbersome. It also does not completely capture the target scattering matrix and is thus limited.

In this paper the complete scattering matrix of the radar target is captured. The targets are assumed to be made up of one or more point scatterers. The target invariant parameters are then extracted for each of the point scatterers and are then used as the feature vector to identify the target [13]. Up until now no one has used this approach in GPR. In this paper the polarimetric responses from various targets are simply compared - no complex target recognition algorithms are developed.

The objective of this paper is to extract the best feature vector that can be used by some classification algorithm to identify mines. The paper begins by describing a simple model for signal processing based on the underlying electromagnetics. This model is then extended to take into account the stepped frequency continuous wave (SFCW) radar modulation. Multi-channel super-resolution techniques are then used to extract the target scattering parameters. The theory of polarimetry is then used to extract the target invariant parameters from the target scattering matrix. Finally real measurements of mine simulants and other possible targets are presented and conclusions are drawn.

MODELLING A SUBSURFACE RADAR

The development of a simple model for signal processing based on the underlying electromagnetics is an area of SFCW GPR research which has been greatly ignored [16, 9]. The models used are based more on signal processing demands and rudimentary free space EM and radar modulation theory. There are occasions where the properties of the subsurface media are addressed (Iizuka [16]), however these are related very simplistically to the signal processing. In this section a signal processing model is developed based on the radar geometry and the electromagnetic propagation in a lossy medium and across a dielectric boundary. The model is shown to be equivalent to the exponential model found in signal processing theory[14, 17].

The model is based on a simple radar geometry as is shown in Figure 1

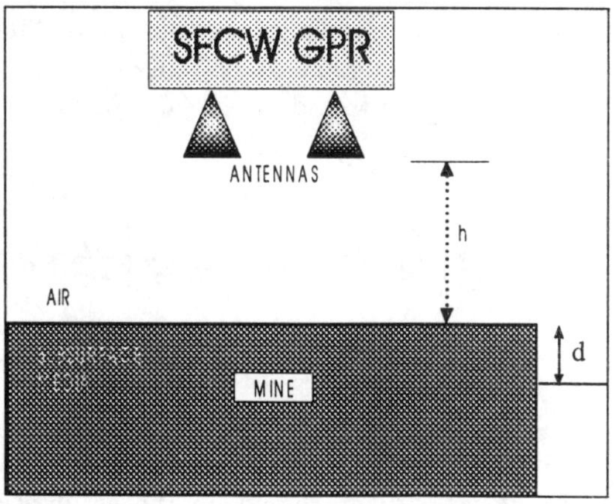

Figure 1 – Geometry of SFCW GPR for mine detection

The antennas are mounted on a vehicle a distance h above the surface of the medium. This distance

"Detection of abandoned land mines", 7-9 October 1996, Conference Publication No. 431 © IEE 1996

is chosen such that the ground surface is in the far-field of the antennas. The ground is assumed to be a single layered lossy dielectric.

The simple GPR model can be developed by considering a TEM plane wave propagating through the air, across a dielectric boundary, through a lossy dielectric, reflecting off a buried mine and propagating back again.

In the next section the elements of this model will be discussed in detail.

Theory of wave propagation in lossy media

Consider a plane wave propagating through a linear, homogeneous and non-dispersive medium which can be characterised by its complex permittivity ε $(\frac{F}{m})$ and complex conductivity σ $(\frac{s}{m})$. The electric far field $\mathbf{E}$ $(\frac{v}{m})$ at a distance r meters from the source is given by Equation 1.

$$\mathbf{E}(\mathbf{r}) = \mathrm{Re}\left[\frac{\mathbf{E}_o}{|\mathbf{r}|} e^{j\left(\omega t - k_m \hat{\mathbf{k}}\cdot\mathbf{r}\right)}\right] \left[\frac{v}{m}\right] \quad (1)$$

Where,

- $\mathbf{r} = \hat{\mathbf{x}}x + \hat{\mathbf{y}}y + \hat{\mathbf{z}}z$ is the displacement vector and $r = |\mathbf{r}| = \sqrt{x^2 + y^2 + z^2}m$
- $\hat{\mathbf{k}}$ is the unit direction vector of the wavenumber k_m,where $k_m = \frac{\omega}{v}\sqrt{1 - j\tan\delta}$
- v is the velocity of the electromagnetic energy in the medium and $\tan\delta$ is the loss factor.

Typically one can separate k_m into a real and imaginary component i.e. $k_m = \beta_m - j\alpha_m$, where α_m ($\frac{Np}{m}$) is attenuation factor and β_m ($\frac{rad}{m}$) is the propagation factor . These parameters as is shown by Balanis[2], are generally complex functions of the medium parameters and frequency.

At frequencies greater than $100MHz$ it is possible to derive a simple yet general approximation to α_m and β_m using the Maclaurin series expansion . This allows one to reduce α_m and $\beta_m$to simple linear functions

$$\alpha_m = \alpha_{m1}\frac{\omega}{c} + \alpha_{m2}\left[\frac{Np}{m}\right] \quad (2a)$$

$$\beta_m = \beta_{m1}\frac{\omega}{c} + \beta_{m2}\left[\frac{rad}{m}\right] \quad (2b)$$

Where $\alpha_{m1}, \alpha_{m2}, \beta_{m1}$ and β_{m2} are constants dependant on the properties of the medium. Equation 2a and 2b are based on the assumption that $\left(\frac{\sigma}{\epsilon'}\right)^2 \frac{1}{\omega v} << 1$. This is true for even high loss soils. Consider a gray clay loam with 20% water content and $\sigma_s = 0.123\frac{S}{m}$ and $\epsilon_r^{'} = 16$ (data from Hipp[10]). Calculating $\left(\frac{\sigma}{\epsilon'}\right)^2 \frac{1}{\omega v}$ at 500 MHz we have that $\left(\frac{0.123}{16\cdot 8.85\cdot 10^{-12}}\right)^2 \frac{\sqrt{16}}{2\cdot\pi\cdot 500\cdot 10^6\cdot 3\cdot 10^8} = 2.5\cdot 10^{-19} << 1$. At frequencies above 2.5 GHz the dispersion effects due to dielectric relaxation at high moisture ($>$ 10%) contents can destroy the validity of the model.

It is convenient to express the received or transmitted electric field in terms of a local co-ordinate system defined by the position and polarization of the radar antenna system. If we assume that we have linearly polarized antennas lying along the x and y axis of its local Cartesian coordinate system then we can rewrite Equation 1 as

$$\mathbf{E}(\mathbf{r}) = (\hat{\mathbf{a}}_x E_h + \hat{\mathbf{a}}_y E_v)\,\mathrm{Re}\left[\frac{1}{|\mathbf{r}|} e^{j\left(\omega t - k_m \hat{\mathbf{k}}\cdot\mathbf{r}\right)}\right] \left[\frac{v}{m}\right] \quad (3)$$

where

- $\hat{\mathbf{a}}_x$ and $\hat{\mathbf{a}}_y$ are the unit vectors which span the plane of the electric field $\mathbf{E}$. The unit vector $\hat{\mathbf{a}}_x$ is defined such that it is perpendicular to the plane of incidence [2]. The unit vector $\hat{\mathbf{a}}_y$ is define such that it is parallel to the plane of incidence.
- E_h is the electric field in the direction of unit vector $\hat{\mathbf{a}}_x$ and is often referred to as the *perpendicular (horizontal or E)* polarized component of the electric field *or the TE wave.*
- E_v is the electric field in the direction of unit vector $\hat{\mathbf{a}}_y$, and is often referred as the *parallel (vertical or H)* polarized component of the electric field *or the TM wave.*

Scattering from mines and the scattering matrix

An important parameter in scattering analysis is the *echo area or radar cross section* (RCS) of a target. Equation 4 gives its formal definition

$$\boldsymbol{\sigma} = \lim_{|\mathbf{r}|\to\infty} 4\pi\,|\mathbf{r}|^2 \frac{|\mathbf{E}_s|}{|\mathbf{E}_i|} \quad (4)$$

where:

- $\mathbf{E}_s$ and $\mathbf{E}_i$ are the scattered and incident electric fields

Typically σ is a complex function of target size, target composition, polarization of the transmitted and received signals and the angle of incidence and frequency of $\mathbf{E}_i$. Assuming a fixed aspect and a single polarization, Hurst [12] suggests that in most practical cases the RCS can be modelled using a damped exponential

$$\sigma = \sigma_o \exp(-\gamma' k) \quad (5)$$

He demonstrated that the exponential model could be used to locate the corners of a square plate for frequencies such that $0.8 < \frac{a}{\lambda} < 4.2$, where a is the width of the plate. This is a similar ratio of wavelength to target dimensions that occur in a mine detection radar.

The scattering coefficient from the target is related to the RCS by $s = \sqrt{\sigma}$ and is given by Equation 6

$$s = s_o \exp\left(-\gamma_p f\right) \tag{6}$$

Where γ is the target scattering decay factor and s_o is the complex scattering amplitude.

When illuminating the target with an arbritary transmit and receive polarization, the target backscatter can be represented in terms of the scattering matrix $\mathbf{s}_{ab}$ (Cloude [6]),where

$$\mathbf{s}_{ab} = \begin{bmatrix} s_{aa} & s_{ab} \\ s_{ba} & s_{bb} \end{bmatrix} \tag{7}$$

and $\{a, b\}$ is the polarization basis function. The scattered electric field $\mathbf{E}^s$ is hence related to the incident electric field $\mathbf{E}^i$ by

$$\begin{bmatrix} E_a^s \\ E_b^s \end{bmatrix} = \frac{e^{j\left(\omega t - k_m \hat{\mathbf{k}} \cdot \mathbf{r}\right)}}{|r+h|} \mathbf{s} \begin{bmatrix} E_a^i \\ E_b^i \end{bmatrix} \left[\frac{v}{m}\right] \tag{8}$$

GPR model of a radar over a dielectric half space

Consider a target at a distance z beneath the surface as is shown in Figure 1 . The received electric field at the radar origin (a distance d_o above the surface) from the target is given by Equation 9.

$$\mathbf{E}^r(f) = \mathrm{Re}\left\{ \begin{array}{c} \mathbf{E}_o \frac{e^{-\alpha_{m1} \frac{\omega}{c} z - \alpha_{m2} z}}{(z+d_o)^2} \times \\ e^{j\left(\omega t - 2\frac{\omega}{c}(\beta_{m1} z + d_o) - 2\beta_{m2} z\right)} \end{array} \right\} \left[\frac{v}{m}\right] \tag{9}$$

Where:

- $\mathbf{E}_o = \hat{\mathbf{a}}_x S_h E_{h0} + \hat{\mathbf{a}}_y S_v E_{v0}$
- $S_h = T_b^h s_h T_f^h$ and $S_v = T_b^v s_v T_f^v$
- s_h and s_v are the radar scattering coefficients from a target buried at a depth of z meters below the surface for *horizontal* and *vertical* polarized incident electric fields.The parameters s_h and s_v are characterised by an amplitude s_{ho} and s_{vo} and an exponential factor γ_h and γ_v.
- T_f^h and T_f^v are the horizontally and vertically polarized transmission Fresnel coefficients when the electric field moves in the direction from free-space to the soil medium.
- T_b^h and T_b^v are the horizontally and vertically polarized transmission Fresnel coefficients when the electric field moves in the direction from the soil medium to free-space.

From Equation 9 it can be seen that to determine the RCS of the target one needs to determine T_b and T_f for both the horizontal and vertical polarized electric fields. These unknowns cannot be measured directly, but can be obtained from the measured reflection coefficients (R^h and R^v). It is simple to show that the Fresnel reflection coefficients are related to the transmission coefficients by invoking the boundary conditions at the interface of the media[2]. This yields Equation 10

$$1 + R^p = T^p \tag{10}$$

where $p = \{h, v\}$

SFCW MODULATION

The model developed in the previous section is a general propagation model that is independent of the radar system that generates or receives the signals. In this section this model will be extended to include the SFCW modulation scheme.

A SFCW radar determines the distance to a target by sequentially measuring the coherent electromagnetic returns from a target for a number of stepped operating frequencies (Langman [15]).

Conceptually the radar is made up of three main modules:

- *A step-frequency oscillator:* This module is responsible for generating the N frequency steps, Δf, starting at f_o and extending over the required measurement bandwidth.
- *The coherent receiver:* This module mixes the required received signal down to baseband and then measures the amplitude of the reflected target backscatter. Some SFCW radars use quadrature modulation schemes to determine both the in-phase (I) and quadrature (Q) components of the reflected target energy [16].
- *A range extractor:* This module transforms the raw data from the frequency domain to the spatial domain.

Single polarization

Assume we transmit and receive a single horizontal polarization (h). Assume also that there are P point targets. Applying the theory of SFCW modulation to the GPR model we can show that the voltage at the input of the radar processor can be expressed by an analytic signal $E_{hh}[n]$ given by Equation 11

$$E_{hh}[n] = \sum_{m=0}^{P-1} (b_m)_{hh} z_m^n + w[n][V] \tag{11}$$

Where $w[n]$ is zero mean white gaussian noise and

$$b_m = A_m \exp(j\theta_m) \tag{12}$$
$$z_m = \exp\left[(\lambda_m + j2\pi d_m)\Delta r\right] \tag{13}$$

and

$$A_m = \frac{G \cdot |S^{TM}|}{(r+d_o)^2} \exp(\iota_m) \quad (14a)$$

$$\theta_m = \angle S^{TM} + 4\pi \frac{f_o}{c} d_m + \beta_{m2} r_m \quad (14b)$$

$$\lambda_m = \pi(\alpha_{m1} r_m + \gamma \beta_{m2}) \quad (14c)$$

$$d_m = \beta_{m1} r_m + d_o \quad (14d)$$

$$\Delta r = -\frac{2\Delta f}{c} \quad (14e)$$

Where A_m is the target amplitude, θ_m is the target absolute phase, λ_m is the target damping factor, d_m is the free space target position and n is the nth frequency step.

Complete polarization

If we assume that the damping factor λ_m is independent of polarization ($\gamma\beta_{m2} << 1$ or $\gamma\beta_{m2} << \alpha_{m1} r_m$ or γ is polarization independent) then we can extend the single polarization equations as follows

$$\mathbf{E[n]} = \begin{bmatrix} \sum_{m=0}^{P-1} (b_m)_{hh} & \sum_{m=0}^{P-1} (b_m)_{hv} \\ \sum_{m=0}^{P-1} (b_m)_{vh} & \sum_{m=0}^{P-1} (b_m)_{vv} \end{bmatrix} \mathbf{Z} \quad (15)$$

where

$$\mathbf{Z} = \begin{bmatrix} z_m^n & 0 \\ 0 & z_m^n \end{bmatrix} \quad (16)$$

Parameter estimation of b_m and z_m

Numerous algorithms exist which enable one to extract the target parameters b_m and z_m from the raw frequency data $\mathbf{E[n]}$ (Steedly[17]). The algorithm used in this paper is a modified version of the algorithms found in these references and is described briefly below

1. The target is illuminated with an impulse of circular polarized radiation yielding $\begin{bmatrix} E_{hl}[n] \\ E_{vl}[n] \end{bmatrix}$ (Steedly [17]). From this set of data the pole (z_m) is found using the generalised matrix pencil method (Hua [11]).
2. $\begin{bmatrix} (b_m)_{hh} & (b_m)_{hv} \\ (b_m)_{vh} & (b_m)_{vv} \end{bmatrix}$ is then determined using a standard least-squared algorithm.

The data is then scaled to remove the effects of geometric spreading of the EM wave and the Fresnel transmission through the medium, to yield the target scattering matrix $\mathbf{s}$ as a function of position for each target.

TARGET INVARIANT PARAMETERS

It is possible to transform the scattering matrix $\mathbf{s}$ from one polarization basis to another $\mathbf{s}'$ as is shown in Equation 17

$$\mathbf{s}' = \mathbf{T}(\boldsymbol{\rho})^T \mathbf{s} \mathbf{T}(\boldsymbol{\rho}) \quad (17)$$

Where $\mathbf{T}$ is a 2×2 unitary transformation matrix given by

$$\mathbf{T} = \frac{1}{\sqrt{1-\rho\rho^*}} \begin{bmatrix} 1 & -\rho^* \\ \rho & 1 \end{bmatrix} \quad (18)$$

and ρ is the complex polarization ratio[3] of the polarization ellipse which is typically describe in terms of its geometric parameters - its orientation angle (ψ) and its ellipticity (τ).

The invariant target parameters can be determined by transforming the scattering matrix [13] such that

$$\mathbf{s}' = m \begin{bmatrix} e^{j\nu} & 0 \\ 0 & e^{-j\nu} \tan^2 \gamma \end{bmatrix} e^{-j\xi} \quad (19)$$

From Equation 18 and 19 we can deduce the six target invariant parameters

- m is the backscatter amplitude
- ξ is the absolute phase (typically calibrated out).
- γ is the characteristic angle where $0 \leq \gamma \leq 45°$
- ν is the target skip angle where $-45° \leq \nu \leq 45°$
- ρ, the polarization state of maximum power described by τ and ψ

These parameters are often referred to as the Huynen parameters.

EXPERIMENTAL MEASUREMENTS

The results reported in this paper have been obtained using a laboratory setup consisting of a vector network analyser, computer, 2-18GHz ridged horn antennas, a sandpit and some test targets. Measurements were taken over a 3-6GHz frequency range and the data was processed as described above. Table 1 lists the Huynen parameters for a number of target measurement. Note that the mine simulants are cylindrical targets with similar dimensions to anti-personnel mines and are made from metal, teflon and polystyrene.

Where numbers in Table 1 correspond to the following targets:

1. Ground
2. Metal mine simulant at a depth of 5cm
3. Teflon mine simulant at a depth 8cm
4. Polystyrene mine simulant at a depth of 7cm

No	m	ν	γ	τ	ψ	ξ
1	1.68	−2.0°	37.5°	1.7°	1.2°	−86.0°
2	1.76	0.0°	42.0°	2.0°	0.3°	−23.0°
3	0.44	−7.2°	45.0°	6.9°	1.1°	−61.5°
4	0.77	0.8°	40.6°	2.2°	−2.5°	80.0°
5	1.61	0.9°	42.6°	3.1°	−1.1°	−36.43°
6	1.78	5.0°	38.6°	1.5°	10.2°	88.0°

Table 1. Huynen parameters for various targets

5. Metal disk at a depth of 11cm
6. Metal rod lying 30° to the radars horizontal axis on the ground.

It is important to note that due to the lack of antenna facilities at the University no accurate antenna calibration could be performed. This did result in the target parameters varying from theoretical predictions.

CONCLUSIONS

A paper has presented the theory to extract the Huynen target parameters from a SFCW radar and demonstrated the technique in the laboratory. From the results it can be seen that it is possible to discriminate between different targets using this technique.

With further calibration and better radar equipment these results can be improved and provide valuable tool to help in the detecting and identification of landmines.

References

[1] I. Arai and T. Suzuki. Experimental Results on Subsurface Radar with Improved Resolution . *Journal of Electromagnetic Waves and Applications*, 7(11):1479–1495, 1993.

[2] C. A. Balanis. *Advanced Engineering Electromagnetics* . John Wiley and Sons, New York, Chichester, Brisbane,Toronto,Singapore, 1st edition, 1989.

[3] W.-M. Boerner, M. B. El-Arini, C.-Y. Chan, and P. M. Mastoris. Polarization Dependence in Electrogmagnetic Inverse Problems . *IEEE Trans on Antennas and Propagation*, AP-29(2):262–271, March 1981.

[4] L. C. Chan, L. Peters, and D. Moffat. Improved Performance of a Subsurface Radar Target Identification System through Antenna Design. *IEEE Trans on Geosc. and Remote Sensing*, (2):1, March 1981.

[5] L. C. Chan and L. Peters Jr. Characterization of Subsurface Radar Targets . *IEEE*, 67(7):991, 1979.

[6] S. Cloude. Polarimetric Techniques in Radar Signal Processing . *Microwave Journal*, pages 119 – 127, July 1983.

[7] D. Daniels, D. Gunton, H. Scott, and C. Tran. Introduction to Sub-surface Radar . *Institute of Electrical Engineers (United Kingdom)*, 135(4), August 88. Pt.F.

[8] A. P. Freundorfer. *Step Frequency Radar* . PhD thesis, University of Toronto, 1989.

[9] S. hamran, D. T. Gjessing, J. Hjelmstad, and E. Aarholt. Ground Penetrating Synthetic Pulse Radar : Dynamic Range and Modes of Operation. *Elsevier Science Publishers B. V.*, 33:7–, 1995.

[10] J. E. Hipp. Soil Electromagnetic Parameters as Functions of Frequency, Soil Density, and Soil Moisture. In *Proceedings of the IEEE*, volume 62, pages 98–103. IEEE Proceedings, January 74.

[11] Y. Hua and T. K. Sarkar. Matrix Pencil Method for Estimating Parameters of Exponential Damped/Undamped Sinusoids in Noise. *IEEE Transactions on Acoustic and Signal Processing*, 38(5):814–824, May 1990.

[12] M. P. Hurst and R. Mittra. Scattering Center Analysis Via Prony's Method . *IEEE Trans on Antennas and Propagation*, AP-35(8):986, 1987.

[13] J. R. Huynen. Measurement of the Target Scattering Matrix . In *IEEE/Proceedings*, pages 936–946. IEEE Proceedings, August 1965.

[14] S. M. Kay and S. L. Marple Jnr. Spectrum Analysis - A Modern Perspective . *IEEE Proceedings*, 69(11):1380–1419, November 1981.

[15] A. Langman, S. P. Dimaio, B. E. Burns, and M. R. Inggs. Development of a Low Cost SFCW Ground Penetrating Radar . volume 3, pages 430–433, Lincoln, Nebraska, May 1996.

[16] V.-K. Nguyen, K. H. Wu, H. Ogura, K. Iisuka, H. Mori, and A. P. Freundorfer. Step-frequency radar . *J. Appl. Phys.*, 56(9):2572–2583, November 1984.

[17] W. Steedly and R. Moses. High Resolution Exponential Modeling of Fully Polarized Radar Returns. *IEEE Trans on Aerospace and Electronic Systems*, 27(3):459, May 1991.

[18] J. Young and L. Peters. *Applications of subsurface Transient Radar* , chapter 8, pages 296–348. Nostrand Reinhold Company, 88.

UWB SAR DETECTION OF DIELECTRIC TARGETS

S. R. Cloude[1], A. Milne[1], C. Thornhill[2], G. Crisp[2]

[1]Applied Electromagnetics, Scotland, UK [2]DRA Malvern, England, UK

1. INTRODUCTION

In this paper we consider the problem of ultra wide band (UWB) electromagnetic wave scattering by a dielectric cylinder. Our main objective is to study the possibility of using UWB Radar for the detection of objects in noise and clutter. UWB Radars utilise a very wide bandwidth to ensure that the target scatters energy at very long, as well as at very short wavelengths. This provides information about the shape and material structure of the object which can be utilised for detection and classification.

The scattering of EM waves by dielectric cylinders is a classical boundary problem with a well known convergent series solution[1]. Here we examine the details of the various phenomenological wave contributions to this solution, with a view to using very short pulse Radar to isolate the mechanisms in time, amplitude and polarization.

A key element of the wave solution for dielectric bodies is the increased complexity compared to metallic objects. This arises because of internal wave effects. Consequently, the classical Radar Cross Section (RCS) of dielectric targets varies rapidly with frequency, making it unlikely that narrow band systems could be used for reliable detection. However, by employing a broad band signal and by considering its time domain structure, we shall see that we can establish a relatively simple description of such objects.

Two features are important in establishing our approach to this problem: the first is that we concentrate on the early time response i.e. we consider the scattering response over a time interval which is of the order of the wave transit time through the object. Concentrating on this regime is advantageous for detection because it contains the bulk of the scattered energy.

The second key feature of our approach is that we consider only the case of backscatter. This greatly simplifies the analysis and matches closely the main application developments in Radar system design. By restricting ourselves to wave mechanisms which yield large scattered signals only in the backscattered direction, we shall see that we can characterise generic features of the UWB signature for such objects.

2. UWB SCATTERING BY DIELECTRIC CYLINDERS

We consider the problem of wave backscattering from a homogeneous cylinder or radius a, length L and refractive index n as shown in figure 1. We consider firstly the case of normal incidence when L is infinite (i.e. the 2-D cylinder problem). Here we can use a convergent series solution together with a 2-D finite difference time domain (FDTD) solver to examine the structure of the solution. We then consider the more general case of a finite cylinder, using a 3-D FDTD solver to generate a Synthetic Aperture Radar (SAR) image. Finally, we compare this solution with experimental results obtained on the DRA ISAR Range at Pershore[2].

In general, there are six important backscattering mechanisms to consider[1] (figure 1). In the time domain, the Radar equation for backscatter from a dielectric cylinder can then be written in the form

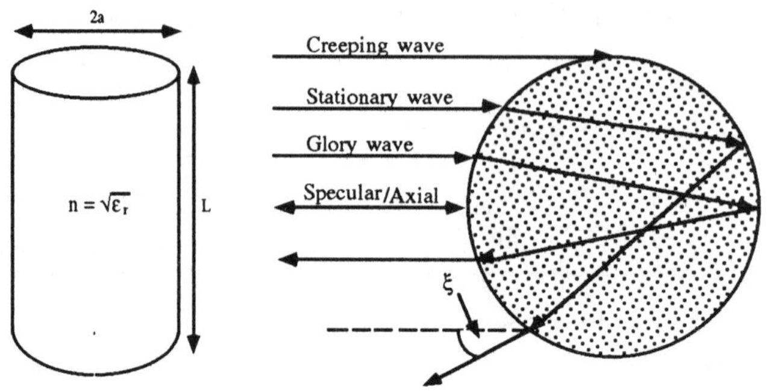

Figure 1: Cylinder Geometry and Dominant Backscatter Mechanisms

$$V_{oc}(t) = \frac{D_r^2 V_{pk}\Gamma(t)}{8\pi^{15} c\tau R^2} \qquad - 1)$$

where R is the range to the target and V_{OC} is the open circuit voltage available at the antenna terminals. We assume the transmitted pulse is a step of rise time τ and peak amplitude V_{pk}, radiated through an antenna with real aperture width D_r. The function Γ is related to the RCS of the target and for a dielectric cylinder can be written in the form shown in equation 4. We now consider each term in this equation (note that for simplicity we assume the material is loss-free and so all β terms expect β_c and β_{cg} are set equal to zero).

2.1 Specular Waves (α^s)

The specular return from the front of the cylinder is taken as the time reference for the signature. It has an amplitude given by the well known reflection formula

$$\alpha^s_{\parallel} = -\alpha^s_{\perp} = \frac{1-n}{1+n} \qquad - 2)$$

This amplitude will be taken as a reference for our discussion of the processing gain of UWB radar (section 4).

2.2 Axial Waves (α^a)

Internal propagation and reflection from the back surface gives rise to a backscatter response with amplitude and timing

$$\alpha^a_{\perp,\parallel} = T^{in}_{\perp,\parallel} R^{in}_{\perp,\parallel} T^{out}_{\perp,\parallel}\left(1 - \frac{2}{n}\right)^{-0.5} \qquad t_a = \frac{4na}{c}$$

$$T^{in}_{\perp,\parallel} = \frac{2}{1+n} \qquad T^{out}_{\perp,\parallel} = \frac{2n}{1+n} \qquad - 3)$$

$$R^{in}_{\parallel} = -R^{in}_{\perp} = \frac{n-1}{n+1}$$

where the last term in equation 3 blows up for $\varepsilon_r = 4$ due to internal focusing by the curved cylinder.

2.3 Glory Waves (α^g)

Waves which are refracted, reflected and then refracted into the backscatter direction are called glory waves. These are often the largest feature in the signature and are polarization dependent, so we consider their amplitude and timing in some detail as shown in equation 5. The most important feature to note about this wave is that it is only significant for material in the range $2 \le \varepsilon_r \le 4$ but over this range it is the dominant backscatter wave mechanism.

2.4 Rainbow Waves (α^r)

While rainbow or stationary waves are well known in optical scattering, it is important to realise that for $\varepsilon_r > 2.5$, significant backscattered energy can still result from these waves, even though their ray paths are not exactly in the backscatter direction. In quantitative terms, this wave is only important for backscatter if $\xi \le 20°$. The angle ξ is defined in figure 1 and the timing for the rainbow wave is given in equation 6

$$\Gamma_{\perp,\parallel}(t) = \frac{\partial}{\partial t}\sqrt{\pi a}\{\alpha^s_{\perp,\parallel}\delta(t) + \alpha^a_{\perp,\parallel}\delta(t - t_a)\, e^{-\beta_a t_a} + \alpha^g_{\perp,\parallel}\delta(t - t_g)\, e^{-\beta_g t_g}$$

$$+ \alpha^r_{\perp,\parallel}\delta(t - t_r)\, e^{-\beta_r t_r} + \alpha^c_{\perp,\parallel}\delta(t - t_c)\, e^{-\beta_c t_c} + \alpha^{cg}_{\perp,\parallel}\delta(t - t_{cg})\, e^{-\beta_{cg} t_{cg}}\} \qquad - 4)$$

$$\alpha^g_{\perp,\parallel} = 2T^{in}_{\perp,\parallel}(\phi)\, R^{in}_{\perp,\parallel}(\psi)\, T^{out}_{\perp,\parallel}(\psi) \qquad t_g = \frac{2a}{c}\left(2 + \frac{\varepsilon_r}{2}\right) \qquad \psi = \frac{\phi}{2} = \cos^{-1}\left(\frac{n}{2}\right)$$

$$T^{in}_{\parallel}(\Omega) = \frac{\frac{2}{n}\sqrt{1 - \frac{\sin^2\Omega}{\varepsilon_r}}}{\cos\Omega + \frac{1}{n}\sqrt{1 - \frac{\sin^2\Omega}{\varepsilon_r}}} \qquad T^{in}_{\perp}(\Omega) = \frac{2\cos\Omega}{\cos\Omega + n\sqrt{1 - \frac{\sin^2\Omega}{\varepsilon_r}}}$$

$$R^{in}_{\parallel}(\Omega) = \frac{\sqrt{1 - \frac{\sin^2\Omega}{\varepsilon_r}} - \frac{\cos\Omega}{n}}{\sqrt{1 - \frac{\sin^2\Omega}{\varepsilon_r}} + \frac{\cos\Omega}{n}} \qquad R^{in}_{\perp}(\Omega) - \frac{n\cos\Omega - \sqrt{1 - \varepsilon_r\sin^2\Omega}}{n\cos\Omega + \sqrt{1 - \varepsilon_r\sin^2\Omega}} \qquad - 5)$$

$$T^{out}_{\parallel}(\Omega) = \frac{2n\sqrt{1 - \varepsilon_r\sin^2\Omega}}{\cos\Omega + n\sqrt{1 - \varepsilon_r\sin^2\Omega}} \qquad T^{out}_{\perp}(\Omega) = \frac{2\cos\Omega}{\cos\Omega + \frac{1}{n}\sqrt{1 - \varepsilon_r\sin^2\Omega}}$$

$$t_r = \frac{a}{c}((1-\cos\psi_0) + 4n\cos\psi_1 + (1-\cos\gamma))$$

$$\cos\psi_0 = \sqrt{\frac{\varepsilon_r - 1}{3}} \qquad \cos\psi_1 = 2\sqrt{\frac{\varepsilon_r - 1}{3\varepsilon_r}} \qquad -6)$$

$$\gamma = 4\psi_1 - \psi_0 \qquad \xi = 4\psi_1 - 2\psi_0$$

2.5 Creeping Waves (α^c)

The creeping wave, guided by the curved surface of the cylinder, is a well known backscattering mechanism in UWB Radar and causes backscattered radiation with a timing

$$t_c = \frac{(2+\pi)a}{c} \qquad -7)$$

However, even for the case of a lossless dielectric, this wave maintains an exponential damping due to radiation loss around the surface of the cylinder. Hence, although it is important in the nonpenetrable target case, its effect on the early time signature of low loss targets is generally negligible compared to other wave mechanisms. It is still important however in the mixed glory/creeping wave to be discussed next.

2.6 Creeping/Glory Wave (α^{cg})

While the direct creeping wave contribution is small and the first glory wave only exists for $\varepsilon_r > 2$, there is always the possibility of a hybrid mechanism causing enhanced backscattering where the wave is backscattered by first being coupled into the cylinder at the critical angle before striking the internal surface, again at the critical angle, after which either a second internal reflection takes place (important for $\varepsilon_r = 2$) or a surface wave is launched to propagate along a *portion* of the curved surface before radiating in the backscatter direction (important for $\varepsilon_r > 4$). The timing of this path can be calculated as

$$t_{cg} = \frac{2a}{c}(1 + n\cos\theta_c + \theta_c) \qquad -8)$$

where θ_c is the critical internal angle for the dielectric ($\sin\theta_c = n^{-1}$).

Figure 2 shows the relative timings for all these wave components as a function of ε_r in the range 2 to 4. Also shown are points derived as the maxima of signatures from FDTD simulations for a dielectric cylinder illuminated with a Gaussian pulse of width 0.1665 cylinder diameters (which matches the experimental results shown in figure 4). We see that across the range the glory/stationary wave combination is dominant.

Figure 3 shows a comparison between theory, computer modelling and experiment for the amplitude ratio of the signal maxima to the front specular return $\frac{\alpha^g}{\alpha^s}$. Both TE and TM are shown and we can see that the ratio is always larger for TE than for TM. Also shown are points obtained from a 2-D FDTD simulation.

We see that there is good general agreement for TE but that the TM values are lower than predicted for small values of ε_r. Also note that there is poor agreement for both TE and TM at $\varepsilon_r = 2$. This can be explained as a failure of the simple model of equation 5 (which assumes zero transmission for tangential rays). Also shown in figure 3 are experimental measurements for $\varepsilon_r = 3$ and 2.5. The data for these points was obtained from an ISAR image obtained at DRA Pershore (see figure 4).

3. UWB-SAR IMAGING OF DIELECTRIC CYLINDERS

We now turn to consider the more general case of a finite dielectric cylinder of length L. Such a dielectric cylinder was modelled using a 3-D FDTD solver and also imaged on the ISAR range operated at DRA Pershore[2]. This range is a time domain short pulse facility transmitting a Gaussian pulse of fwhh = 0.1 ns. The target is located on a straight rail across the aperture and target motion is translated via a k-space Fourier processing method[3] into a high resolution image of the target. Note that an autofocus technique has also been used to focus the image onto the brightest feature in the signature. The experimental configuration used was $\theta = 90^o$ for which this brightest feature is the glory wave. Note that without this autofocus, the glory wave contribution is partially defocused and we obtain poor estimates of the TE and TM amplitude ratios.

Figure 4 shows experimental results for TE and TM polarizations. These time series were obtained directly from the autofocused ISAR images. The early time signatures are clearly apparent. We can see the front specular flash followed some time later by the much larger glory wave contributions. Data from these time series shows good agreement with theory as indicated in figures 2 and 3. We now consider how we might use these observations to design time domain templates for the detection of targets in noise or clutter.

4. DIELECTRIC TARGET DETECTION USING UWB SAR

We now turn to consider how the above observations may be used to establish detection probabilities for dielectric targets in noise and clutter. For the case of detection in noise we note that by using an UWB

SAR, we can improve the signal to noise ratio by a gain factor G_S defined by

$$\left(\frac{S}{N}\right)_{sar} = G_s \left(\frac{S}{N}\right)_{raw} \qquad G_s = \sqrt{\frac{Rc\tau f_p}{2D_r v}} \qquad - 9)$$

which is valid for transmitted pulse widths such that $\frac{c\tau}{D_r} \leq 1$. There are two new important system parameters in equation 9; the platform velocity v and the pulse repetition frequency (PRF) of the source f_p. The ratio of these two is constrained by the sampling theorem so that the distance travelled between samples must be less than half the real aperture width so that

$$\frac{f_p}{v} \geq \frac{2}{D_r} \qquad - 10)$$

If the target response were limited to the front specular return, then G_S would be the only gain factor available for improving the signal to noise. However, we have seen that the UWB signature has several contributions so that in simplified form we may model the target backscatter as a finite impulse response signature as shown in figure 5 (taking only the first three terms in equation 4 i.e. specular, axial and glory wave contributions). Here we express the template as a doublet and triplet separated by the glory wave transit time (compare with figure 4).

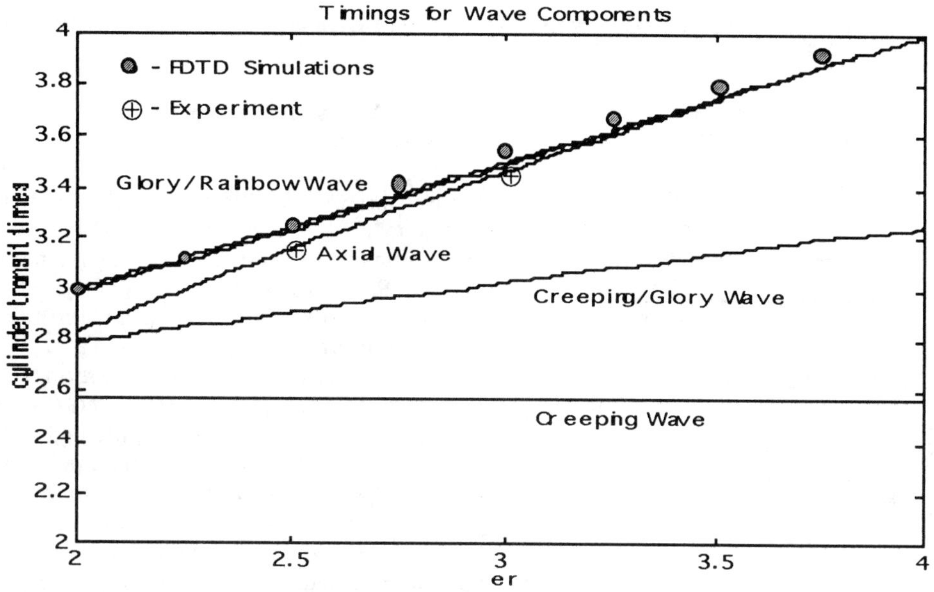

Figure 2: Timings for Wave Components

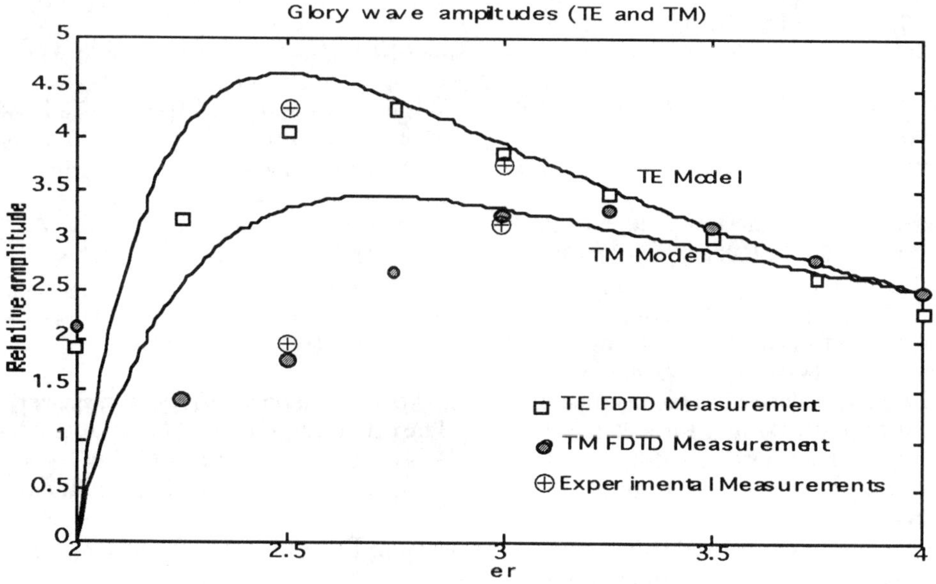

Figure 3: Relative amplitudes (α^g/α^s) for Glory wave

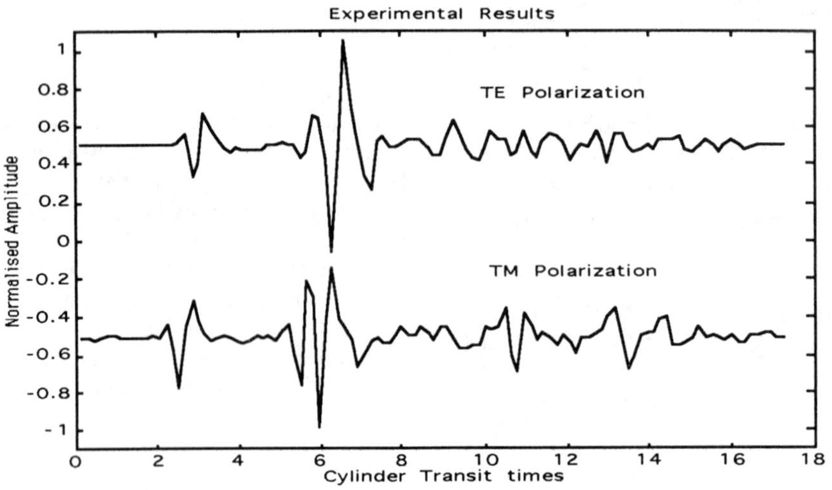

Figure 4: Experimental Data from ISAR Image of Dielectric Cylinder

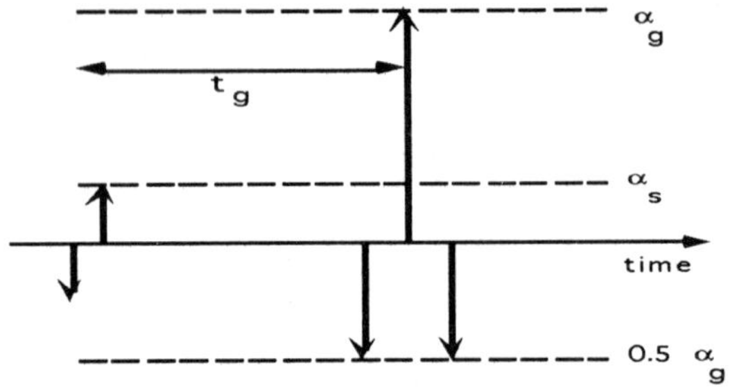

Figure 5: Simplified Template for UWB Cylinder Detection

If we combine templates for both TE and TM polarizations then we obtain a maximum target signature detection gain G_T given by

$$G_T = \sqrt{2 + 1.5\left(\frac{\alpha_{g,\parallel}}{\alpha_s}\right)^2 + 1.5\left(\frac{\alpha_{g,\perp}}{\alpha_s}\right)^2} \qquad - 11)$$

where we have taken α_s as the reference for 0dB gain and have assumed that the Radar uses a step fed aperture antenna system.

The final processing gain for the UWB detection problem is given by $G = G_t G_s$. Calculation of these gain terms enables a quantitative performance analysis to be carried out of CFAR detection based on polarized broad band signals. The main conclusion from our analysis is that G_t can be a significant factor when assessing the detection performance of UWB Radar against dielectric targets.

5. CONCLUSIONS

In this paper we have considered in detail the early time scattering behaviour of dielectric cylindrical targets. We have shown that the backscatter response, although complicated in the frequency domain, is characterised by only a few parameters in the time domain. We have used a mixture of series solutions, 2-D and 3-D FDTD numerical predictions and experimental measurements made at DRA Pershore to study the relative amplitudes, timings and polarization dependence of the various wave scattering mechanisms involved.

In particular, we have demonstrated the importance of the glory wave in UWB dielectric target backscatter problems and illustrated how it may be used to design a FIR template for enhanced detection performance. It is important to bear such gain terms in mind when comparing the performance of UWB Radar with conventional narrow band detection systems. We have shown that through a combination of SAR imaging and multi-polarization template design, significant gains can be achieved for target detection.

REFERENCES

[1] G T Ruck, D E Barrick, W D Stuart C K Krichbaum "Radar Cross Section Handbook", , Plenum Press, 1970, Chapter 4

[2] S R Cloude, P D Smith, A Milne, D Parkes, K Trafford, "Analysis of Time Domain Ultra Wide Band Radar Signals", SPIE Volume 1631, Pulse Engineering, January 1992, pp 111-122

[3] S. R. Cloude, P.D. Smith, A. Milne, C. Thornhill, G Crisp, "K-Space Imaging Algorithms applied to UWB SAR", Proceedings of 1st IEEE International Conference on Image Processing (ICIP), Austin, Texas, pp 13-16 November 1994

MODELLING AND PROCESSING OF ULTRA WIDE BAND SCATTERING OF BURIED TARGETS

M Jaureguy, P Borderies

Centre d'Etudes et Recherches de Toulouse - Office National d'Etudes et de Recherches Aérospatiales

FRANCE

INTRODUCTION :

Among the techniques which may be proposed for detection and reconnaissance of abandoned mines, very short pulse scattering analysis should be very helpful, by itself or combined with other sensors information. It is well known that the field scattered by a target illuminated by a ultra wide band (UWB) pulse contains a lot of information : high resolution imaging (location and strength of diffracting centers), dispersive phenomena and resonances, to which could be added polarimetric characteristics for example, which should be very useful to detect and identify abandoned mines. Apart from data acquisition, simulation of electromagnetic behaviour of such scenarii is essential to determine what is, on a given impulse response, the impact of the numerous parameters involved in the scenario description and by the way help and orient the subsequent signal processing. Among these parameters should be included those related to the medium (surface roughness, volume heterogeneities, frequency dispersion...) as well as those related to the measurement system and in particular the pulse shape. The point is then to be able to extract the supposed discriminating features from the received scattered signals, and computations should be accurate enough to highlight these features without introducing numerical artefacts. This paper deals with Finite Difference in Time Domain (FDTD) scattering modelling, results and processing of such scenarii.

ELECTROMAGNETIC MODELLING OF UWB DIFFRACTION :

FDTD looks to be the most appropriate method to model electromagnetic diffraction of a buried object by a UWB pulse. Indeed, it may model the soil properties in detail (surface roughness, volume heterogeneities) as well as take into account arbitrary constitution and shape of the scatterer. Another advantage of FDTD is its ability to take into account large computational volumes due to its computational power.
Let us recall that our goal is here to model the Impulse Response (IR) with a dynamic range compatible with possible measurements, typically 60-80 dB, which is not straightforward using UWB FDTD. We are going to adress this point first for targets in free space, before considering it for buried objects.

UWB SCATTERING IN FREE SPACE :

We implemented Yee's leapfrog algorithm with Mur second order absorbing conditions (1) and Taflove's corner ones (2). Dielectric and magnetic losses of materials are taken into account conventionally through real electric and magnetic conductivities. FDTD with UWB pulses and scatterers including dielectric materials yield specific problems :

The high sampling rate in space and time, imposed by the shortest significant wavelengths included in the pulse spectrum, is also inversely proportional to the optical index of the material and need to be fine enough to conform to boundaries. Furthermore use of uniform structured mesh is necessary to avoid numerical dispersion effects ,which becomes very shortcoming in terms of memory storage and time computation. To handle this point, we have implemented local mesh refinements by power of 2 by extending Yee's subgridding technique (3) based on interpolation in space and time for which, to our knowledge, free space results had not been previously reported. We could check that such an interpolation introduced low level spurious reflections with respect to the incident plane pulse at their crossing which of course depend on the size of the subgridded zone (4). Fig 1 reports the levels of these spurious reflections induced by a subgridding by 4 over nearly the entire computational space and it is important to notice that these levels are identical to those induced by a mere subgridding by 2. They are compatible with the dynamic range we wish.

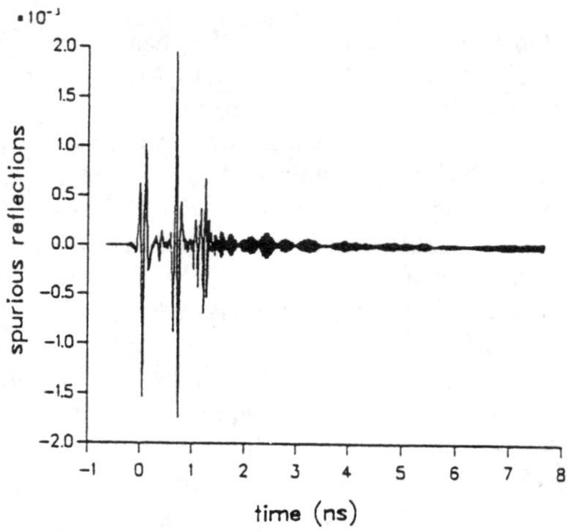

Fig. 1

"Detection of abandoned land mines", 7-9 October 1996, Conference Publication No. 431 © IEE 1996

frequency dispersion in materials : it has to be considered here because for the pulses which are able to penetrate most significantly soils, complex dielectric permittivities vary substantially. Therefore we have introduced the possibility of taking into account the presence of such dispersive materials with either Debye or Lorentz models (5) .

Huygens surface : We perform an initialisation of the incident field in which it is updated by the Yee's algorithm in exactly the same way as the total field (6), which eliminates approximation errors and prevents any leakage of the incident pulse out of the inner volume and gives reliable results. The classical approach which uses analytical incident field was found to fail in many cases where the pulse length is short with respect to the grid dimensions, which generates significant numerical dispersion.

UWB SCATTERING OF BURIED OBJECTS :

Previous techniques were accommodated to the situation depicted by Fig. 2, ie a computational volume composed of an upper free space zone separated from the host medium , representative of soils, by a rough or flat interface and where a plane pulse or a beam impigges the soil.

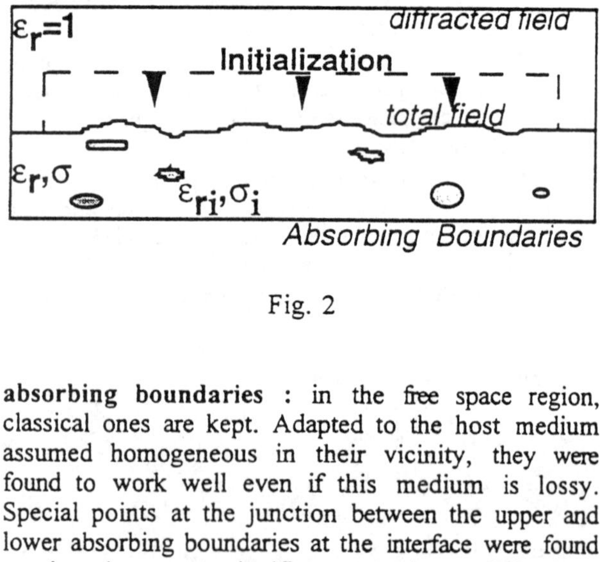

Fig. 2

absorbing boundaries : in the free space region, classical ones are kept. Adapted to the host medium assumed homogeneous in their vicinity, they were found to work well even if this medium is lossy. Special points at the junction between the upper and lower absorbing boundaries at the interface were found to introduce no significant spurious diffractions. However, when a plane wave hits a flat interface and enters the host medium, it may be very energetic when the wave reaches the lower absorbing boundary which, as well known, is not perfect. When the host material is lossy enough, this natural absorption combined with MUR's ones are sufficient to insure a reflection inferior to say -80 dB. When it was not the case, we added a rectangular slice of broadband radar absorbing material of intrinsic impedance that of the host medium and dielectric and magnetic loss angles following within the slice a gaussian law which insured the same performances.

Frequency dispersion : Fig 3 shows matching of a 4 pole Lorentz model to a 5% water content clay soil. Fig 4 shows the differences between consideration of the dispersion inside the medium together with the use of a mean conductivity over the frequency band of interest. In this case, need of accurately model the dispersive character of soil is evident.

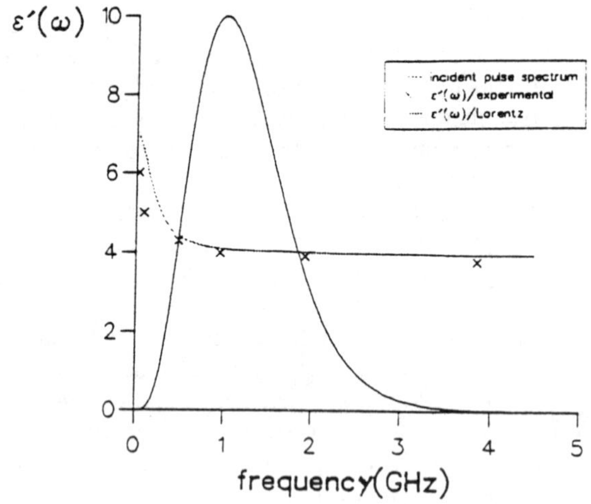

Fig.3

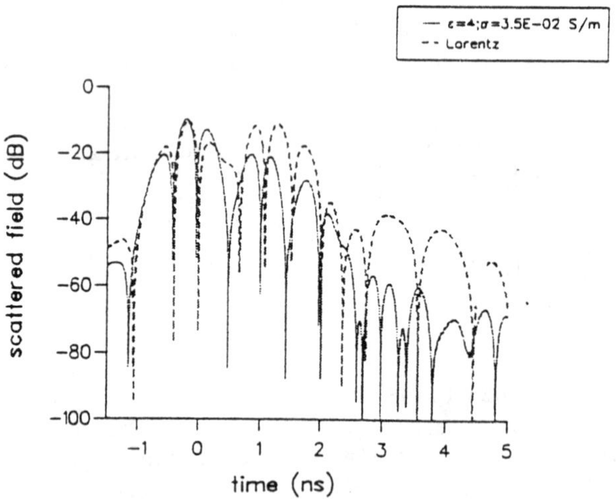

Fig.4

Subgridding is here extremely useful because of the high variety of permittivities which may be encountered in the volume. Let us recall that according to the moisture content, that of argile range from 4 to 16 and water inclusions to 80. For example Fig. 5 shows scattering of dielectric cylinder of relative permittivity 80 with mesh refinement compared with analytical solution.

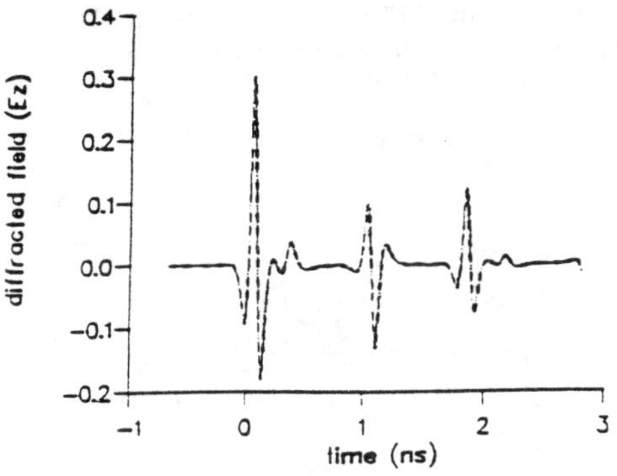

Fig. 5

of creeping waves around the scatterer. Not represented here is the response in the orthogonal polarization which behaves quite differently in the sense that the top reflections are embedded among the creeping waves echoes. When dealing with a dielectric cylinder, both polarizations behave more similarly but the top and the bottom echoes are clearly separated, the second one being stronger than the first. These impulse responses exhibit characteristic features of each situation which can be used for classification, and unless losses are too high in the host medium, the late time response can be used in the identification process, its complexity being synthesized through resonances extraction.

Huygens surface : when considering free space scattering, Huygens surface concept is a very natural one because it clearly corresponds to actual situations where the scatterer is wholly surrounded by a plane wave. Here, in practical cases, a portion of soil can be illuminated by a plane wave, the truncation of this soil sample being done by time gating or by SAR processing for example. If the emitting antenna is close to the interface, the incoming wave is rather a beam including natural apodisation. In the first case, the concept of Huygens surface to separate the total field zone from the scattered field one has been reported (7) in the case of flat interface and stratified media by using analytical reflection and transmission of multilayered media. Though this technique permits eliminating reflection of the incident plane wave on the absorbing boundaries, it is not suited to the case of rough interface and apparently is computationally intensive. After implementing several combinations of absorbing boundaries (MUR's, broadband RAM, Berenger's) with the Huygens surface, we obtained the best trade-off spurious reflections-computational volume with launching the incident wave on the interface and absorbing it further by the above technique.

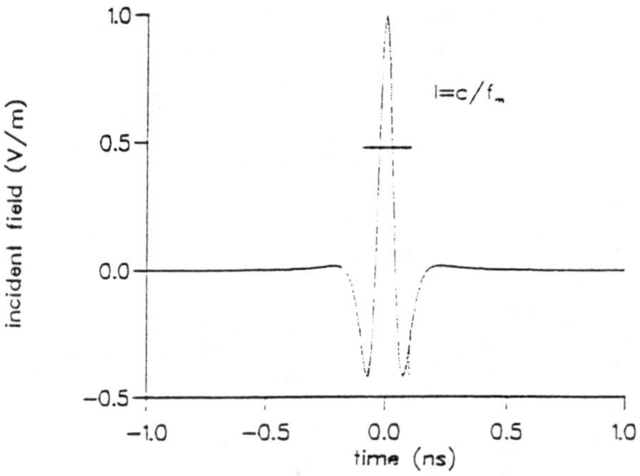

Fig. 6

RESULTS

The emitted pulses (Fig 6) we use are 2 l long, l being the wavelength most energetic in the pulse spectrum window (l=c/fm) and where fm parametrizes the pulse. Here are some bidimensional examples relative to cylindrical geometries.

Flat interface

Fig 7 represents the field scattered by a metallic cylinder of a diameter of 3 cm buried at 30 cm below the interface of an homogeneous medium for three values of increasing losses and for TM polarization. One can easily identify the successive diffractions due to the interface, then to the top of the cylinder and later successive multiple reflections with in between the effect

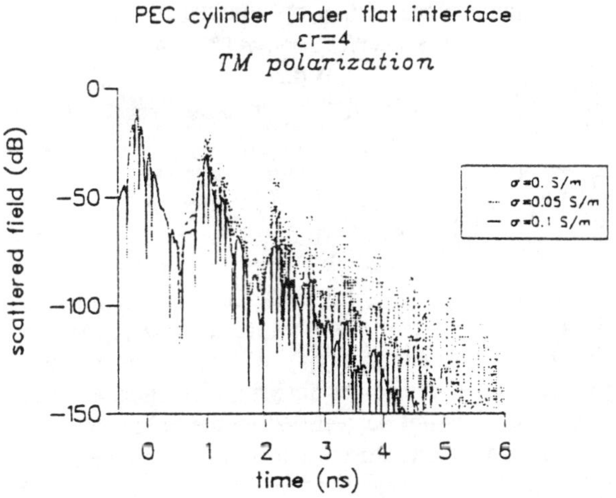

Fig. 7

Rough interface

Similar cylindrical targets have been considered when located under a rough interface. Fig 8 and 9 shows the response of respectively a dielectric and a metallic one buried in a homogeneous medium of relative permitivity 4. The interface is here a sine of amplitude 5 cm and period 10 cm. The dielectric one appears mixed up with the soil echo, whereas the metallic one very clearly stands out. The well known polarisation properties of the metallic cylinder are stressed on Fig 9 and 10 at lower frequencies (fm=500 MHz) and where the resolution is no more sufficient to separate the interface reflection from the target one.

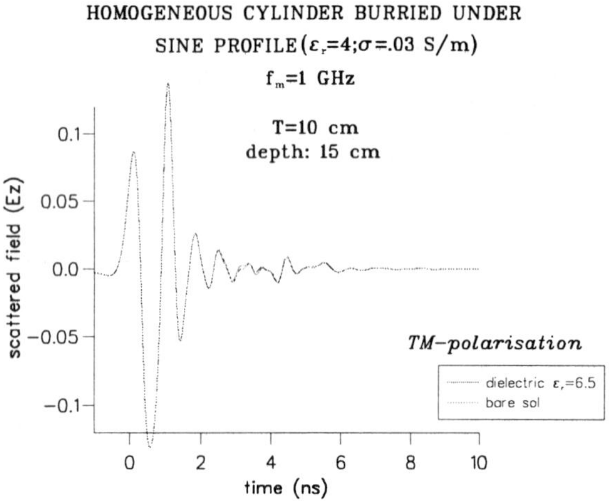

Fig 8

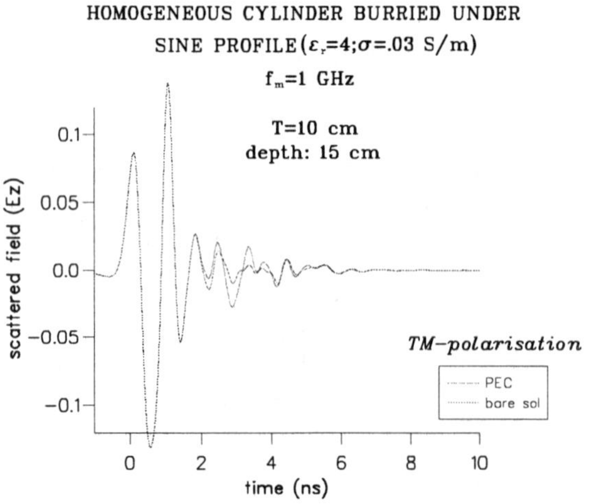

Fig 9

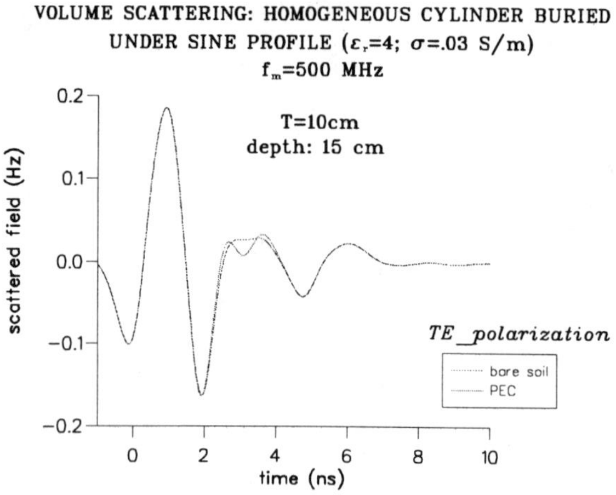

Fig 10

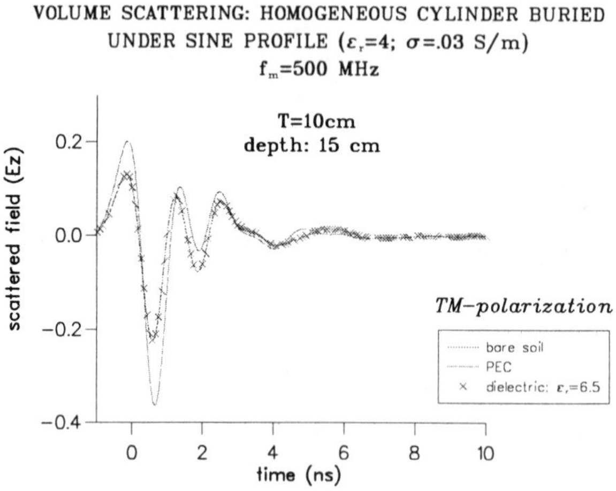

Fig 11

PROCESSING

As already stressed, classification of buried targets, whatever the algorithm used, implies first a phase of selection of the discriminant parameters of the target. Those are of course related to the intimate properties of the target searched (polarimetric characteristics, strongly resonant, characteristic succession of diffracting points) and of the local clutter. So addressing this question needs to particularise the problem, and one can project the use of several processing techniques such as time-frequency filtering, imaging, resonances extraction according to the relative properties of target and clutter. We discuss here combined exploitation of early and late times response.

For long, target resonances extraction is considered as a very characteristic feature of a scatterer because the resonant frequencies are aspect-independent. Prony method (8) and its extensions, including least squares Prony method (LSPM) (9), have been extensively used in order to extract these resonances from the signal tail after their turn-on time. The scattering centers give as well a lot of information about the target and may lead to radar imaging with SAR techniques. Consequently, exploitation of energetic resonances and scattering centers give very useful results with complementary information contents.

Although scattering centers appear rather at the early times of the IR and resonances at late times, they may be mixed up. In this case, presence of SC inside the signal is a major problem to extract correctly the poles of resonance by conventional methods such as LSPM and if one desires to make imaging, presence of resonances introduce wrong scattering center localizations.

Separate out scattering centers and resonances contributions is then a challenge to fully benefit from both of them especially in the very short pulse regime. A recent paper (10) has addressed this problem by using conventional LSPM over several segments in both time and frequency domains followed by a weighted averaging process. We have implemented a different approach (4) based on the time domain data and an unconventional use of LSPM.

Prony method being well documented, we will just recall that it was designed to deal exclusively with signals consisting in a finite serie of damped sinusoids, which is the case of the late time response of a free space scatterer, and that it can extract poles and residues.

Whatever the signal input in a LSPM routine, the solution usually reconstructs the input data provided that the number of poles is chosen sufficiently great but the poles may have or not relation with true resonances. When applied to the impulse response of a scatterer in a complex environment which is the case of buried targets, some poles will account for true resonances whereas lots of them will account for isolated diffracting points, but with a damping factor stronger whatever the residue value. Also, at late times only remain the true poles. Based on these considerations, one can build an iterative process which can separate both contributions and take fully benefit of both of them by working with a great number of poles and playing with time windows translations. This process was applied to some canonical examples. By the way, reliability of resonances extraction was addressed and we showed that to be efficient this time window must cover at least 3 periods of the lowest resonant frequency involved in the IR, if not resonant peaks are considered as distinct scattering centers and generate only non physical poles.

In order to illustrate this, fig 12 shows the impulse response of a water pipe (ε_r=80) whose diameter is 3 cm long buried at a depth of 25 cm under the flat interface of a dry soil (ε_r=4). Incident pulse is here characterised by fm=500 MHz. We can distinguish the interface reflection, then the cylinder top mixed up with resonances. Applying the above analysis make appear a single pole of resonance at 823 MHz which by itself accounts for the late times of the IR and which is very characteristic of such a scatterer in this soil, and all this information can now be injected in a classification algorithm.

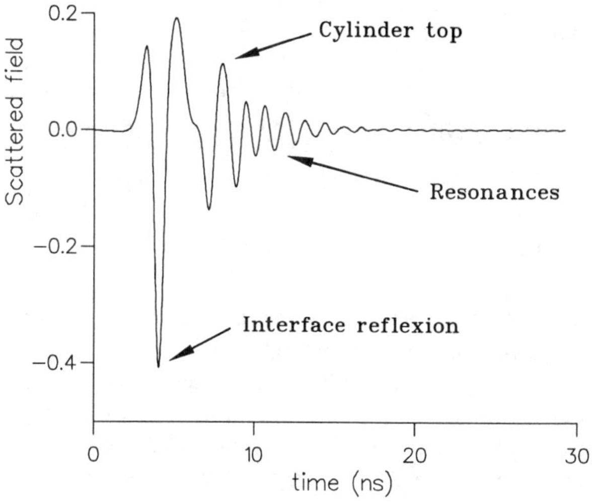

Fig 12

REFERENCES

1. Mur G. 1981. IEEE Trans. Electromagn. Compat., EMC-23, 377-382

2. Taflove A. and K.R.Umashankar K. R. PIER 2. « Finite Element and Finite Difference Methods in Electromagnetic Scattering », ch.8

3. Zivanovic S. S., Yee K. S. and Mei. K. K., 1991. IEEE Trans. on microwave theory and techniques, vol.39, n°.3

4. Jauréguy M. 1995. thèse de l'ENSAE. « Etude de la diffraction par impulsions électromagnétiques très courtes d'objets en espace libre ou enfouis : modélisation numérique et extraction des parametres caracteristiques ».

5. Luebbers R., Steich D. and Kuntz K., 1993. IEEE Tr. AP , Vol.41, n°9

6. Jauréguy M. and Borderies P., 1995. IEEE-URSI Symposium

7. Demarest D., Plumb R. and Huang Z., 1995. IEEE Tr. AP , Vol.43, n°10

8. Van Blaricum M.L.and Mittra R., 1975. IEEE Tr AP, vol AP-23, p. 777-781

9. Poggio A.J. and al. 1978. IEEE Tr AP, vol AP-26, p. 165-173

10. Moore J. and Ling A., IEEE Tr AP, vol AP-43, p.623-626

MICROWAVE IMAGING ALGORITHMS FOR ARBITRARY SPACE AND TIME INCIDENT WAVEFORMS USING ULTRAWIDE BANDWIDTH GPR TECHNIQUE

C Dourthe†, Ch Pichot††

† CERMICS/INRIA France, †† LEAT/UNSA/CNRS France

Introduction

This paper deals with a reconstruction algorithm for microwave tomography of buried mines. This technique can use a synthetic pulse (multifrequency or swept frequency mode) as well as a real pulse (impulse mode) transmitted through a broadband antenna in a reflecting mode configuration. The backscattered electric field generated by the buried mines is mesured at various locations with a GPR technique.

Development

The detection and identification of buried non metallic mines using electromagnetic waves are areas of current importance. Step or multi-frequency radars are one of possible means for such a goal [Clarricoats (1) - Daniels and Dittmer (8)]. Recent developments in GPR techniques and systems can be found in [GPR'94 (9)]. Different signal processing methods have been investigated for identifying buried objects [Daniels et al (2)]. Among them, one may mention methods based on target dependent methods, target resonance, target polarization properties, microwave holographic synthetic aperture imaging method, diffraction stack migration, Kirchhoff methods.

Imaging techniques are of common use in many areas of physics. In most methods, diffraction, which occurs when the illuminating wave interacts with the object to be imaged, is neglected. In the microwave resonance domain, diffraction effects are important and cannot be neglected without decreasing the quality of the image. The approach described here is based on diffraction tomography or active microwave imaging [Chommeloux et al (10) - Bolomey and Pichot (12)]. This technique processes the backscattered field signal generated by the buried inhomogeneities and measured over a probing area for computing a tomographic image of the ground. The method originally developed for multifrequency and plane illumination (10) has been extended, recently, for time pulse illumination and for arbitrary spatial waveform of the emitting antenna. The main advantages of such a method is the use of explicit formulas (based on the Fourier Diffraction theorem) for solving the imaging problem and to take profit of Fast Fourier transforms to implement the reconstruction algorithms on PC's allowing rapid image display. Presence of stratifications can be taken into account in the reconstruction algorithm when a *priori* information on them is available.

For instance, this imaging technique can use a synthetic pulse (multifrequency or swept frequency mode) as well as a real pulse (impulse mode) transmitted through a broadband antenna in a reflecting mode configuration. The backscattered electric field generated by the buried inhomogeneities is measured at various locations with a receiving antenna for different positions of the emitting antenna (Fig. 1). For detecting various types of mines (antitank, antipersonnel...) the frequency used in the applications is from 500 MHz up to a few GHz.

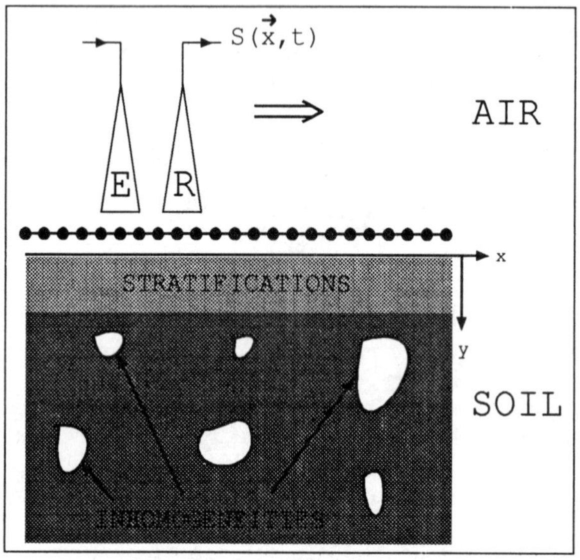

Figure 1: Experimental configuration

A one-to-one relationship via Fourier transforms relates the scattered field $\widetilde{E}_z^D$ measured at various locations x_S and at a height y_1 above the interface air/soil to the object complex permittivity within the Born's approximation, using the so-called Fourier Diffraction theorem (12) which is a generalization of the Radon Projection theorem used in X-rays tomography. This relationship can be extended, for the general case, to the mean equivalent current distribution K inside the inhomogeneities. For each angular frequency ω we can define

$$\left\{\begin{array}{l} \alpha = \nu - \eta \\ \beta = \frac{-1}{2\pi}(\widetilde{\gamma}_3 + \widetilde{\xi}_3) \\ \widetilde{\gamma}_3 = \sqrt{\omega_3^2 \mu_0 \epsilon_3 - 4\pi^2 \nu^2} \\ \widetilde{\xi}_3 = \sqrt{\omega_3^2 \mu_0 \epsilon_3 - 4\pi^2 \eta^2} \end{array}\right. \qquad (1)$$

and then we obtain

$$\widehat{K}(\alpha,\beta) = \mathcal{F}(\nu,\eta,y_1)\,\widehat{\widetilde{E}}_z^D(\nu,\eta,y_1) \qquad (2)$$

where $\widehat{K}(\alpha,\beta)$ and $\widehat{\widetilde{E}}_z^D(\nu,\eta,y_1)$ denote the 2D spatial Fourier transforms of $K(x,y)$ and $\widetilde{E}_z^D(x,x_S,y_1)$, respectively, while $\mathcal{F}$ represents a frequency domain filter depending directly on the plane wave spectrum $A_1(\eta,\omega)$ of the emitting antenna in the air and on the associated transmission coefficient $T_\perp(\eta,\omega)$ of each plane wave in the soil taking into account that the soil is a homogeneous or multilayer medium.
The reconstruction algorithm is based on this relationship (2) mapping the 2D Fourier transform of the measured backscattered field data onto arcs defined in the Fourier space by

$$(\alpha+\eta)^2 + (\beta + \frac{\widetilde{\xi}_3}{2\pi})^2 = (\frac{k_3}{2\pi})^2 \qquad (3)$$

where the spectral variables are satisfying the following conditions

$$\nu\,,\eta \in \left[-\frac{\mathcal{R}e(k_1)}{2\pi};\frac{\mathcal{R}e(k_1)}{2\pi}\right] \quad \nu\,,\eta \text{ real} \qquad (4)$$

Tomograms (2D slices of the ground) are obtained after taking a 2D spatial inverse Fourier transform (Fig. 2). This method could be generalized to a surface probing of the backscattered field for displaying 3D images of the ground.

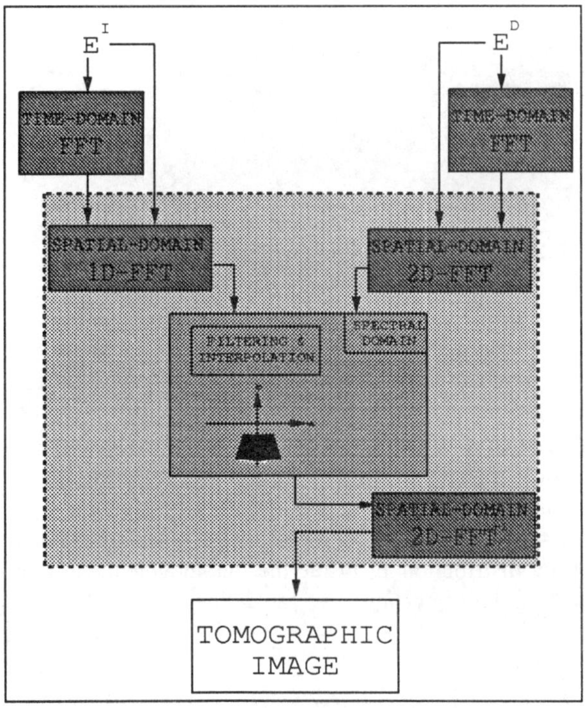

Figure 2: Flow chart of imaging algorithm

Numerical results

We present here some tomographic reconstructions from synthetic data for two different types of buried mines.
Let us first consider (Figs. 3 and 5) an antitank mine (permittivity $\epsilon_r = 3$, conductivity $\sigma = 0\ Sm^{-1}$) of size 30 $cm \times 14$ cm embedded in a dry soil ($\epsilon_r = 3.7$, $\sigma = 4.10^{-3}\ Sm^{-1}$) at a depth of 30 cm for the frequency band $[0.5 - 2.5]$ GHz. The second mine (Figs. 4 and 6) is an antipersonnel mine ($\epsilon_r = 3$, $\sigma = 0\ Sm^{-1}$) of size 11 $cm \times 5.5$ cm buried in the same soil at a depth of 5 cm for the frequency band $[2 - 4]$ GHz. We used in these simulations a space and time gaussian pulse for the incident wave. Contour plots of polarization currents induced in the buried are given in the reconstructed images. We note that a source variation guarantees a good description of upper and lower sides of each mine. But on the other hand, sides perpendicular to air/soil interface are not reconstructed with this bidimensional imaging procedure. However, in Figures 5 and 6 where mines are surrounded by inhomogeneities ($\epsilon_r = 1, \sigma = 0$ Sm^{-1}) we obtain a sufficient resolution allowing us to characterize the mine among the voids. Nevertheless, we note a light enhancement of the lower sidefor each reconstucted mine.

Conclusion

This method benefits from the use of fast numerical algorithms as FFTs to implement the reconstruction algorithms on PC's and to allow rapid image display. However, some approximations are introduced in the spectral formulations in order to deal with usual Fourier transforms. The available spectral information is restricted to a certain domain leading to a spatial resolution limit of the order of $\frac{\lambda}{2}$ (Rayleigh criterion). This reconstruction method is related to the induced currents inside the inhomogeneities and then to a qualitative imaging reconstruction. Further work will be devoted to methods allowing the quantitative reconstruction of a physical parameter as the complex permittivity profile of the buried inhomogeneities. Nevertheless, some extensions of the method has been carried out in order to incorporate the spatial wavefront of the emitting antennas or to work as well with time-domain signals allowing to process GPR system data. These extensions will be also useful for improving the antenna design with the aim of a better object illumination and reconstructed image.

References

1. Clarricoats P.J.B., 1977, Radar '77 IEE Int. Conf., 547-551

2. Daniels D.J., Gunton D.J. and Scott H.F., 1988, IEE Proc.-F, 135, 4, 278-321

3. Iizuka K. and Freundofer A.P., 1983, Proc. IEEE, 71, 276-279

4. Chan L.C., Moffatt D.L. and Peters L., 1979, Proc. IEEE,67, 991-1000

5. Mast J.E. and Johansson E.M., 1994, SPIE, 2275

6. Azevedo SG., Gavel D.T., Mast J.E. and Warhus J.P., 1995, Proc. WAPM'95, 48-515

7. Chignell R.J., 1995, Proc. WAPM'95, 52-61

8. Daniels D.J. and Dittmer J., 1995 Proc. WAPM'95, 62-70

9. 1994, Proc. Fifth Int. Conf. on GPR, London

10. Chommeloux L., Pichot Ch. and Bolomey J.Ch., 1986,IEEE Trans., MTT-34, 10, 1064-1076

11. Tabbara W., Duchêne B., Pichot Ch., Lesselier D., Chommeloux L. and Joachimowicz N., 1988, Inverse Problems, 4, 305-331

12. Bolomey J.Ch. and Pichot Ch., 1990, Int. J. Imag. Syst. Techn., 2, 144-156

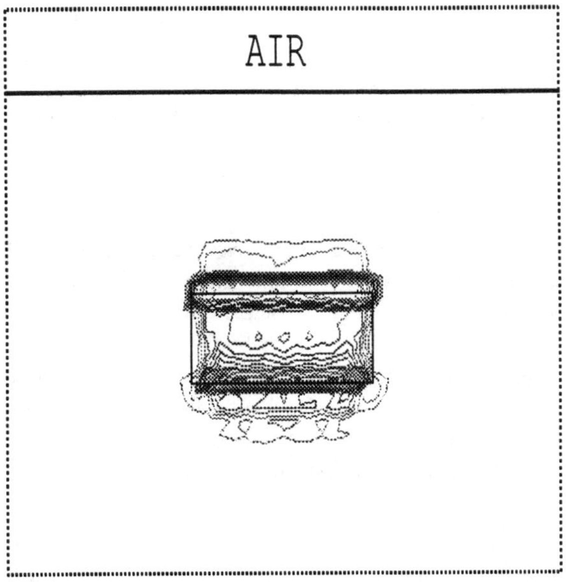

Figure 3: Reconstruction of a buried antitank mine

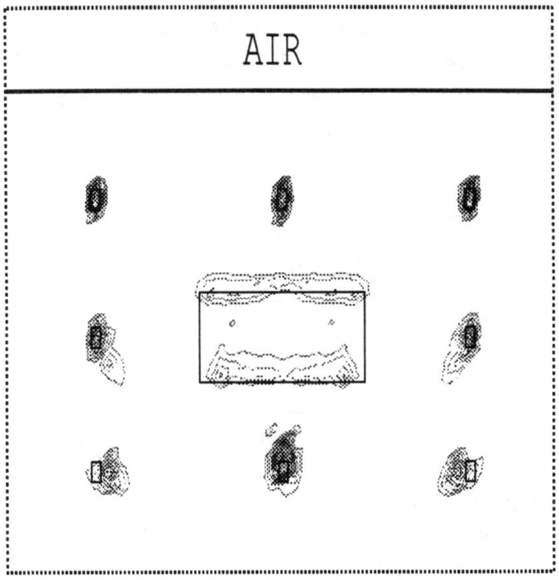

Figure 5: Reconstruction of a buried antitank mine surrounded by inhomogeneities (voids)

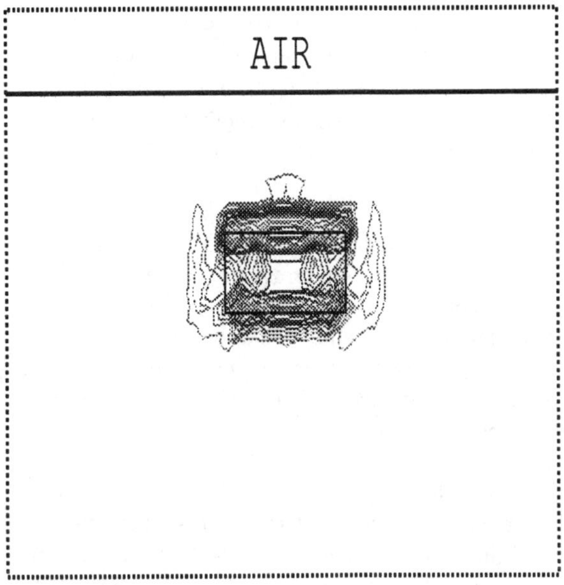

Figure 4: Reconstruction of a buried antipersonnel mine

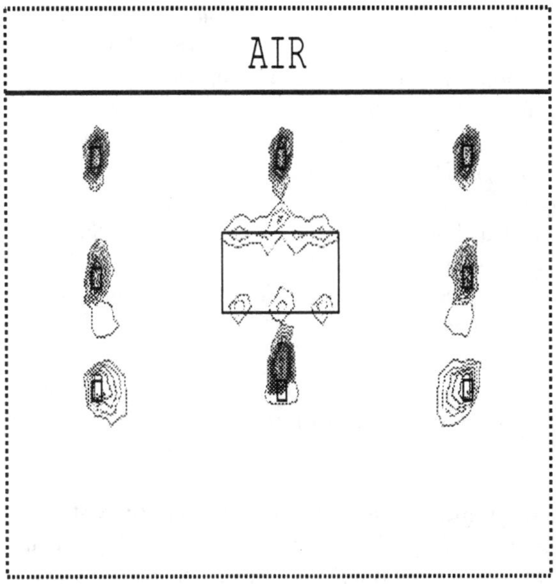

Figure 6: Reconstruction of a buried antipersonnel mine surrounded by inhomogeneities (voids)

NEAR-FIELD SPOT-FOCUSED MICROWAVE SENSING FOR THE DETECTION OF BURIED LAND-MINES

R. Benjamin.

Bristol University, UK.

1. Introduction

The most promising technique to date for detecting buried mines, ultra-wide-band ground-penetrating radar [1], is still quite limited as regards:-

- depth of penetration of attenuating ground
- false alarm rate
- missed detections
- need for skilled operation and interpretation
- area covered per unit time.

We hope to achieve significant improvements in all these respects by the use of near-field focused radar. Spot-focusing of the transmitted and received radiation enhances the resolution, and the focusing convergence gain counterbalances the path-loss, in a highly attenuating medium such as moist soil. Furthermore, unwanted reflecting objects are unfocused, thus substantially reducing obscuration of wanted targets by "clutter". This is of special importance in an attenuating medium, where a shorter unwanted path would benefit from reduced path loss. These gains in sensitivity should also enable us to cope with the extra attenuation of a somewhat shorter wavelength, thus improving the compromise between resolution and depth of penetration. Several novel variants of this technique will be considered.

2. The Resolution Achievable.

A. Horizontal Resolution. Antenna reciprocity states that the coupling between two antennas is independent of which is transmitting and which receiving. When an antenna is focused onto a point in the near field then, by the definition of focusing, that point is coherently irradiated. The resulting focal spot can then be regarded as a receiving antenna. By antenna reciprocity, its diameter is therefore equal to that antenna aperture whose beam width would just encompass the source antenna, see Fig. 1. Thus the spot diameter is $D_1 = \lambda/\theta$, where θ is the convergence angle and λ the wavelength. Beyond $\pm D_1/2$, the received phase will no longer support coherent integration of the signal impinging on the focal point. For mono-static coherent reception, we are concerned with the signal returned from the focal spot back to the transmitting antenna. Thus we are concerned with the range of phase changes experienced in the *two-way* path from and to the antenna, and so the relevant coherence-length is halved to $D_2 = \lambda/2\theta$. In a bistatic configuration, the sensitivity of the phase, observed from the two-way path, to lateral displacement depends on γ, the angle between the transmitted and received rays, and so $D_3 = \lambda/2\gamma$.

B. Vertical Resolution. Since we are concerned with resolution in 3D, we are equally interested in the depth of focus. By the Rayleigh principle, an antenna does not form a beam of width θ originating from its centre, but an essentially parallel beam, of diameter equal to its aperture D, up to the Rayleigh distance where the "ideal" divergent beam would have spread out to width D. Hence the Rayleigh distance is

$$R_d = \frac{D}{2}\cot\frac{\theta}{2}. \quad (1)$$

However, the radiation converges towards the focal spot and then diverges again, thus maintaining a near-constant width D within length $\pm R_d$. See Fig. 2. Hence the depth of focus is

$$F = D\cot\frac{\theta}{2} = \frac{\lambda}{2\theta}\cot\frac{\theta}{2}. \quad (2)$$

The approximation $\cot\theta/2 \cong 100/\theta'$ is accurate within ± 10% for $90° \geq \theta' \geq 10°$, where θ' is now measured in degrees. In our form of near-field focusing the convergence angles will be fairly large, and so $D \cong \lambda$, and $\lambda \leq F \leq 3\lambda$. Here λ is the wavelength in the propagation medium. For moist soil this is between 1/3 and 1/8 of the free-space wavelength.

Note that our focal depth is shorter than any practical pulse-length for a phased array. Hence we have to cope with non-coherent clutter from a thicker zone, bracketing the focal range. (This contrasts with synthetic-aperture radar, where the convergence angles are small, and the focal depth is larger than the effective pulse length.)

Inhomogeneities in the ground will perturb the equi-phase condition at the focal spot, and so they reduce - indeed they can ultimately destroy - the focusing gain. Hence λ must be kept large enough, compared to the scale of such inhomogeneities, even at the expense of reduced resolution.

C. 3D Sidelobes. As in a conventional far-field focused array antenna, the focused beam is flanked by a pattern of sidelobes, due to the non-coherent combination of the element patterns. In our scheme, the viewing directions differ from element to element, and so the clutter arcs seen by the antenna elements do not coincide except close to the focal cell, thus reducing any coherence between these clutter contributions. However, since practical pulse lengths will be substantially longer than the dimensions of the focal cell, we are also subject to a similar process, due to focusing in the range dimension. Hence, our focal cell is surrounded by sidelobes in all three dimensions, with some resultant impairment of our signal/clutter ratio, and some blurring of our resolution.

3. Ground Matching.

Our design options are severely affected the constraints imposed by the medium, and hence these must now be considered:-

A. Index Matching. The propagation speed - and hence the wavelength - in soil is typically 3 to 8 times smaller than in free space. Hence any phased-array antenna would have to be in effective contact with, and matched to, the propagation medium. But the size of the elements and their spacing define the wavelength with a tolerance of, at most, about ±20%. Hence we propose to change the frequency to cover the appropriate range of soil conditions, thus maintaining a substantially constant wavelength, and hence both a constant phased-array configuration and a constant resolution in the medium.

In a good conductor, the wavelength varies inversely with the square root of the frequency and of the conductivity, and the attenuation *per wavelength in the medium* is independent of the frequency or conductivity. Although our medium is not a pure conductor, the path loss should vary relatively slowly, with changing soil conditions, and so wavelength matching should also substantially reduce the variations of path loss.

Frequency selection will however have to accompanied by the appropriate impedance matching in the phased array. It is proposed to face the antenna always with the appropriate one of range of compliant sheets, matched to the locally pertaining soil impedance. Depending on the operating conditions, this sheet may be in physical contact with the ground or separated from it by half a wavelength in air (or an integral number of half wavelengths), so that the reflections from the two interfaces cancel

B. Loss Matching. In a lossy medium, ray-to-ray variations in path-length can seriously distort the pattern of relative powers of the rays converging from different directions onto the near-field focal spot. For maximum focusing gain and resolution, we would wish to use the largest practical antenna array. Its useful size is however restricted by the fact that the elements furthest from the near-field focal spot have a substantially longer path length - and hence experience substantially more attenuation - then the elements nearest the focal spot, and so the potential contribution of the more distant elements is limited. If we want the true resolution benefit of a given aperture for focusing (which, for very near objects, may not always be the full available aperture), we have to normalise gains by attenuating the signals transmitted from the nearer elements of the array, to equalise their lower path-losses. Hence there is a maximum antenna diameter (roughly 5 times the electrical "skin depth" of the medium, at the given frequency) beyond which the extra aperture gain is more than counterbalanced by the extra normalising loss. In mine-search, safety considerations may require the antenna to be laterally offset from the vertical through the search-point. This increases the path lengths and losses, but mitigates slightly the problem of differential path losses.

C. Ground Contour Matching. If the ground is overgrown, the sensor, on a boom ahead of the search vehicle, may have to be preceded by cutting and clearance tool on a further forward extension of the sensor boom. Since the ground will normally be rough, good contact may require the front-face of the antenna to be faced with a soft mat of the appropriate index-matching material, as already indicated. For such contact, we would have to place the antenna onto the ground and take a series of measurements, and then lift the antenna and move it to the next, adjacent ground position. The ground is also likely to be uneven on a larger scale. After considering a number of alternatives, we have opted for the arrangement shown in Fig. 3. Here a T-shaped boom projects forward from the operating vehicle, far enough to ensure the safety of the vehicle in an accidental mine explosion. Six (say) disc arrays (with padded faces) are mounted at the front-end crosspiece of this boom. Each of these discs can adapt to the gross variations of the ground, both vertically and by tilting about the roll and pitch axes, with sensors measuring these displacements, so that the excitation can be adapted accordingly

D. Ultra-Sonic Variant. With in-contact discs, an ultra-sonic acoustic version of the scheme could also be considered. This would perform best in wet ground, the very conditions where a microwave system is severely limited. The echoing delay times would of course be enormously larger than in the electro-magnetic variant. However, as shown in the next section (for the electro-magnetic scheme), any in-contact scheme the search rate is limited entirely - and quite severely - by non-sensor considerations.

E. Ground-Contact Search Sequence. To cover a search volume of 2 m (width) x 1.5 m (forwards) x 0.6m (depth), at 5 cm intervals in 3D, would involve the transmit and receive beams focusing sequentially onto 14,400 distinct resolution cells. Even for the slowest propagation speed (refractive index of 8), the average echoing time, would be under 0.03μs. Even allowing a total of 0.7 μs per resolution cell, to make generous provision for beam forming and for any "chirp" pulse-spreading, the 14,400 spot searches would only take a total of 0.01 s. Thus the search time would be governed virtually exclusively by mechanical processes. For instance we might envisage a cyclic search scheme somewhat as follows:-

1.	fit discs to ground	3 s
2.	sequential beam forming and search	-
3.	advance boom by 1.5m (vehicle remaining stationary)	6 s
4.	repeat 1 - 3	9 s
5.	repeat 1 - 3	9 s
6.	repeat 1 - 2	3 s
7.	advance vehicle 6m, whilst retracting boom by 4.5m	15 s
TOTAL DURATION		45 s

Subsequent cycles would repeat steps 1-7, above.

This scheme is designed to minimise the time taken by the mechanical processes, which dominate the search rate. These arbitrary but not unrealistic figures give a total of 45 s for searching 2 m (front) x 6 m (advance) x 0.6m (depth), - i.e. an average advance of 1 m every 7.5 seconds.

D Non-Contact Search. The same equipment could be made to operate in a non-contact mode whenever the ground conditions are suitable, thus permitting a far higher search rate to be achieved. For non-contact search, the electronic scan could be limited to a single resolution cell in the direction of vehicle travel, the third dimension being provided by the vehicle's own advance. In some circumstances it might however be expedient to search over one mine diameter in the forward direction. In either case, there would be no significant electronic limit to the search rate. The limit would be set by the requirement, on detection of a mine to stop, say, within 5m at a retardation of 0.25G. This would correspond to a forward speed of 5m/s = 18 km/hr. Any higher speed would require a further boom extension and/or a higher braking rate.

In the non-contact configuration, it should be easier to focus onto voxels close to the surface. Even so, the achievement of practical convergence angles may have to invoke a special antenna-operating mode, using part of a single disc for transmission and another part for reception.

4. Possible Radar Configurations

A. Bistatic Operation. The very close ranges involved in mine-search pose considerable duplexing problems, when using the same antenna for transmission and reception, which are avoided by bistatic operation, with distinct transmit and receive antennas. In each of the two triplets of discs shown in Fig. 3, the centre disc is assumed to be the transmitting array and the outer ones (used one at a time) the receiving ones. As already mentioned, such a bistatic configuration also enhances the resolution by benefiting from the combined convergence angle between the two antennas. Bistatic operation entails some increase in path lengths, and hence in losses, but the propagation loss due to the combined go and return paths varies much less drastically with focal-cell location than in a monostatic configuration. This reduces the system's dynamic range, and it reduces the risk of reflections from nearer scatterers impairing detection of more distant ones. Most important of all, the beams from the two antennas (or at least their main lobes) coincide only at the selected focal cell. Hence any reflecting object, outside the common focal cell of the transmit and receive antennas, cannot be in the main beams of both antennas, even in an unfocused state. This should drastically reduce any obscuration of the wanted target by "clutter".

B. The Potential Heterodyne Variant. For optimum resolution, the operating frequency will be chosen so that echoes from the maximum required depth of penetration will just be detectable against the background of clutter and/or thermal noise. This involves an attenuation of many tens of dB in both the 'go' and 'return' paths. However, in a possible alternative operating mode, both antennas of the bistatic scheme might transmit, at frequencies differing by a small pre-determined offset (say 10% of the carrier frequency). The transmission would then comprise very short pulses of high peak power, but with a low "duty cycle", to maintain the same mean power. (The risk of radiation setting off a mine detonator is primarily a function of mean power.)

The two short, high-power pulses, converging on the focal spot at the same instant, would then excite any exploitable dielectric non-linearity of the substance at that focal spot, thus making it operate as a heterodyne mixer, producing a a signal at the given low difference frequency. This difference-frequency signal would be transmitted, with negligible loss, to a separate antenna and receiver on the surface. (Crossed dipoles might be used to cope with the uncertain polarisation.) If one or more harmonics of the difference frequency are also received, the relative power of these components may also be of diagnostic value, for classifying the object irradiated.

If indeed there is an exploitable dielectric non-linearity in land mines - which remains to be confirmed - this heterodyne operation, with negligible return-path loss, would permit a significant increase in the depth of penetration, at a given wavelength, even after allowing for the low efficiency of the conversion from the incident radiations into this difference-frequency signal. In addition, this mechanism would give a virtual guarantee that the received signal can only come from the focal spot where:

- both incident signals arise at the same point,
- and at the same instant in time,
- and at a high power level, localised in time (by pulsing) and in space (by focusing).

Any such heterodyne detection of a mine would of course depend on the existence of a noticeable dielectric non-linearity, manifesting itself at a system-wise practicable field-strength at the target, and on this dielectric non-linearity being different from (probably higher than) that of the surrounding medium. It is thought likely that a non-linearity may manifest itself as the molecular dipoles attempt to align themselves to the incident electro-magnetic field. The large molecules of a plastic mine-casing, or indeed of the mine filling, should have a much larger electric dipole moment than the surrounding earth. Hence, if dielectric non-linearities are present and detectable at all, the prospects of a significant *differential* non-linearity seem fairly good.

The use of short pulses, essential for heterodyne operation, is also desirable to limit the range extend of the volume of side-lobe clutter straddling the focal cell. Furthermore, pulse lengths shorter than the time-constant of resonant absorption mechanisms can reduce absorption losses. (In this respect ultra-short pulse radars can gain an even bigger benefit.)

5. Conclusions

1. Near-field focusing promises to give a substantial novel capability of high-resolution 3D microwave imaging, even in lossy media.
2. A bistatic configuration seems to offer significant practical advantages in this context.
3. Where the substances of interest exhibit sufficient differential dielectric non-linearity, the heterodyne variant of the scheme is particularly attractive.
4. The simple 6-disc scheme should provide adequate adaptation to the shape of the surface.
5. Changing frequency, with a given array, can adapt adequately to the nature of the soil.
6. In very wet soil, the in-contact ultra-sonic variant seems promising.
7. Otherwise, the non-contact microwave mode of operation with its far superior search-rate, is clearly preferable to the in-contact mode, wherever conditions permit.
8. The system will be substantially larger, more complex, and more expensive than existing ground-penetrating radars. However, its expected greater depth of penetration, greater certainty of identification, independence of skilled interpretation, enormously greater search rate, and greater immunity to mine explosions, should also make it more effective - and cost-effective.
9. In view of its promise, the scheme has been patented by Bristol University

6. References.

1. Daniels, D.J. Dec. 1994, Microw. Jnl. pp. 68 - 82.

2. Benjamin R., Sept. 1995, Microw. Jnl. pp. 68- 81.

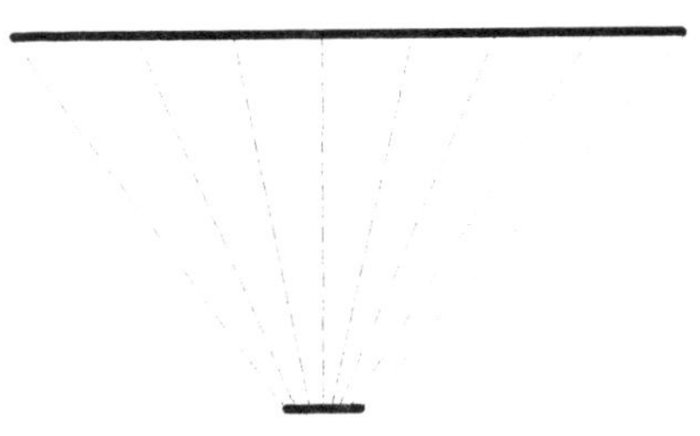

Fig. 1 Mutually Reciprocal Antennas

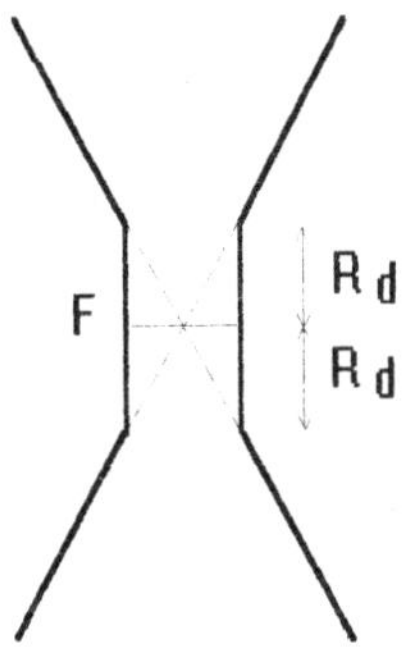

Fig. 2. Depth of Focus

F = focal plane at beam waist

R_d = Rayleigh distance

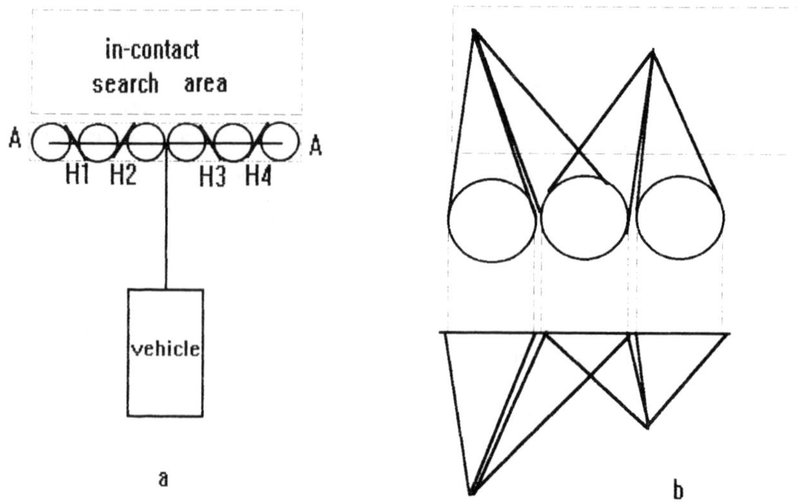

Fig. 3. The 6-disc Configuration

a full picture
30 cm dia. discs, with 4 cm gaps
H1, H2, H3, H4 = receive dipoles
for heterodyne variant
A - A = non-contact mode search area

b enlargement of search area
plan view and front elevation
of two sample search points

Synthetic, Post-Reception Focusing in Near-Field Radar.

R. Benjamin.

Bristol University, UK

1. Introduction

A separate paper [1] has developed the concept of searching for buried mines by means of a (probably bistatic) radar, near-field focusing onto all relevant 3D resolution cells (voxels) in turn. A specific example has been postulated, where each of two triplets of disc antennas searches a volume of 1m x 1.5m x 0.6m with 5 cm 3D resolution, by focusing sequentially onto all 7,200 resolution cells so defined (a total of 14,400 for the two triplets). This a realistic practical solution, within the state of the current art. However, it goes against the grain to have so many overlapping transmissions into the same volume, each restricted to examining a single voxel. Hence we are here putting forward a proposal for using the transmitted power much more efficiently. This "synthetic focusing" radar is believed to be a new concept, although it shares some features with "synthetic aperture" radar [2]. It also has applications to some conventional, far-field radar scenarios.

2. The Concept of Post-Reception Synthetic Focusing.

With "conventional" spot-focused operation, in our postulated application, each of 7,200 transmissions from each of the transmit antennas would excite all elements of that transmitting antenna. In the basic form of our new proposal we reduce the aggregate power transmitted 7,200-fold, for the same performance, by using all transmissions from all antenna elements to contribute in full to the examination of all 7,200 resolution cells, examined by that antenna. This is achieved as follows:-

Assume transmit and receive antennas of N (say 150) elements each. Each of these elements, individually, must then have a wide enough radiation pattern to encompass all the volume of interest. In "conventional" spot focusing, we adjust the relative timings and phases of the outputs from the network feeding the N transmit elements to compensate for the fixed element-to-element variation in path length from feed to element, and for the element and focal-cell dependent variation in path length from the given element to the selected focal cell. This ensures that the transmit signals from all N elements arrive at the selected focal cell in time and phase coincidence. Similarly, we adjust the relative timings and phases of the inputs to the network collecting the signals from the N receive elements so that the signals re-radiated from the selected focal cell will be combined in time and phase coincidence. This processing of the received signals can be accomplished in real time, by means of appropriately switched delay elements, or in "reel time", by independently receiving the relevant signals at all receive elements and (digitally) recording them for subsequent processing. The total process involves N single-element focusing operations at the transmit array and a further N at the receive array, for each of the 14,400 voxels examined.

However, fundamentally, we are only concerned that the signals from all N transmit elements, as re-radiated from the selected resolution cell, are combined coherently at the output from all N receive elements. If we energised the N transmit elements sequentially, one at a time, we could receive and record the resulting returns from the selected resolution cell independently, at all receive elements. For subsequent off-line near-field focusing we could apply to each of the N signals at each receive element the appropriate correction, so as to equalise the combined total delays and phase shifts for the two-way path from the given transmit element via the selected focal cell to the given receive element. Furthermore, we could apply to this signal that amplitude scaling for path-loss equalisation, and possibly also for aperture shading, that would have been applied to the relevant transmit element plus that which would have been applied to the given receive element, when both are focused onto the given resolution cell. See Fig. 1. Thus, after this equalisation, the N separate contributions noted in each receive element (one per transmit element), summed for each receive element and then summed over the N receive elements, will jointly comprise the full pattern of contributions otherwise received as a result of independent, concurrent transmit and receive focusing.

Hence we can combine the N^2 signals for the given focal cell, representing the N^2 transmit-element/receive-element pairings, to reconstitute a 2-way focused signal which is essentially identical with that from conventional operation with off-line processing. The only difference arises because any path-length equalising or aperture shading power reduction at a transmit element is replaced by an equivalent attenuation of the eventual received signal. This then reduces both signal and noise equally, so resulting in a superior eventual combined signal/noise ratio. Note also that the single focusing operation for the two-way path is subject to the combined tolerances otherwise applied to the separate transmit and receive focusing operations.

However, in place of recording the signals corresponding to a single 3D resolution cell, we use the transmission from each of the N transmit elements to record, at each of the N receive elements, the full range of signals received:

- within the time bracket of relevant 2-way path delays
- and resolvable with the given pulse length.

These N^2 sets of records therefore provide all the data required to reconstitute the two-way focused signals *from all relevant 3D resolution cells*. Note also that this information is collected by N (i.e. 150) parallel receivers from N (i.e. 150) single-pulse transmissions, as compared to (in our example) 7,200 transmissions, each in parallel from all N transmit elements. Of course, when focusing onto one space cell (voxel), each receive element has to make N separate contributions, one for each transmit element. Thus two-way focusing involves the coherent combination of N^2 contributions, at the receive array, in place of N at the transmit and N at the receive arrays. This N/2-fold increase in processing - and independent reception at all receive elements - are the price paid for the benefits of

- N-fold reduction in transmitted peak power
- 7,200-fold reduction in transmitted mean power
- 7,200/N = 48-fold reduction in total transmission time (not of importance in our application).

To reconstruct the data for any given 3D cell, we compute - or look up - the path delays from each transmit-antenna element via that cell to each receive-antenna element, and extract the corresponding signal from our stored data. Each of these contributions is then scaled in power to normalise the losses in all 2-way paths to match that of the longest 2-way path involved in focusing onto the given voxel, and is further weighted for any taper, for side-lobe reduction, associated with the relevant transmit and/or receive antenna element.

Synthetic post-detection focusing, when one element at a time is transmitting, all he other ones are free to receive. Thus we can combine the freedom from duplexing problems characteristic of bistatic operation with the full exploitation of the all the aperture characteristic of monostatic operation. In such a quasi-monostatic implementation, each of the N elements of the array has to transmit in turn. However, for full information, we need only record the signals received in half of the other N-1 elements:- half the two-way paths will be observed with element i transmitting and element j receiving, and half with j transmitting and i receiving. All elements must transmit for 1/N of the time, as well as receiving for half the time: we cannot observe a path between two transmit-only elements or two receive-only elements. Quasi-monostatic operation also permits us to select the data from that subset of elements which is optimum in relation to the horizontal position and depth of any given focal cell.

Irrespective of bistatic or quasi-monostatic operation, or indeed of "conventional" near-field focusing or post-reception synthetic focusing, each activation of a transmit element will cause each receiving element to accept the resultant of the reflections from all inhomogeneities within the volume defined by:-

- the directional pattern of the transmitting element,
- the directional pattern of the receiving element,
- the leading edge of the received pulse,
- the trailing edge of the received pulse.

See Fig. 2. The loci of constant two-way path length, defining the leading and trailing edges of the received pulse, are ellipsoids of revolution, with the relevant transmit and receive elements as their foci. (They reduce to a constant-range sphere when a single element receives the return from its own transmission.) The N^2 pairings of transmit and receive elements generate N^2 such zones, differing in lateral position and curvature. See Fig. 3. Even were these clutter zones overlap, their returns do not combine coherently, except only in the selected resolution cell, whose size is determined by the wavelength and by the focusing and bistatic convergence angles. Only in that voxel do the signals received by all N elements of the receiving array from all N elements of the transmit array combine

- at the same time,
- at the same phase,
- at the same power level.

Exactly the same selection, focusing and sidelobe generating mechanisms operate in conventional quasi-optical focusing.

3. Implementation of Synthetic Focusing.

Our postulated transmit and receive arrays would have about 150 elements each, and the difference between the longest and shortest likely two-way 3D path lengths postulated is approximately 3 m. With a pulse length equivalent to, say, 30 cm of path delay, each receive element need only record at most 20 discrete complex signals per transmit element to provide a full continuous record of all resolvable single-element signals. However, in focusing, we are concerned with the size of our resolution cell of say 5 cm, and so we have to centre a time gate onto the reconstructed continuous received signals to a fraction of a pulse length. (All pulse lengths and phase delays relate of course to the phase velocity *in the medium*.) Hence "I" and "Q" correlaters, associated with each receive element, would have to be sampled up to 20 times for each transmit-element activation. Thus the data collection per transmitted single-element pulse, in its basic form, would need to perform 6,000 correlations:- 150 (arrays elements) x 20 (discriminable pulse positions) x 2 ("I" and "Q" phases). In processing these data, we would have to compute the sum of 150 x150 = 22,500 complex contributions per focal cell, for each of a total of 14,400 resolution cells. This comes to a total of 6.48×10^8 additions.

We postulate that mechanical limitations would impose a minimum time of 9 seconds to advance the search area 1.5 m forward, for the next search scan, corresponding to a computation rate of 7.2×10^7 additions per second, a fairly trivial task.
It would be possible to drive each transmit element k times, so that 1/k the number of receive correlators can be time-shared, at the expense of a k-fold reduction in the power saving achieved. The number of receive correlators could be reduced by yet further factor, again at the cost of a proportionate increase in the total transmitted power, by increasing the size of the transmit antenna and reducing that of the receive antenna, by a common ratio, thus keeping the number of path-length combinations constant. Both the preceding measures would increase the data-collection time. However, in no circumstances will the electronic observation time impose any constraints on the system design.

In the non-contact mode of operation, the range dimension of the 3D scan is provided primarily or indeed entirely by the mechanical forward movement of the search vehicle. Hence the range-wise dimension of the volume to be searched electronically, using the data from the 150 x 150 path combinations, would be reduced from 150 cm (30 voxels) possibly to 30 cm (6 voxels) i.e. one disc diameter, but probably to 1 voxel (5 cm). This would however have to be repeated for every 30 cm or 5 cm of advance of the vehicle, respectively. The number of (i, j) pairs to be covered would of course be unchanged. However, the difference between the maximum and minimum 2-way path delays to be covered would be reduced to only about 90 cm, as compared to 3 m, thus permitting a pro rata reduction in the number of correlators. Here, too, the electronic search rate will not limit the overall search rate achievable.

Since the components of a focused spot are obtained sequentially, the technique (in all its variants) is best suited to observing static situations as in mine search (or tumour diagnosis). However, specific hypotheses of target movement can also be covered.

4 Synthetic, Post-Reception Beam Forming for Far-Field Radars.

Synthetic focusing can also be applied to far-field radars. In that case, our previous range-dependent 3D beam-focusing, onto one voxel at a time, is reduced to range-independent 2D beam-forming, and steering onto one resolvable angular direction at a time. Hence the number of discreet phase delays required for scanning the array is drastically reduced, and indeed these delays can be implemented as phase shifts, if desired. Range is then resolved purely by pulse arrival time, and the data recorded at each receive antenna element must cover the full echoing time to maximum range. For any given beam direction, of 2D phase slope θ, we then apply to the signals from transmit element i, as received and recorded at receive element j over the full inter-pulse period, an incremental phase delay

$$\Delta t(\theta) = \Delta t(\theta, i) + \Delta t(\theta, j)$$

equal to the sum of those arising at elements i and j from applying phase slope θ separately to conventional transmit and receive antennas. See Fig. 4. The resultant from summing these N^2 signals is then the signal of the desired steered beam. Since the number of independent beams, which can be formed in a phased array, is equal to the number of elements, the mean power is reduced N^2-fold (and the peak power N-fold), and the data collection time, with sequential transmission from all antenna elements in turn, is the same as for a conventional phased array performing a sequential search of all distinct beam directions. Like a normal phased array, the synthesised far-field scanning system could save both search-time and processing, when searching a more restricted angular range. For our synthetic far-field beam forming, this would be achieved by configuring the transmit and receive antennas as sets of sub-arrays, with their (common) beamwidth and "squint" angle matched to the desired angular coverage. However, any resultant reduction in "N" would reduce the power saving *pro rata*.

Because synthetic far-field beam forming depends on the coherent combination of data collected sequentially, it is best matched to scenarios which can be regarded as static during the data-collection time frame, but where we wish to observe frame-to-frame changes - with appropriate inter-frame intervals. This condition would be less restrictive in single-dimensional synthetic beam forming and steering: For instance, for a vertical receiver-synthesised scan, 20 horizontal lines of transmit elements, within a 300-element circular transmit array, might be energised sequentially, one complete line at a time. Thus the minimum target coherence time required would be reduced in the ratio 20/300 = 1/15. Hence there should be further applications where, for instance, vertical synthetic beam forming and steering could be combined with rotational or phased-array scanning in the horizontal plane.

6. Conclusions.

1. Post-detection synthetic focusing or beam forming is essentially restricted to near- or far-field scenarios which are static or change only slowly.

2. However, it then offers major- or indeed dramatic - benefits where:-

- Radiated power is limited by safety considerations.
- Radiated power is limited by considerations of hardware procurement cost.
- Mains power is limited by logistic considerations.
- Radiated power is limited to reduce the likelihood of interference with "friends".
- Radiated power is limited to reduce the likelihood of reception by "enemies".

3. All these conditions could apply to some extend to the mine-search problem, but none are likely to be of overriding importance.

4. Significant technological development will be required to implement the scheme.

5. However, it might eventually become the most cost-effective solution to mine-search - and to some other problems.

6. In view of its promise, this scheme is being patented by Bristol University.

7. References.

1. Benjamin, R. Near-field spot-focused microwave sensing for the detection of buried land-mines, EUREL/IEE Conference on the detection of abandoned land mines, Edinburgh, Oct. 1996.

2. Benjamin, R., 1995, Microw. Jnl. Sept. 1995, pp. 68-81.

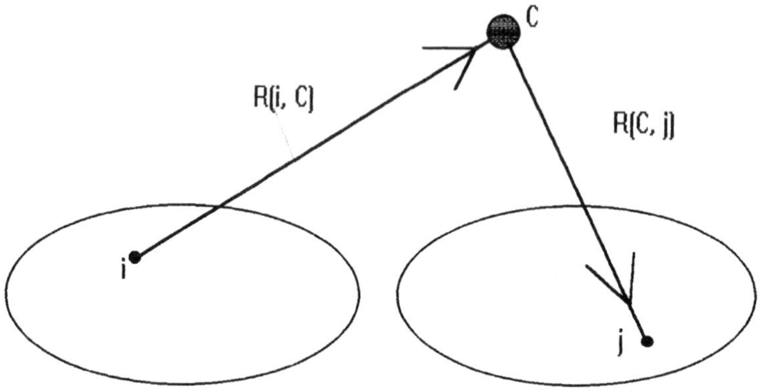

Fig. 1. Equalisation of two-way Path

i = transmit antenna element C = selected resolution cell J = receive antenna element

delay equalisation = $\{[R(i, C) + R[C, j)]_{max} - [R(i, C) + R[C, j)]\}/c$

where c = velocity of propagation

path loss equalisation = $L\{[R(i, C) + R[C, j)]_{max} - [R(i, C) + R[C, j)]\}$

where L = path loss in dB/m

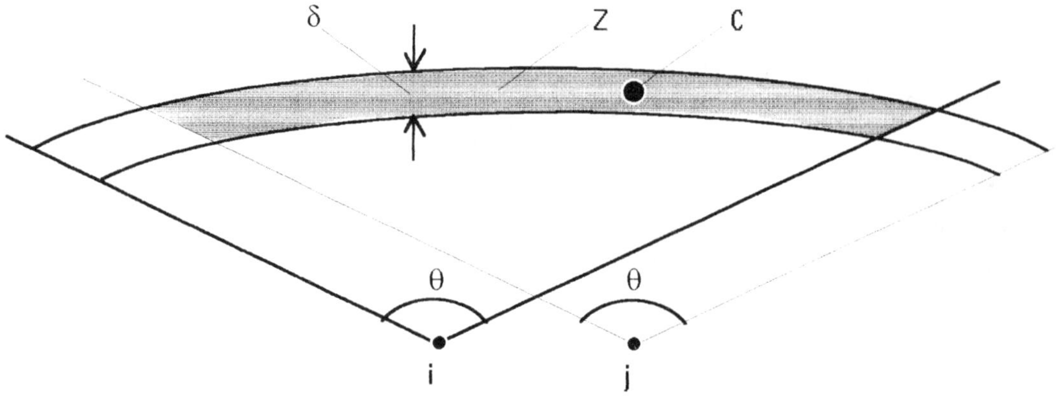

Fig. 2. Limits to Clutter Volume for Element Pair

i = transmit element j = receive element

θ = width of element pattern δ = path-length increment due to pulse length

C = selected resolution cell Z = clutter zone (shaded area)

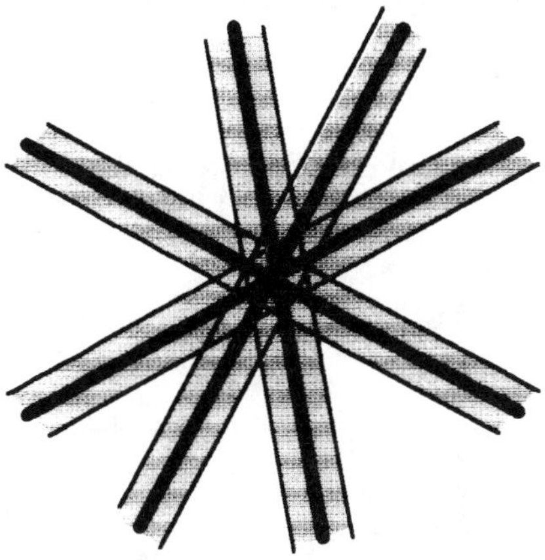

Fig. 3. Overlapping Clutter Zones around Focal Cell

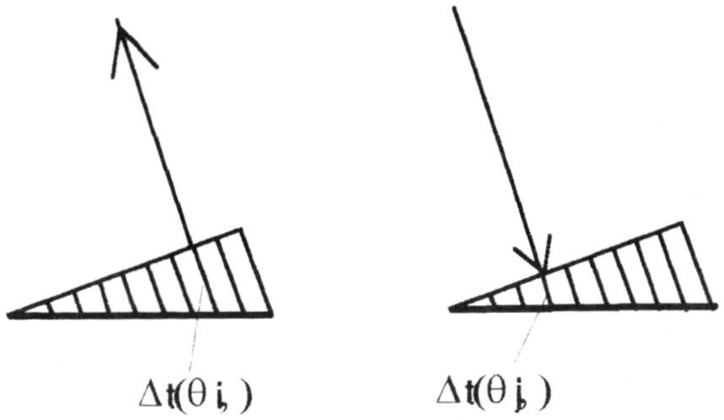

Fig, 4. Pairing of Transmit and Receive Elements in Far-Field Synthetic Focusing

OVERVIEW OF CURRENT RADAR LAND MINE DETECTION RESEARCH AT THE DEFENCE SCIENCE AND TECHNOLOGY ORGANISATION, SALISBURY, SOUTH AUSTRALIA.

I J Chant & A R Rye

Microwave Radar Division, Defence Science & Technology Organisation, Australia.

ABSTRACT

This paper describes the Australian Defence Science and Technology Organisation's (DSTO) efforts in buried land mine detection by Ultra Wide Band (UWB) radar technology. Detection of low-metallic content land mines with reduced probability of false detection is a high priority. Present emphasis is on the development of a hand-held system for demining operations. Vehicle-mounted land mine warning systems are also being investigated.

An effective hand-held system is likely to combine a Ground Probing Radar (GPR) with a metal detector. An initial version of such a system could be a cheap, light-weight GPR attachment for current mine detectors. This work is also being extended to cover close-in vehicle-mounted systems for road clearance. Experimental work has commenced on stand-off UWB radar systems for vehicle-mounted anti-vehicle mine warning systems.

The current thrust is in the application of 1 to 3 GHz bandwidth impulse radar systems. Later development will include integrated multiple-sensor systems incorporating automated target recognition and sensor fusion.

INTRODUCTION

Australia's requirements for land mine detection derive mainly from our commitment to support United Nations (UN) peacekeeping activities. One need is for humanitarian demining, such as that being undertaken by Australian troops working with indigenous demining teams in Cambodia and Afghanistan. The second requirement is for vehicle-mounted systems to allow safe transit over unsealed roads.

The humanitarian demining requirement is essentially for a hand-held system to detect, and identify, land mines. The priority is for improved target discrimination to reduce the number of false targets found by current metal detectors. Development emphasis the detection of low metal-content mines and detection of non-metallic mines will be a target for later development.

The most promising technology to fulfil the hand-held detector requirement is a combination of an inductive metal detector with GPR. Thermal imagery might also be incorporated to cue the operator on to potential targets.

Operational mobility is essential in low-level conflict such as UN peacekeeping missions. Land mines can severely curtail the ability of peacekeeping troops to provide aid and respond to crises. A vehicle-mounted land mine detection or warning capability is required to ensure mobility. The technologies being tested for this purpose are extensions of the hand-held GPR detection system.

A stand-off radar sensor combined with other stand-off sensors would allow a mine-hardened vehicle to survive anti-personnel (AP) mine explosions and to avoid anti-vehicle mines. The most likely sensors for stand-off detection are optical/multi-spectral and/or thermal imaging combined with Ultra Wide Band (UWB) radar operating at ground penetrating frequencies.

HAND HELD GROUND PROBING RADAR

Two GPR systems developed in Australia were trialed under contract to determine the feasibility of GPR for land mine detection. One system was a Step Frequency GPR under development at the University of Queensland, the other an Impulse GPR system developed by the Commonwealth Scientific and Industrial Research Organisation (CSIRO). Both systems were trialed on sand-filled test-beds and on a local test site in both wet and dry soils. The targets were inert land mines filled with paraffin wax.

Impulse Ground Probing Radar

The impulse system was the CSIRO FM127 GPR. The system is a 127 kHz sequential sampling device acquiring 512 samples per sounding in any time interval from 4 ns to 16 ns. This device produced 127 soundings per second of 512 time-domain amplitudes across the time interval. The device stretches 20 ps samples to 70 ns to allow sampling by an oscilloscope or computer with a 250 kHz data acquisition card.

The system uses bistatic, bow-tie antennas with the transmit pulse created by a fast-recovery diode at the transmit antenna. This technique allows the centre frequency and bandwidth of the transmitted pulse to be changed by changing the antenna. The GPR unit is light-weight and of comparable size to existing mine detectors. The acquisition and display unit can be a notebook sized computer or portable oscilloscope.

DSTO is testing various antenna designs with this system. A set of differential, or balanced bridge, antennas has so far proved the most effective configuration for small target detection. These antennas difference the input from two separate receive antennas, one on either side of the transmit antenna. The configuration was effective in reducing the ground return and other constant-time features. Other configurations under test include separated-aperture, cross-polar and spiral, circularly-polarised antennas. Currently the impulse GPR has been tested with 1.4 GHz and 0.8 GHz band-centre, 90% bandwidth antennas. A 3 GHz antenna is under development.

The antenna configurations above are for use in close proximity to the ground. The system will also be tested with broad-band, horn antennas suitable for a close-in vehicle-mounted system.

Initial tests where undertaken with acquisition software designed for geophysical use. This software allowed post-processing in the form of background subtraction, migration and filtering in one and two dimensions, as well as phase reconstruction. It is unsuitable for demining applications which require real-time display of filtered images.

CSIRO have been contracted to develop real time acquisition and analysis software suitable for a GPR land mine detector. This software incorporates a running average background subtraction as well as a number of display features specified by DSTO. Dual-sensor, acquisition and display software to accommodate joint operation with metal detectors will be added in the next few months. This software package can be extended to include more sophisticated signal processing such as target identification if it is needed for an operational system.

The impulse GPR was the more mature of the two systems tested and was able to detect low-metal content AP mines as well as the larger mines. Only the smallest non-metallic mine tested proved difficult to detect (Chant (1)). Figure 1 shows the impulse GPR image of a PMN AP mine common in Cambodia and Afghanistan.

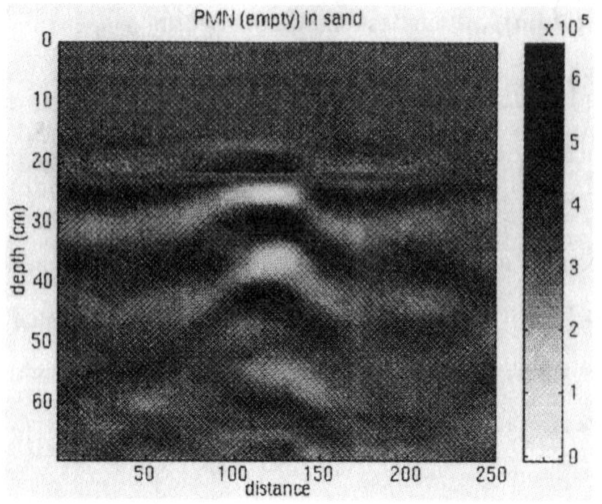

Figure 1: Impulse GPR image of a PMN mine in sand.

Step Frequency GPR

A Step Frequency GPR system being developed at the University of Queensland was used to examine the feasibility of using this technique, as opposed to the time-domain impulse technique, to detect buried land mines. This system is a 1 to 2 GHz (approx.) stepped frequency device using separated aperture bow-tie antennas. The device is being developed to trace shallow mineral layers. It was not optimised for the detecting small targets, such as the AP mines, but could detect the larger mines such as the plastic anti-vehicle (AV) mines. Redesign would be necessary to optimise this equipment for mine detection (Chant (1)).

ULTRA WIDE BAND RADAR

A stand-off vehicle-mounted detection system is required to alert drivers to mines in roads. To fulfil this requirement DSTO is experimenting with UWB radar in combination with other sensors. Both impulse and frequency-domain radar systems are being tested.

Impulse UWB Radar

The impulse radar experiments conducted at DSTO used a 4 kV, 100 ps rise-time impulse from a Kentech impulse generator to drive UWB horns in a bistatic configuration. This gave an effective bandwidth of 3 GHz at approximately a 1.6 GHz band centre. The data were sampled at 6 GHz using a successive sampling digital oscilloscope triggered by the signal source. Initial experiments were conducted with the antennas at 3 m separation, 0.9 m above the ground, with the targets 5 m distant. Targets were alternately placed on top of the ground and then buried under 2 cm of soil in the same location. Figure 2 shows the signature of a plastic AV mine on (upper trace) and on (lower trace) the ground.

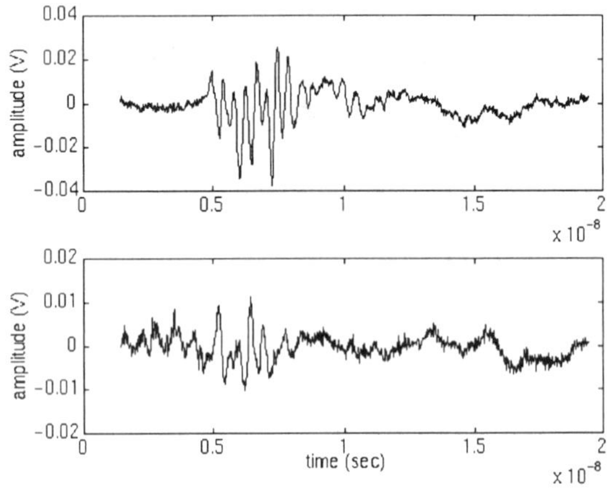

Figure 2: UWB radar 3 GHz response from a MISAR AV mine.

The results of this experiment indicate that AV mines can be detected by this method both on the surface and buried. AP mines are more difficult to detect and the smaller AP mines are unlikely to be detected (Chant et al (2)). The technique would provide an AV mine warning system which could be used aboard vehicles which are hardened against AP mines.

This work was undertaken in an area of grass stubble on clay soil. Further experiments will examine the clutter characteristics of various vegetation and soil types as well as the effect of soil moisture content. Both side-looking and forward-looking SAR imaging are planned.

Step Frequency UWB Radar

The Step Frequency GPR developed by the University of Queensland was used with separated aperture, horn antennas to provide stand-off forward and side looking SAR imaging of buried objects. This work was undertaken for an honours thesis (Radcliffe (3)) and was partially funded by DSTO. The system was used with the antennas 1.2 m high and 1.2 m from targets buried in a flat area of dry sand. A 12 cm diameter steel rod and a 450 hp diesel engine piston were successfully imaged but a 1.25 litre plastic bottle filled with water was poorly resolved.

TARGET IDENTIFICATION

Target identification is a crucial part of any new land mine detection system. False alarm rates in Cambodia exceed 10,000 to 1 for current mine detectors. To address this problem DSTO is examining the following detection/identification options for GPR and UWB SAR.

Resonance Based Methods

Radar targets which have a physical size within 10 (approx.) wavelengths of the wavelength of the illuminating radiation have a natural resonance. It has been documented that buried plastic mines have an internal resonance while buried metal mines have an external resonance (Daniels et al (4)). These signatures can be used to identify targets of the approximate size, shape and material characteristics of land mines.

DSTO is examining methods of target identification using the resonant response of land mine targets. Initial work is under way using Prony resonance pole extraction methods to isolate the first few poles of a target's resonance. Other techniques to be investigated include matched filter and neural network methods.

Both numerical modelling and experimental measurements will be used to determine the effect of clutter conditions on the ability to detect and identify a land mine using its resonance signature. It is also essential to determine how the signature changes with differences in manufacture, soil and vegetation conditions.

Imaging Methods

Two and three-dimensional imaging methods are being examined to determine if targets can be identified using pattern recognition. This work will be carried out with both focused and raw image data. Imaging techniques are likely to be most effective for vehicle-mounted systems where the spacing of GPR soundings across a target is well controlled.

Data Fusion

DSTO is experimenting with the combination of the CSIRO impulse GPR and in-service inductive metal detectors to enhance detection and recognition of mines with some metal content. These types of mines are currently the most widely deployed. Initial results indicate that the current detectors provide a valuable cue for the GPR system while the GPR provides extra information for target identification.

We expect that the combination of sensors will allow the induction detector to be used at high sensitivity while the GPR is used to reject false targets. This approach would allow the GPR system to be used as an attachment to current mine detectors. Initially the operator will make the target recognition decisions but this will latter be extended to automated or semi-automatic target recognition. Thermal or optical imaging might also be useful as a cuing device for the hand-held systems.

We propose a similar approach for stand-off UWB SAR using optical or thermal imaging as the second sensor, and have placed an initial contract with the Cooperative Research Centre for Sensor Signal and Information Processing (CSSIP) to determine data fusion techniques for both hand-held and stand-off systems.

SOIL CHARACTERISATION

While a major thrust of DSTO's land mine detection program is humanitarian demining, there is also a requirement to know how effective this equipment would be in Australian conditions. To determine the reliability of GPR and UWB radar mine detection in Australia we need to determine the expected range of soil environments. DSTO has commissioned CSIRO to provide a coarse soil map of Australia and recommend a range of Australian soils to test. The report will characterise soils in terms of their electrical characteristics and relate these properties to conventional soil classifications.

This will involve the measurement of the dielectric constants and attenuation coefficients for a range of soils as a function of frequency and moisture content. These measurements will be used to generate soil dielectric models (Peplinski et al (5)). If successful we hope to extend these studies to countries where Australian peacekeeping forces are operating.

SURROGATE TARGETS

DSTO has begun to develop a set of surrogate land mine targets to overcome the difficulty of obtaining large numbers of inert land mines. Previous experiments were carried out with inert mines filled with paraffin wax which simulates the electrical properties of explosive fillers. Simple surrogate targets have been manufactured from solid PVC cylinders and PVC pipe filled with paraffin wax.

For sensor fusion DSTO requires surrogates with the same electrical, thermal and magnetic properties as the target land mines. To this end we are developing PVC surrogates having a small metal insert with the same shape and orientation as genuine land mines. These surrogates are likely to filled with teflon which has electrical and thermal properties similar to explosive fillers.

CONCLUSION

DSTO will continue its work on radar land mine detection for at least the next two years. GPR and UWB radar target signatures in various environmental conditions will be collected, and surrogate land mines with appropriate multi-sensor signatures developed. Test sites using surrogate targets and inert mines will be developed to determine the detection and false-alarm rates of various detector systems. These will supplement our current sites which are designed for signature measurement.

Investigations into optimal GPR antenna configurations, signal processing, target identification and sensor fusion methods will continue. An UWB SAR capability will be developed to test the utility of stand-off imaging techniques. Target signature and clutter measurements for various environments will be generated for both GPR and UWB SAR.

REFERENCES

1. Chant, I. J., 1996, "Microwave Detection and Neutralisation of Land Mines", DSTO Technical Report DSTO-0110.
2. Chant, I. J. and Staines, G. W.., 1996, "Near Distance Detection of Land Mines using a 3 GHz Bandwidth UWB Radar Phase 1", DSTO Technical Report, to be published
3. Radcliffe, M. J. 1995, Synthetic Aperture Radar Detection of Shallow Buried Objects", Honours Thesis, University of Queensland
4. Daniels, D. J., Gunton, D. J., and Scott, H., F., 1988, IEE Proceedings, 135, 278-320
5. Peplinski, N. R., Ulaby, F. T., and Dobson, M. C., 1995, IEEE Trans. Geosci. Remote Sensing, 33, 800-807

A HIGH RESOLUTION DIFFERENTIAL RADAR FOR MINE DETECTION

W Murray, C J Williams & J T A Pollock
CSIRO Applied Physics Australia

ABSTRACT

The performance of mine detecting radar systems depends critically upon the quality of the raw data prior to processing. Hand held, high resolution radar hardware for the centre frequency range 800 MHz - 3.5 GHz has been developed and used in a series of field trials targeting buried plastic land mine equivalents. Hardware details and encouraging trial results are presented.

1. INTRODUCTION

Ground Penetrating Radar (GPR) systems are used to detect structures below the ground surface. Although the technology started with the defence industry, it has been adopted most enthusiastically by geophysicists to describe delineation zones in land form structures. In addition, GPR is finding general use for the detection of buried objects such as pipes, oil drums and mines. Plastic objects are generally harder to detect than metallic ones, as the dielectric constant difference with the ground is usually too small to provide significant reflection of the radar signal.

Geophysical radars (1, 2) tend to rely heavily on post-acquisition processing, using elaborate software algorithms to extract sometimes unexpected information from an unwanted background. Sometimes data from many different "runs" are combined at a later date to provide an overall picture of a ground structure, such as an ore body or old mine workings.

The requirements of a civil engineering radar may be quite different. The approximate position and type of target may be known. However the highest resolution attainable for the given target range may be required to determine the precise location and condition of, for example, a corroding reinforcing member or perhaps the position and size of a void in a pavement. Also, the real time data acquired should be of a quality which makes immediate decision making a distinct possibility.

We entered the radar field in 1990 with a civil engineering perspective. Impulse radar systems with centre frequencies of about 700 MHz were designed for characterising road structures, finding buried objects such as pipes, tunnels and optical fibre cables, and detecting concrete cancer in concrete structures. Penetration depths are about 1-10 metres depending, as usual, on the centre frequency and the dielectric characteristics of the material through which the transmission passes. Over the past three years, we have focused on higher centre frequencies, 800 MHz - 3.5 GHz - our goal being the highest resolution possible for the task. These radars, embodying novel sampler and transmitter designs, together with bow-tie antennas, form the basis of instruments constructed for applications in the Australian Coal Industry. A bore-hole radar and machine guidance radar have been field trialed (3, 4).

Since late 1994, we have configured these high resolution radars for the detection of plastic landmines. In this paper, we describe novel sampler and time base transmitter designs and variants in antenna design employed. We also present data captured with plastic mine equivalents (5) as target objects.

2. ELECTRONICS HARDWARE

A high resolution, portable, hand held radar for mine detection has very demanding hardware specifications because of the short time-base range, the real time display required and other factors such as battery life, weight and likely manufacturing cost. Some commercial geophysical radars produce a rather unsteady return waveform display, requiring a considerable amount of waveform stacking to smooth this jitter and reduce white noise. However, for a hand held radar scanning the ground at near walking speed, stacking is not an option unless very high transmitter pulse repetition frequencies (PRFs) are used, with increased battery drain leading to reduced battery life.

In summary, for plastic mine detection, a rock steady return waveform from the target, as viewed on an oscilloscope, is a key specification. Any time or amplitude fluctuations must degrade the real time processed final display. Electronic design factors affecting the final display are discussed briefly below, with reference to the hardware we have developed and field trialed.

2.1. Timing

For a hand held mine detection radar to provide good ground cover and adequate resolution, a scanning rate of about 100 complete clean waveform returns per second is required. In our design, with 127 returns per second, the return waveform, measured on an oscilloscope, appears free of time and amplitude jitter. The wave form is reconstructed from 512 sample points, hence the pulse repetition frequency (PRF) is 512x127 Hz, or 65 KHz. The return signal time range is 6 ns, for detecting mines to 360 mm depth in ground with $\in_r = 6$, giving an equivalent sample time resolution of 6/512, or 12 ps. To maintain these sampling points distinct, the system jitter must be less than 6 ps. This includes contributions from the time-base, transmitter and sampler. Since this jitter is some 2 orders shorter than the delays and rise times of the typical, off the shelf, integrated logic device, considerable care needs to be paid to design detail. Power supply rail noise and interference from asynchronous switching power supplies, for example, can cause unacceptable time and amplitude jitter.

Direct measurements of such effects on a given radar are difficult because of the high speeds involved, problems in defining what is being measured and the reality that timing jitter can lead to amplitude jitter in the recovered and filtered waveform. The radar is more easily evaluated by measuring jitter of the wave form from a strongly reflecting static target. Time and amplitude jitter of less than 0.5% in the waveform recovered from a buried strong reflector (a reinforcing bar 100 mm below the surface in a concrete slab) are achieved with our radar. This is near the limit of what is observable on a typical analog oscilloscope display.

2.2. Samplers

Repetitive samplers are normally used as receivers in high frequency impulse radars (1). Geophysical radars require very high dynamic range and low noise and we have developed both bridge and balanced dual-gate MOSFET samplers for these longer range applications. However, for a mine detection radar, noise and dynamic range are perhaps less important considerations than linearity. Poor receiver linearity, especially with shallow targets and high amplitude return signals, can alter the return waveform or "signature" of a mine and make subsequent processing more difficult. Non-feedback samplers, although simpler to design and implement, do not handle large return signals well. In a diode based non-feedback sampler, diode on-resistance and on-time vary with signal amplitude, the latter because of inevitable slope on the edges of the switching signal.

Feedback samplers overcome these problems by effectively keeping the diode bridge near balance during sampling. This is the design approach we have taken.

For plastic mine detection, we use a half-bridge sampler employing a hybrid containing the sampling diodes and pulse generator. This sampler is capable of good high frequency performance and easily meets the timing requirements set out in 2.1. An integral loop keeps the bridge near balance. Since a diode sampler may only be optimised for a particular frequency range, due to factors such as diode resistance, capacitance, frequency response and sampling pulse width, we have used two different hybrids containing the sampling components. One or the other of these may be inserted on a single PC board to suit the frequency ranges required.

3. ANTENNAS

3.1 Choice of frequency

If a small plastic mine in isolation is radiated with a swept frequency EM field, a small increase in reflected energy is noticeable around 2.5-3 GHz. This can be readily demonstrated on a network analyser as shown in Fig. 1. This weak response peak is dependent on the orientation of the mine and drops considerably if the mine is in a medium of similar dielectric properties. Because detection of a buried mine also depends on ground propagation loss, frequencies much lower than 2.5 - 3 GHz may offer better results.

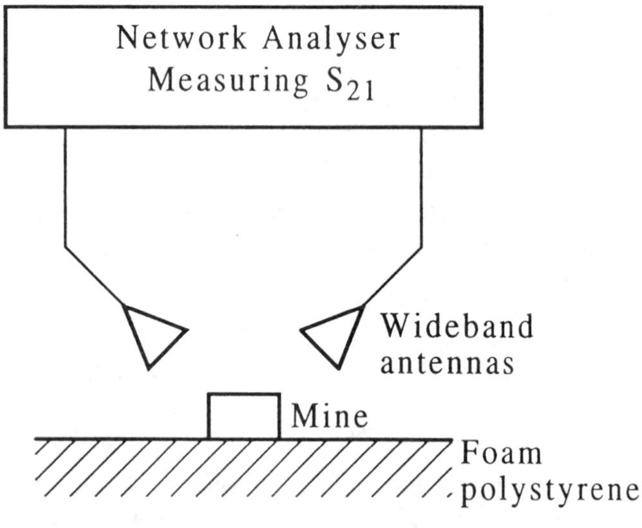

Figure 1. Measuring reflected energy vs frequency

We have used impulse radar centre frequencies of 800 MHz to 3 GHz for mine detection. The requirement to detect at depths to 250 mm (5) is very difficult to meet, if the soil is lossy or there are other scattering centres present, such as small stones. Under such conditions a lower frequency, for example 800 MHz, gives better results than 3 GHz. Thus a mine detector which offers the choice of a higher or lower centre frequency will be more versatile than one operating at a single frequency only

3.2 Antenna types

Resistive dipoles, bow-ties, horns and, to a lesser extent, circularly polarised spirals are the main types of antenna used in GPR where a wide frequency response and constant phase centre are important. Reference 1 provides a detailed review of various antenna types.

Two goals of antenna design are to reduce the transmitter-receiver coupling and the top surface return so that mines near the ground surface may be more readily distinguished. Configurations which provide isolation between transmit and receive antennas also tend to reduce surface reflections use either polarisation or differential techniques/ geometries and include:

(a) cross polarised resistive dipoles
(b) circularly polarised antennas such as spirals
(c) differential horns, bow-ties, etc.

Since systems using (a) are commercially available, we have concentrated on antenna designs based on (b) and (c).

3.2.1. Circularly polarised antennas. Circularly polarised antennas generally tend to "chirp" or disperse the pulse. This can be corrected in software or hardware. Spiral antennas also tend to be more complex and generally use a damped cavity backing to absorb the reflected radiation, which otherwise causes spurious signal returns.

We have developed simple PCB spiral antennas using flat metallic backing, employing ferrite tiles and a thin conducting rubber sheet to reduce the back radiator. The transmit antenna has a step recovery diode (SRD) transmitter mounted on its feed structure and both antennas have leads 1.7 m long, so that any reflections from the cables fall outside the time range.

3.2.2. Differential antennas. Differential antennas are described in the literature for lower frequency applications (6) however, they have not been exploited at higher radar frequencies. Two types of hardware differential operations are possible: firstly, post-sampler subtraction of the recovered or strobed down waveforms and, secondly, pre-sampler subtraction of the RF return signals at the antennas. The latter offers greater benefits in dynamic range and linearity, since the sampler does not have to deal with the direct high level transmit pulse (which may contain close-in mine return signals). Cancellation of the common mode signal (transmit pulse and surface layers) at higher frequencies (up to 3 GHz) requires considerable care. The cancellation of the direct returns allows for the use of low gain preamps (15 db) in the 1.4 - 3 GHz range even in lossy ground conditions. Based upon these considerations, we have fabricated differential bow-ties and horn antennas.

Shielded and terminated bow-ties are simple wideband antennas which operate well when spaced above the ground at distance $\lambda\backslash 10$, where λ is the in ground wave-length (7) In addition, their low overall profile allows for easy operation in confined areas. The ground clearance requirement is convenient at lower frequencies (~ 20 mm at 800 MHz), but the very close clearance (~5mm) required at 3 GHz for good performance is more of a problem. Signals are combined either with resistors or in a transformer, the former often having the better frequency and phase response. The two receive antennas are phased oppositely to subtract the common mode signals. The cable length from antenna to combiner needs to be at least 1.7 metres (for solid teflon dielectric and a 6 ns time base), if the reflected signals due to imperfect matching are not to appear in the amplified far field. This length is convenient for a hand held wand unit.

When designing, the antenna configuration and bow tie angle can be varied. Since for shallow mine detection it is desirable to keep the distance between adjacent antennas about equal to a mine diameter (say 80 mm), only certain geometries are applicable. Two of these are shown in Fig. 2.

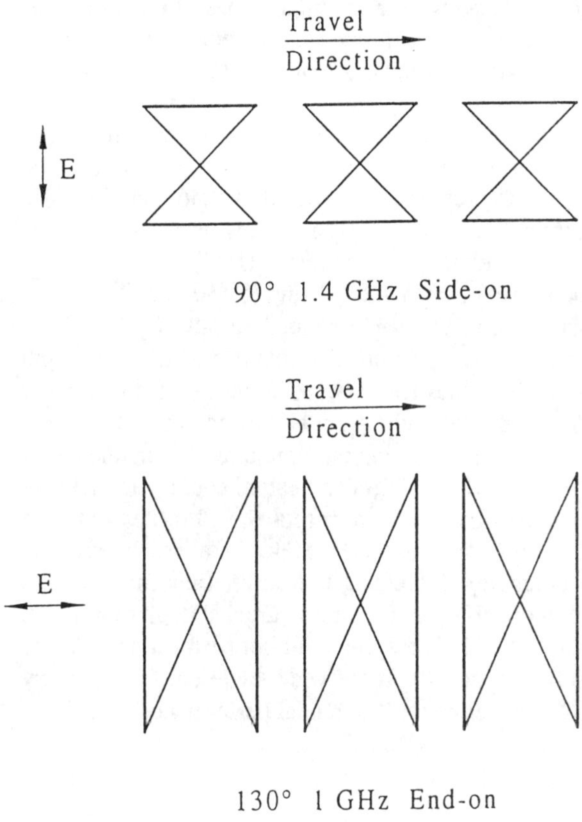

Figure 2. Differential bow tie antenna geometries

The following antenna geometries were used,

(i) 60° side-on, 800 MHz bow-ties
(ii) 90° side-on, 1.4 GHz bow ties
(iii) 130° end-on, 1 GHz bow-ties.

The two configurations, side-on and end-on, have different detection characteristics, A metal rod or a tree root orthogonal to the direction of motion will be detected by (i) and (ii), but not by (iii), since the E field is in the direction of the motion in (iii).

Differential horn antennas were trialed, as they were judged to have some advantages. In particular, their relatively high gain allows them to be held further from the ground than bow-ties. Both ridged and TEM horn antennas were constructed. In general, horn operating characteristics improve with length, allowing easier optimisation and reduction in internal reflections. Final test geometries were determined by the chosen 6 ns time base, with overall lengths in the range 900 - 450 mm. The shorter length was only possible with dielectric loading of $\epsilon_r = 4$ and an associated weight penalty.

4. RESULTS

Results with circularly polarised antennas were disappointing and are not reported. Poor results were also obtained with the horn antennas. In addition, the overall weight and profile of long horn antennas makes them unsuited to hand held mine detection.

Results with the differential bow-tie antennas were encouraging. Fig. 3 shows typical unprocessed detection profiles taken with a 90° side-on, 1.4 GHz differential bow-tie. The targets were four plastic anti-personnel mine equivalents (5) containing no metal parts buried at depths up to 150 mm. As shown, strong profiles were obtained at all depths. Close examination of the profiles obtained at 150 mm depth (Fig. 3c) indicates that, as a result of the rapidly increasing far field gain employed, the M14 mine (3c(i)) displays a ringing signature at a frequency of 2.3 GHz. This ringing signature could potentially be used for identification purposes. Profiles obtained with 60° side-on, 800 MHz antennas were also encouraging, indicating that lower frequency systems are more effective for mines buried at depths up to 250 mm. The 1 GHz system did not perform particularly well. It appears that the wide angle end-on geometry gives less effective differential performance.

5. CONCLUSIONS

Attention to electronic hardware design can improve the raw data returns and considerably ease software processing requirements. An appropriate specification for overall displayed jitter is 0.5% of the recovered waveform from a strongly reflecting buried target.

Antenna design is very important. Differential and circularly polarised antennas can remove the static returns from the direct transmit pulse and the early surface ground reflections, to give a much clearer and more easily interpreted raw data display allowing even shallow targets to be clearly detected. The properties and detection characteristics of the antennas discussed suggest that a hand held mine detector could benefit from a combination of at least two frequencies, for example 800 MHz and 1400 MHz.

6. REFERENCES

1. Daniels, D.J., Gunton, D.J., and Scott, H.F., 1988, 'Introduction to subsurface radar', IEE Proceedings - F, 135, 278-320
2. Fisher, S.C., Stewart, R.S., Harry. M.J., 1994, 'Processing ground penetrating radar data', Proc. 5th Inter. Confer. on Ground Penetrating Radar, Kitchener, Canada.
3. Hatherley, P., Murray, W., Stickley, G., Eisler. P., and Techen, T., 1995, 'Development of in-seam borehole logging tools for detection of outburst prone structures', International Symposium on Management and Control of High Gas Emissions and Outbursts in Underground Coal Mines, ed. Lama, R.D., 93-100.
4. Murray, W., Williams, C.J., Pollock, J.T.A., Hatherly, P., and Hainsworth, D., 1996, 'Ground probing radar applications in the coal industry' To be published, GPR '96 Proceedings, Sendai, Japan.
5. Gans, W.L., Geyer, R.G., and Klemperer, W.K., 1991, 'Quantifying The Performance of Electromagnetic-based Mine Detectors', NISTIR 3982, USA.
6. Barnes, M., Poirier, M., and Peteres Jr., L., 'Novel GPR antennas', to be published in a special issue of J.Appl. Geophys, 1997.
7. Smith, G.S, 1984, 'Directive properties of antennas for transmission into a material half space', IEE Trans. on antennas and propagation, 32, 232-246

7. ACKNOWLEDGMENTS

We are pleased to acknowledge the help and assistance of Mr Roy Bird and Dr Ian Chant of the Australian Defence Science and Technology Organisation.

(i) (ii) (iii) (iv)

3(a) Top of mine near surface (4ns time base)

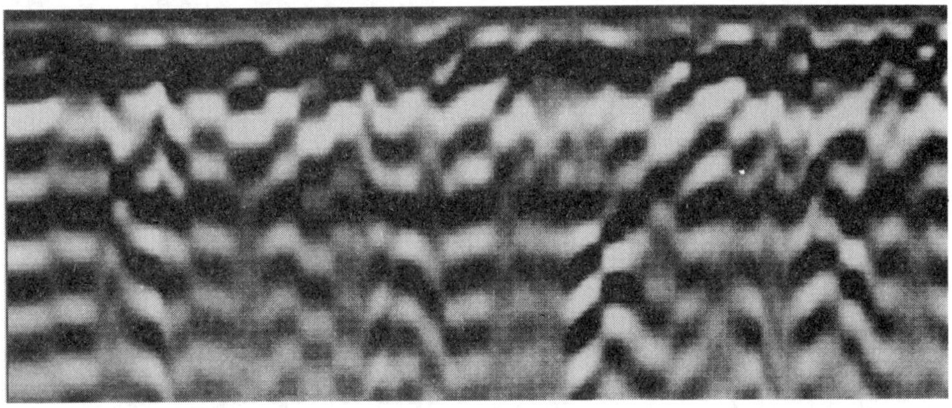

(i) (ii) (iii) (iv)

3(b) Top of mine 100 mm below surface (6ns time base)

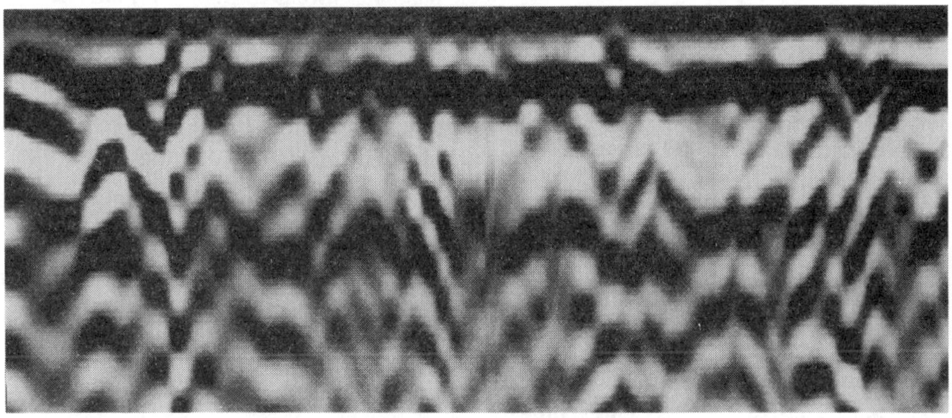

(i) (ii) (iii) (iv)

3(c) Top of mine 150 mm below surface (6ns time base)

Figure 3. Unprocessed tests results obtained with a 1.4 GHz, 90° side-on, differential bow-tie System. Four plastic anti-personnel mine equivalents were used;

(i) nylon M 14 equivalent (69 mm dia)

(ii) polyethylene M14 equivalent (69 mm dia)

(iii) polyethylene type 1, class B equivalent (76 mm dia)

(iv) nylon type 1, class B equivalent (76 mm dia)

IMPULSE RADAR MINE DETECTION

P Mallinson, D J Daniels

ERA Technology Ltd, UK

1 INTRODUCTION

Mines are weapons which are used in the military role to frustrate and slow down the manoeuvrability of enemy forces. There are internationally agreed procedures for laying down and recording the position of minefields and individual mines. In the most recent conflict in the Gulf, minefields were laid by the Iraqi Armed Forces according to agreed protocol. The problem of mines appears to be associated with operations in regions where agreed military procedures are ignored.

Where mines have been used or scattered indiscriminately, the civilian population suffers the consequences of such irresponsibility. It is estimated that there are some 100 million uncleared mines in the world, Williams (1). Mines are still being laid at the rate of 2 million per annum, Jackson (2) and the rate of clearance is painfully slow. In Afghanistan it has been estimated that it would take 4300 years to clear mines manually from just 20% of the country, Williams (1). Eblagh (3) reports that in the period 1989 to 1994, 34 sq. kilometres have been cleared of 63,000 mines at a cost of US$1.15M per 1 sq. kilometre.

According to the Mines Clearance Planning Agency in Afghanistan over a fifteen year period an estimated 20,000 civilians have been killed and 400,000 wounded by landmines in that country. For those wounded the cost of treatment and rehabilitation averages US$5000 per victim. The vast majority of civilian casualties are caused by anti-personnel mines which fall into the categories of blast and fragmentation. Blast mines cause traumatic amputation to feet or legs and the lower abdomen whilst fragmentation mines, which shoot out metal fragments at a height of typically 1 metre in an arc reaching up to 50 metres, are evidently lethal.

Gray (4) notes that the International Committee of the Red Cross found that the casualty rates in North West Cambodia were such that 58% of injuries were suffered by non-combatants. In general women and children were most likely to be injured while cultivating fields or gathering firewood.

Mine injury is particularly difficult to deal with, in that most of the wounds are so heavily contaminated with dirt and debris that extensive amputation is necessary to avoid the risk of further infection. Typically 60% of victims require up to three operations. In third world countries medical care suffers from the lack of medical resources such as adequate blood banks.

The problem is immense and all means to alleviate the problem should be considered and applied. Conventional mine detectors detect the metal content of mine but the plastic mine remains largely undetectable by current technology. For this problem new solutions are sought and this paper describes one approach to plastic mine detection.

2 DETECTING PLASTIC OR NON METALLIC MINES

It has been estimated that the numbers of mines of all types are as follows in the following countries: Afghanistan (10M), Angola (60M), Bosnia/ Croatia (1M), Cambodia (4.5M), Falklands, Iraqi-Kurdistan (1M), Kuwait (1M), Mozambique (2M), Nicaragua (0.2M), Somalia (1M), Vietnam (4M). Many of these countries contain many earlier generation metallic mines, but some proportion of plastic or non metallic mines can be found in a variety of countries and terrains.

Mine clearance involves massive effort and resources and is dangerous for mine clearers. On average one expert is killed and two injured for every 5000 mines cleared. On that basis some 20,000 deaths among the deminers could be anticipated during the course of the clearance of the remaining mines in the world.

An alternative means of mine clearing using machines would appear to be a much safer and attractive procedure. One possible method would be to use a robotic vehicle which is programmed to search the ground ahead and indicate the presence of mines. However, there still is the need for a hand held detector in situations where access for a robotic vehicle is not possible.

A short list of attributes for a universal mine detector would be

1 Operation in hot, wet and dry climates
2 Ability to find mines of different size and type in close proximity
3 Find mines in damp or wet soil and earth materials
4 Find mines in salt water
5 Find mines with small or no metal content
6 Operate in various soil conditions particularly those with high magnetic or ferrous content
7 Able to achieve discrimination in favour of wanted targets
8 Able to work through vegetation
9 Simple indication of the presence of a target
10 Low cost
11 High reliability
12 Automatic ground marking
13 Operation at depths of up to 0.3 metre against AP mines and 1 m against AT mines

Field experience in Afghanistan has shown that the priorities for clearance are usually as follows as reported by Keblagh (5):

A - Roads from mines
B - Villages from mines
C - Canals or water courses from mines
D - Agricultural land from mines
E - Grazing land from mines

3 POSSIBLE SOLUTIONS

The basic challenge for a mine detector is to overcome the problem that most techniques will suffer from a high level of indications from objects other than mines. For example, the metal detector probably will encounter from between 500 to 1000 indications before a mine (metal) is detected. Similarly, explosive detectors may also detect traces of explosive on spent ordnance. It is reasonable to suggest that whatever the physical principle, the number of unwanted indications will be high and in practice a ratio of several hundred to one will be encountered.

There is a school of thought among some representatives of the non governmental organisations, who have accumulated considerable field experience, that suggests that unless a future mine detector is low cost, capable of working in all environments and provides a 100% detection rate, then it is not suitable for serious consideration and hence research and development is not justifiable.

This view possible reflects the frustration of those at the sharp end with the slow progress in finding a technical solution and a reluctance to see scarce resources applied to research and development. However, unless some real drive is made towards a concerted R&D programme it is unlikely that improved equipment will ever be available to the deminer.

New physical methods of mine detection are being researched and techniques such as explosive detection by chemical means; plastic detection by means of radar; anomaly detection by means of infrared or microwave radiometry are the currently regarded most promising means capable of detecting mines. Various conferences have been held and discussion of possible technical means have been evaluated, technical details can be found in Daniels (6,7 and 8).

An obvious technical solution is to use a multi-sensor approach coupled to a robot. Apart from the cost and logistics this offers a considerably improved detection capability at minimum risk to the deminer. This is one approach that we are currently investigating, albeit at present with a single sensor. A photograph of a robot (Alvis "Wheelbarrow") carrying an impulse radar antenna is shown in figure 1.

4 IMPULSE RADAR

Impulse radar has been investigated over nearly a decade for the detection of plastic mines and the first real attempt to exploit the technology was by the UK Ministry of Defence in the early 1980's when it contracted ERA Technology to research the technique and construct and evaluate prototype equipment against trial minefields in the Falklands Islands, in which technically promising results were indicated.

The end of the Falklands conflict removed the urgency from the programme and since then ERA has developed the technology for civil applications achieving notable successes in forensic work, notably the detection of bodies in the basement of 25 Cromwell Street. However, in the intervening period we have assisted other organisations with the supply of special experimental equipment and organisations in Sweden, Switzerland, Australia and Germany have all carried out research programmes based on our technology.

The latest commercial product is a time domain impulse radar. This type of radar radiates a short pulse of electromagnetic energy into the ground and detects the backscattered energy from the buried target.

The buried target must be different in electromagnetic terms from the surrounding soil and this means that to a first order its dielectric constant should be significantly lesser or greater than the host soil. Typically, the aggregate relative dielectric constant of a mine lies between 2 and 3 whereas most soils exhibit a relative dielectric constant which ranges between 2 to 25. Fresh water has a relative dielectric constant of approximately 80, which means that mines can be detected in fresh water.

The antenna of a surface penetrating radar is moved on the surface of the ground in either a straight line or in an arc over the target and as the reflection from the target is detected it is displayed on the screen of a computer as a cross sectional image. The key feature of a radar is the quality of the resulting image which should provide a clear and uncluttered image of the target. The choice of the characteristics of the impulse and the performance of the antennas are absolutely critical to achieving optimum image quality.

As a result of an extensive in-house development programme, we have developed a new generation of designs of antenna and radar transmitter-receiver which provides the highest quality of data. The new designs have been incorporated into a range of products and these have been used to investigate the type of image that can be presented with the latest generation of impulse radars. However, it should be noted that for humanitarian demining, a commercial radar system designed for civil engineering applications could usefully be repackaged and modified for the different application. At present the design is most appropriate for research and evaluation purposes as a test benchmark evaluation.

5 TEST RESULTS

The equipment used comprised a standard SPRscan surface penetrating radar using a P1000 antenna. The results shown in this section were gathered at a test facility at ERA Technology. In all of the results shown the data are processed by correction for attenuation and spreading loss and by removal of the average background.

A number of inert mines were loaned to ERA Technology Ltd by Bactec International and these comprised AR409 (AP), P2 (AP), Type 72 (AP), VS-2.2 (AT) and DATS (AT) types. Their characteristics are listed by the US Department of Defence as types PRB BAC H-28 (the PRB 409 predecessor), Type 72 and P2 Mk2 AP mines and the VS-2.2 AT mine.

The first set of results were taken on the three anti-personnel mines and these were buried just below the surface (approximately 5 mm to 10 mm) of a moist sand test pit. Mines buried just below the surface are generally difficult to detect with radar because the reflection from the mine is mixed up with the reflection from the front surface. As can be seen in figure 2, three separate images can be seen as a result of the mines. The image to the right also exhibits a secondary image to its right. This is due to a lump of wet clay inadvertently left in the site. However, the polarity of its reflection is inverted compared with the mine and can thus be distinguished as a non-mine like target in this soil. The photograph shows the depth of one mine for reference.

We believe that testing radar detection in soils like sand is only useful as a very primitive indicator of image performance. A more realistic material should correspond to the types of unmade road material encountered in third world countries. Accordingly, what is known in the UK as sub-base road material was used as a host material. This comprises a mixture of soil, sand, gravel, small pebbles and larger stones and was laid to a depth of 400 mm over a reinforced concrete base. Some of the larger stones approach the size of anti-personnel mines.

Radar cross sectional images of a type P2 anti-personnel mine are shown in figure 3, together with a photograph of the soil material. Note that although there is evidence of reflection from the stones the image of the mine is the most prominent feature.

For reference, an image of an anti-tank mine type VS-2.2 is shown in figure 4. It should be stressed that for all of these mines all metallic parts had been removed and the mine could be classified as truly non-metallic. As an aside, radar will also detect metallic mines and the images gathered to date show a stronger image in the case of metal mines than non-metal mines.

Finally, a careful area search was carried out over the type DATS anti-tank mine. The resultant plan image is shown in figure 5 and the radar image of the mine can be clearly seen. While this type of image can be gathered using a hand-held system, it is a logical proposal to use a pre-programmed robot which will always gather properly registered area scans.

6 CONCLUSIONS

We have shown that the latest generation of commercial surface penetrating impulse radar can be used to detect plastic mines in various soil materials, including what in the past was considered as a difficult type, i.e. the unmade road material. Impulse radar could be developed along two lines; as a hand held detector or as a robot mounted detector. The probability of detection would be considerably improved by adopting the radar as part of a multi-sensor detection system.

Given the scale of the problem and the predicted consequences of present approaches, we believe that a concerted international approach to the future research and development of new and improved means of detection is long overdue.

Impulse radar is one prime candidate technology for proper support to move beyond the stage of demonstration of feasibility to field trials of engineered prototypes. We look forward to being able to contribute to future programmes of work.

7 ACKNOWLEDGEMENTS

We would like to thank the Directors of ERA Technology for their support of the work that enabled these tests to be carried out. We would like to thank Alvis Logistics (particularly John Spence) for their assistance with experiments with the "Wheelbarrow". Finally we would like to thank Bactec International, (Mr G Lucas) for the loan of the mines.

8 REFERENCES

1 Williams J., April 1993, International Committee of the Red Cross, Montreux, pp 69-79

2 Jackson P., October 1994, Microwave Engineering European Editorial

3 Eblagh K., April 1993, International Committee of the Red Cross, Montreux, pp 153-173

4 Gray R., April 1993, International Committee of the Red Cross, Montreux, pp 63-67

5 Eblagh K., June 1994, UN/FOA International Workshop of Technical Experts on Ordnance Recovery in the Framework of International Demining Operations

6 Daniels D.J., June 1994, UN/FOA International Workshop of Technical Experts on Ordnance Recovery in the Framework of International Demining Operations, pp 120-128

7 Daniels D.J., Nov 1994, Workshop on detection of anti-personnel landmines, ISPRA, Italy, pp AN1-30

8 Daniels D.J., Dittmer, J.K., June 1995, Design challenges for ultra-wideband radar for AP mine detection, EPFL Workshop, Lausanne, pp AN5-25

Fig.1: Photograph of Alvis "Wheelbarrow" with P1000 radar antenna

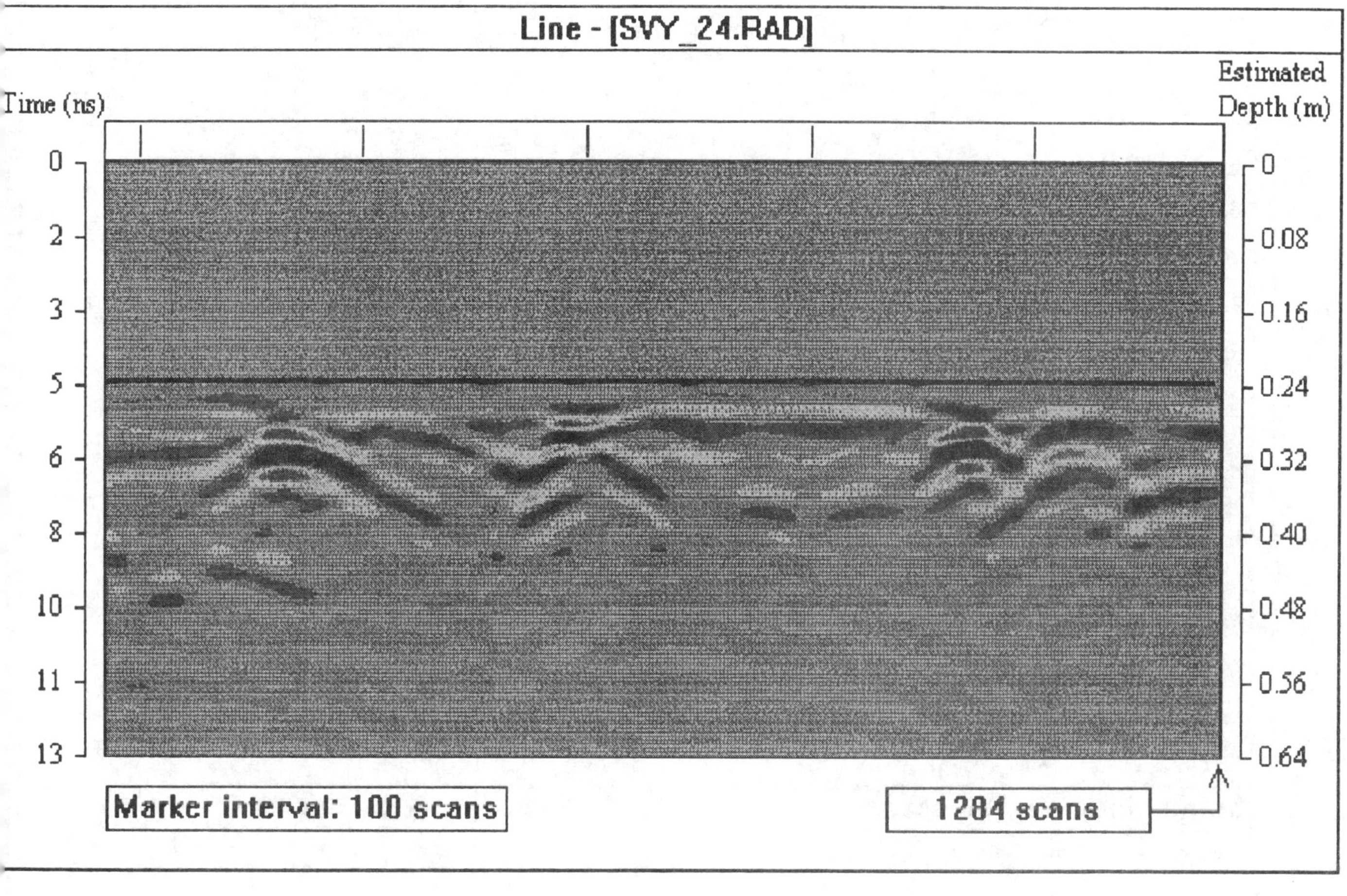

Fig.2: Radar cross sectional image of three anti-personnel mines buried in sand

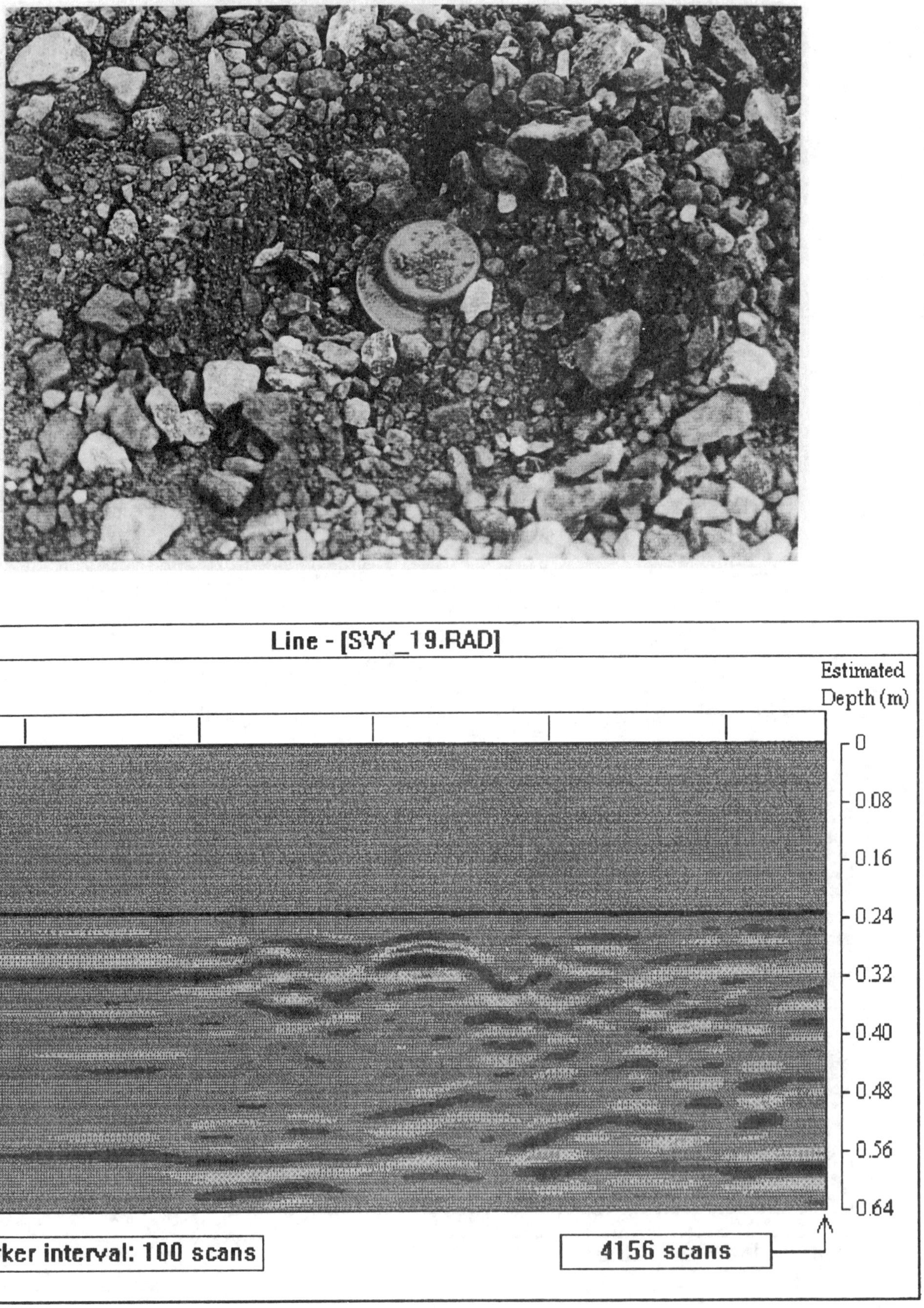

Fig.3: Radar cross sectional image of an anti-personnel mine buried in unmade road material

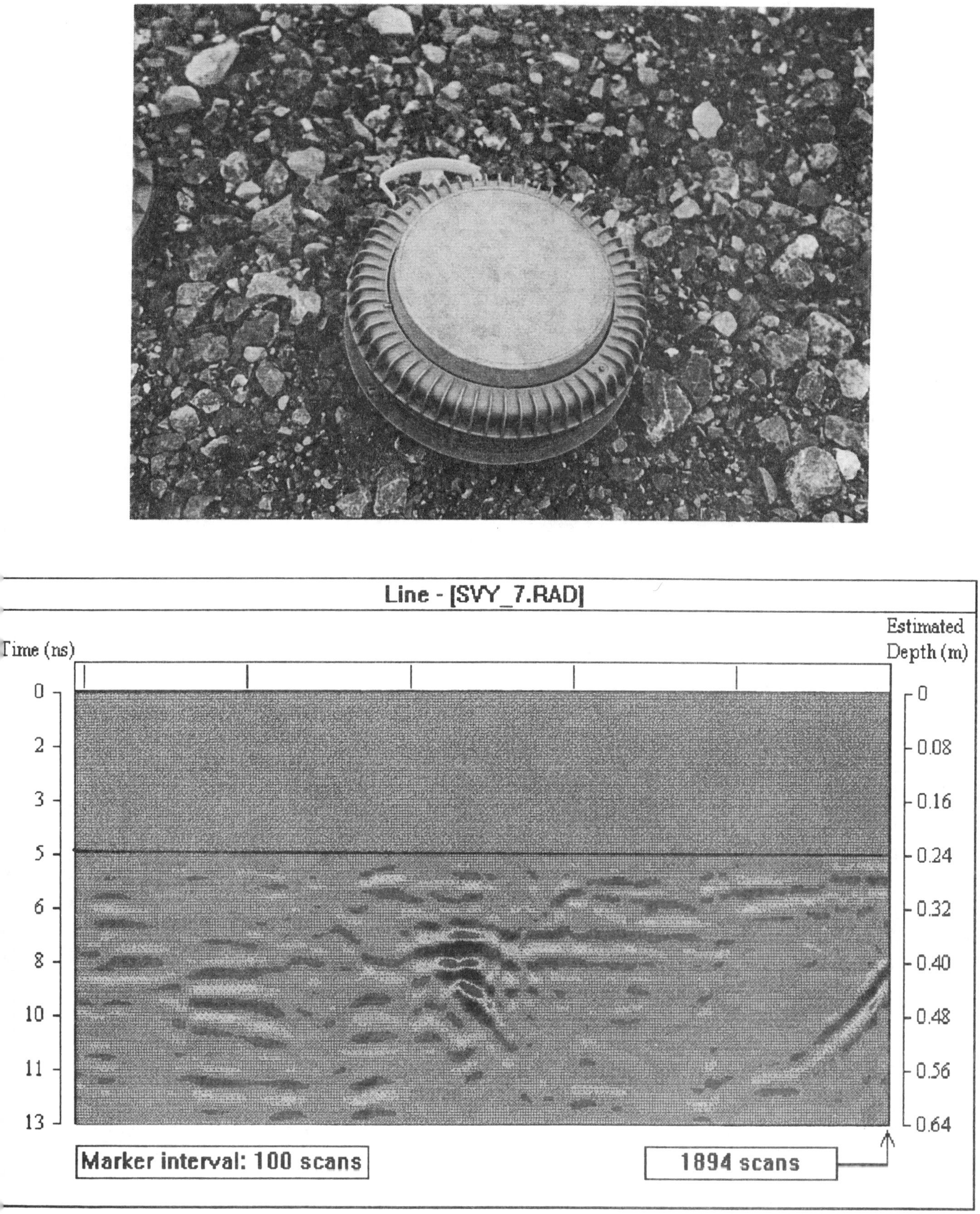

Fig.4: Radar plan image of an anti-tank mine buried in unmade road material

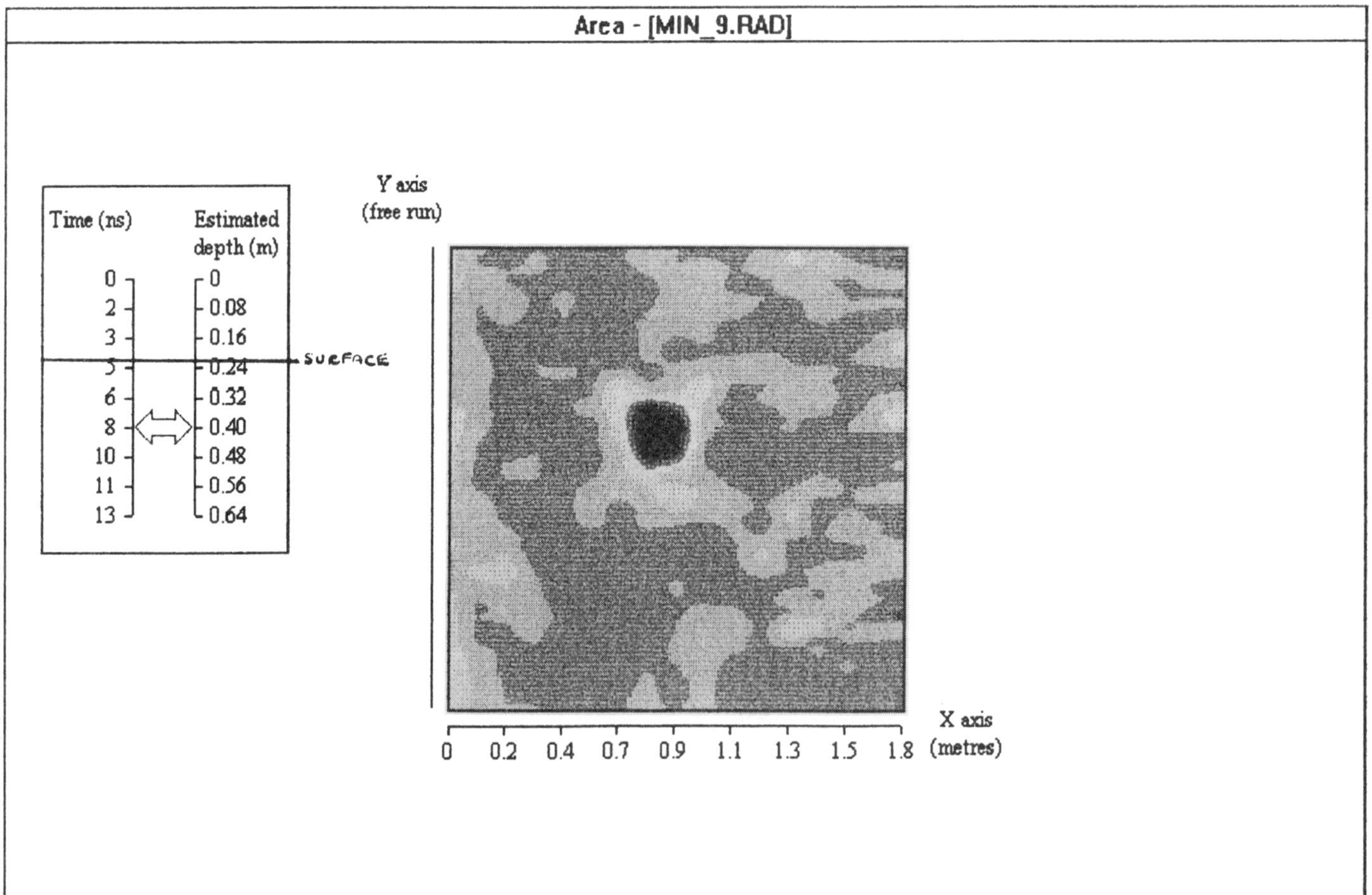

Fig.5: Radar plan image of an anti-tank mine buried in unmade road material

GROUND BASED S.A.R. IMAGING TOOL FOR THE DESIGN OF BURIED MINE DETECTORS

P Millot , A Berges
Centre d'Études et de Recherches de Toulouse - Office National d'Études et de Recherches Aérospatiales
FRANCE

INTRODUCTION

The detection of mines buried in the soil or covered by vegetation has become a topic of major importance. For this purpose, we propose to use electromagnetic waves in the U.H.F. (Ultra High Frequency) range that have the ability to penetrate the soil or the vegetation. This technique is now well known as G.P.R. ("Ground Penetrating Radar"). Our contribution to the improvement of G.P.R. is to use a synthetic aperture antenna in order to achieve azimuth (or cross-range) resolution. In "stand-off" detection, the large antenna footprint would give a very bad resolution to a classical G.P.R. With a synthetic aperture antenna, it is expected that a low frequency G.P.R. can deliver fine images of soil surfaces as well as subsurfaces. Furthermore, at low incidence angles, surface soil clutter level is very low. This is why, in order to help to design such future radars, an imaging tool has been developed at ONERA/CERT at the microwave laboratory. It is thus possible to study the detection of buried targets in various conditions of soils, humidity ... The system versatility allows to test the influence of the frequency bandwidth, the looking angle, the polarisation of the waves as well as to evaluate signal processing algorithms.

SYSTEM DESCRIPTION

This system can be considered as an imaging tool that can be moved easily close to objects simulating land mines (figure 1). It is mainly composed of a rail of 3 meters length carrying one or many antennas that illuminate an area of soil in side-looking mode.
This rail is at 2.5 meter above the soil and the antennas look the area to be imaged with an incidence angle of 10 to 60 degrees. A vector analyzer is used to emit and receive the signals. Antennas move step by step along the rail. Each step is equal to a quarter wavelength at the mean frequency in order to avoid aliasing in angle domain. At each step, a frequency sweeping between two frequencies selected by the operator is performed by the vector analyser. Limit frequencies can be chosen between 100 MHz and 3 GHz. Data resulting from coherent detection are sent in a P.C. by the IEEE bus. The complete data form an array of dimension (N_f x N_a), N_f being the number of discrete frequencies emitted in the frequency sweeping and N_a the number of antenna points. This array of data (typical amount 25,000 points) is automatically processed to show the level of reflection at any point inside a selected surface in the horizontal plane centred at a distance that can vary from 2.5 to 12 meters from the rail. Direct visualisation of the image is made on the P.C. screen.
Our ultimate goal is to use an ultra large bandwidth ranging from 100 MHz to 3 GHz. In fact, we are limited in low frequency to 400 MHz because of the cut-off frequency of used antennas. We use rather small antennas: one is a folded dipole in the range 400-800 MHz, the other one a ridged horn in the range 800 MHz-3 GHz. Even if two antennas operation is better especially in case of horn antennas, we have demonstrated that one can work with the same antenna to emit and receive waves at the same time. So it is possible to cover the full frequency range with two antennas in a single measurement, using the two channels of the vector analyser. The kind of antenna have been carefully selected. Many antennas that are often used in the U.H.F. range bring large distortions in the pulse response that could not be eliminated, even by signal processing.

SIGNAL PROCESSING

High resolution processing

In one dimension (range processing), processing is equivalent to digital range compression and is achieved by FFT (Fast Fourier Transform) on frequency data. So the distance resolution in vacuum is given by:

$$R_d = k\frac{c}{2B}$$

B is the frequency bandwidth, c the speed of light and k a factor depending on signal processing (from 1.2 to 2). The highest frequency may sometimes be attenuated in wet soils. If we assume an effective bandwidth of 1 GHz in every case, the resolution will be roughly 20 cm.
Cross-range processing is achieved as in airborne or spaceborne S.A.R. by correlation with a set of replica. But in our case, the length of antennas displacement is limited to L=3m. Furthermore, the target is in the near field (Fraunhofer's zone) of the synthetic aperture. So a special processing has been developed, which is highly geometry dependant, using the matched filter approach see Mensa (1). The azimuth resolution is given by:

$$R_a = k\frac{cD}{2F_m L}$$

Here F_m is the mean frequency, D the distance antenna to target. For F_m=1 GHz, R_a=22 cm at D=3m and 45 cm at 6m.
This signal processing coded in MATLAB on a PENTIUM PC is very fast. Of course, it depends on the number of pixels of the image. Typically, with an area of 3x3 m^2 and a pixel of 10 cm, computation time takes only a few seconds including calibration

processing and visualisation.

Superresolution processing

It is well known that linear prediction methods such as the maximum entropy method can lead to an increase of resolution in domains that look like spectral analysis. On the other hand, a major drawback of these methods is that the level of the received signals is not well kept normally, because it is not matched filtering.
Superresolution methods are now well known for D.O.A. (direction of arrival) estimation (2) but their application to our case of S.A.R. imaging is not obvious. Superresolution processing is not normally two-dimensional. So, one can think to separate the processing in two one-dimension processing. But there is the necessity is to preserve the phase of signals after range processing. A.R. (autoregressive) spectrum estimation has been used instead of classical Fourier transform for range processing. But the result is a pseudo-spectrum with a loss of phase information. So we have processed in a non conventional way: we have kept the existing data and extended them by linear prediction using the coefficients of the A.R. filter, identified with the help of the original data. (This corresponds to an extension of the above array along the columns). Burg's method and the modified covariance method have been used with success for A.R. identification. We have found a new set of data that have been processed by the above known processing (FFT).
A problem in A.R. methods is the choice of the model order. Popular criteria such as Akaike's, M.D.C. ... appear to be of relative small use in practice. We used to select the best order by simulations of many realistic cases (typical order value of 20). Nevertheless, improvements in range resolution is effective as expected. It is very interesting in case of natural bandwidth limitation due to strong soil absorption.
Superresolution in azimuth or cross-range is also achieved as in range processing by data extension. One must extend (back and forth) the extracted vectors corresponding to a migration track and not the array lines, which is an original way to do. Superresolution in cross-range can overcome the problem of the limited length of the rail. In particular, we have found that an increase of a factor of 2 or 3 in data length and so in resolution gives highly resolved images with a respect of amplitude levels, that could have been obtained by a rail of length 6 or 9 meters.
Other known methods based on the calculation of the covariance matrix such as the maximum likelihood, MUSIC (Multiple Signal Classification) ... have also been applied only in azimuth processing, but the application of such methods remains here tricky. Problems arise due to the special S.A.R. geometry (migration). There is also the difficulty to process correlated signals and so to apply spatial smoothing. Because of these problems, we have found that this kind of processing can be used efficiently in two cases: if the target is very far from the rail (at least 10 meters) or if one takes only few measurement points on the rail. Despite of the large computation time and uncertainties in the obtained amplitudes, resolution is excellent (more than 10 times classical value) and so this processing is not yet abandoned and is still considered for mine detection, in addition to a generalised Prony-S.V.D. (Singular Value Decomposition) method that preserves rather well the relative amplitude. We expect in the future to apply in a second time only on interesting parts of the images superresolution methods in order to improve identification of mines.

PRESENTATION OF A RESULT

We present one measurement result obtained in a difficult situation. We have buried a flat large metallic cylinder (diameter 20 cm height 10 cm) inside a clay soil containing many stones covered with thick grass and large tufts (depth : 20 cm). The soil is rather dry due to an absence of rain during one week. A metallic anti-personal mine from World War II (height 15 cm diameter 8 cm) in vertical position has been hidden in a tuft and was not visible for the eye. The used frequency band was 0.7 to 2.7 GHz and the polarisation was linear vertical. Figure 2 presents the result of the backscattering level (on a linear scale) at a distance from the rail varying from 4 to 8 meters. One can distinguish the buried target in **2**, the hidden mine in **3** and even a previous hole in **1** that remained from an another measurement.

CONCLUSION

A ground based S.A.R. imaging system has been developed and has given preliminary results. It can deliver images of surface or subsurface targets that represent the strength of radar cross section targets for the incidence angle and allows target detection by choosing a threshold. Sophisticated and original signal processing algorithms have been tested. Furthermore, superesolution algorithms can be applied in the special geometry of S.A.R. imaging. Preliminary results have been obtained, a campaign of systematic measurements is planned in a near future on a well controlled soil in order to know exactly the influence of the used frequency, the angle of incidence and the polarisation of waves. This tool will also lead us in the future to test mine identification algorithms by the highly resolved radar signatures, obtained in realistic environments.

REFERENCES

1. Mensa, D. L.,1991,"High Resolution Microwave Imaging",Artech House,Norwood,USA

2. Johnson, R. L., Miner, G. E.,1986,"Comparison of Superresolution Algorithms for Radio Direction Finding", IEEE T AES,22, ,432-441

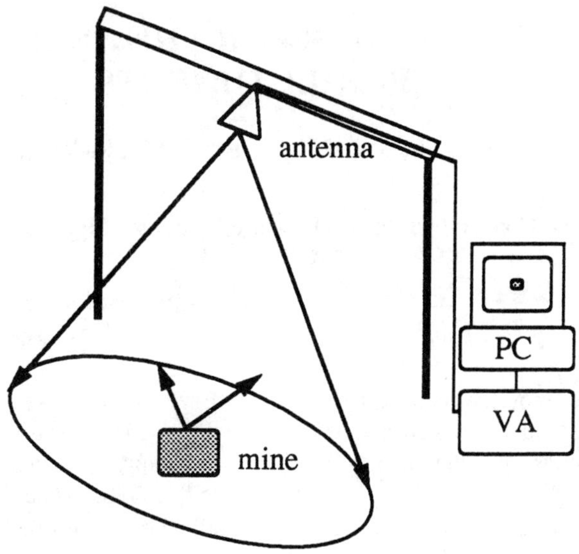

figure 1 : diagram of the imaging system

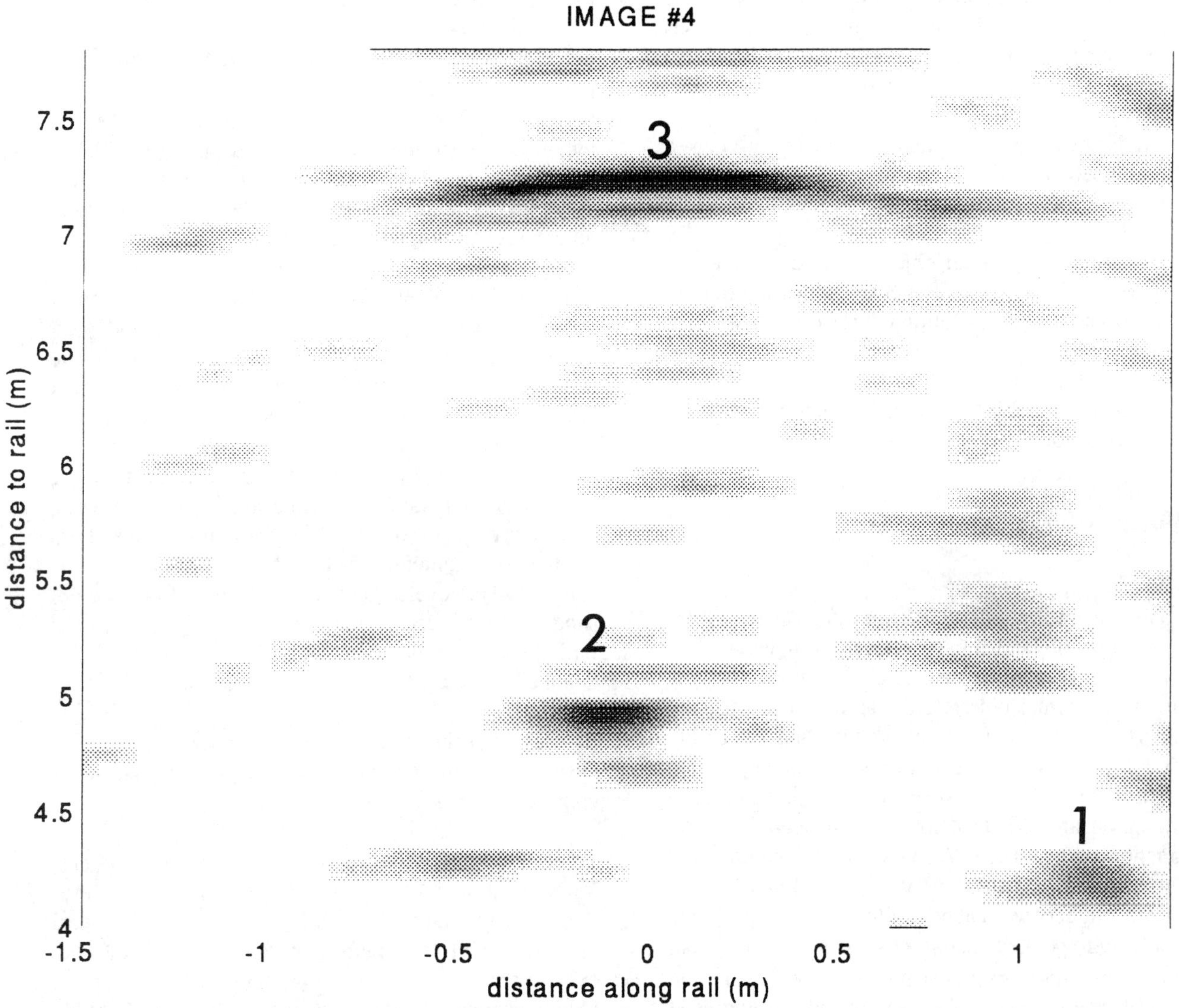

figure 2 : a typical result

SYNTHETIC APERTURE RADAR FOR THE DETECTION OF SHALLOW BURIED OBJECTS

G.F. Stickley, I.D. Longstaff, M.J. Radcliffe

University of Queensland, and Cooperative Research Centre for Sensor Signal and Information Processing (CSSIP), AUSTRALIA

ABSTRACT

This paper illustrates the use of Synthetic Aperture (SAR) processing techniques applied to step-frequency ground penetrating radar (GPR) for the detection of small buried objects.

Conventionally GPR systems operate with the antennas held against the surface of the ground for good coupling and depth of penetration; clearly a mode of operation in which the antennas do not make direct contact with the ground is preferable for mine detection. We have shown that small buried objects can be detected with the antennas elevated off the ground and tilted sideways to the direction of motion.

The radar covered the band 1-2GHz, giving a resolution cell size of 5 to 15 cm which is suitable for detecting objects similar in size to anti-tank mines.

Results showed that small objects can be more easily distinguished from clutter after SAR processing than in the unprocessed high resolution range profile history.

A non-metallic object with diameter of 24cm was clearly distinguishable from clutter when buried about 6cm in dry sand.

INTRODUCTION

Ground penetrating radar (GPR) has great potential for the detection of land mines as even completely non-metallic mines can be detected. To maximise the return from buried objects GPR antennas are usually held in close proximity to the ground and preferably moved directly over the region in which the objects are thought to be buried. However if a GPR is being used to sweep an area for land mines such operation risks detonation of a mine. We have instead investigated using antennas in a stand-off mode. In this mode the transmit and receive antennas are elevated above the ground and pointed about 20° below horizontal. Synthetic aperture processing provides good resolution parallel to the direction of motion of the antennas and also improves signal to noise and signal to clutter ratios.

EXPERIMENT

Experiments were conducted using a radar originally designed for measuring coal seam thickness. The radar is a stepped-frequency system operating between 1 GHz and 2 GHz. The radar incorporates an odometer to allow range profiles to be collected at regular intervals. A more complete description of the radar can be found in Noon (1).

Two ridged horn antennas were used as transmit and receive antennas. The linearly polarised antennas were operated with the E-field polarised vertically. The horn antennas were mounted on a four wheeled trolley and pointed down an angle of 20° to the horizontal and perpendicular to the along track direction. The two antennas are mounted alongside one another with the centre of both horn's apertures held approximately 70cm above the level of the sand. The centre to centre antenna separation was about 40cm.

The antenna trolley was moved along a straight track parallel to the edge of a sand filled beach volleyball court in which eight spheres of varying size were shallow buried. The spheres were grouped into pairs of approximately the same size. One from each pair was then buried at about 1.2m from the antenna track and the other about 2.0m from the antenna track. The spheres are separated from one another in the along track direction by 0.5m. The position of the spheres relative to the antenna track is shown in Figure 1. The spheres were covered by between 3cm and 6cm of dry sand. The surface of the sand in the vicinity of the buried spheres was raked flat. The sand was dry to a depth of about 10cm.

In each experiment the antenna trolley was wheeled slowly along the antenna track. Range profiles were collected every 2.0cm.

To ensure strong radar returns from the buried spheres initial experiments were conducted with the spheres wrapped in aluminium foil. The spheres were uncovered and the aluminium foil removed for later experiments. In this case the spheres, all of which are hollow, represent a dielectric discontinuity and should still have a considerable radar cross-section.

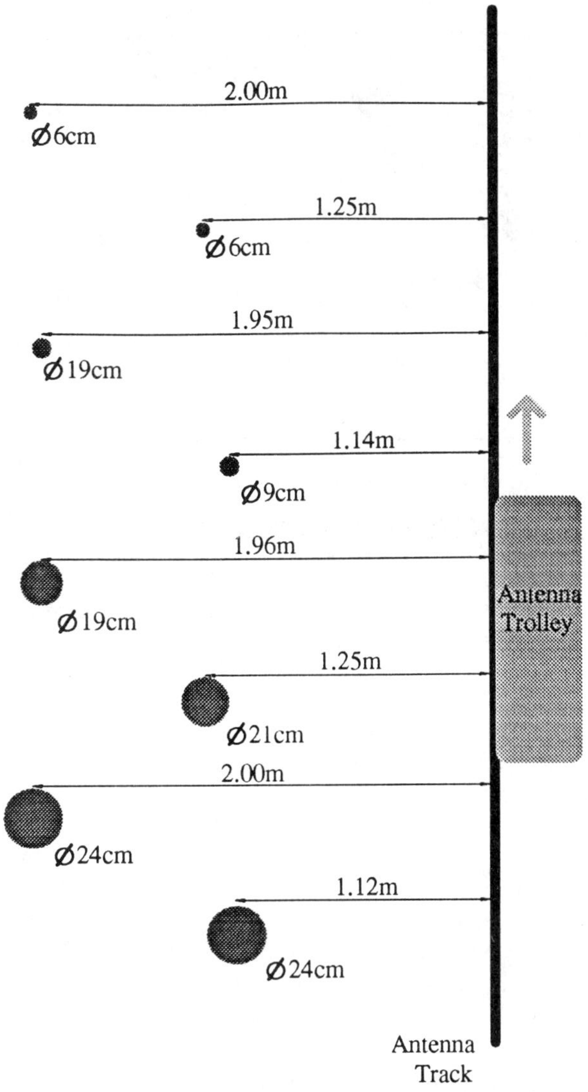

Figure 1- Layout of test site showing the eight buried spheres, their diameters and positions. The top of the spheres were buried about 3 to 6cm below the surface. The experiment was conducted in a sand-filled beach volleyball court. The spheres are separated by approximately 0.5m in the along track direction.

PROCESSING

The frequency domain measurements are converted to the time domain by prepading with the appropriate number of zeros and then applying a Fourier transform. Interpolation in the time domain is achieved by using a FFT length much greater than the number of frequency domain measurements. Sufficient interpolation of the time domain data should be achieved in this step to facilitate the linear interpolation as required in the SAR algorithm.

A simple time delay SAR algorithm has been applied to achieve azimuth compression. As we are solely

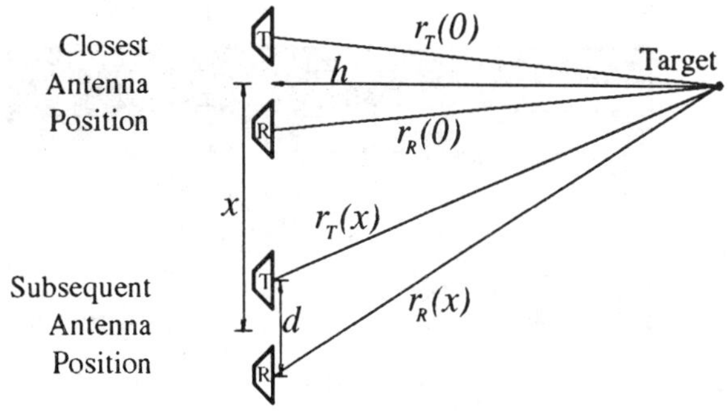

Figure 2- Antenna geometry for point target. The two-way range to the target $r_{2w}(x)$ is just the sum of $r_T(x)$ and $r_R(x)$.

interested in shallow buried targets no account has been taken of refraction at the surface of the sand or of different velocities within the sand. Consider the antenna geometry shown in Figure 2. As the antennas are moved along track past the position of closest approach to the target the two way range to the target is increased. The two-way range to the target, r_{2w}, can be expressed as a function of the distance, x, the antennas have moved along track from the position of closest approach, h the distance of the target from the antenna track, and d the centre-to-centre antenna separation.

$$r_{2w}(x) = r_T(x) + r_R(x)$$
$$= \left(h^2 + \left(x - \frac{d}{2}\right)^2\right)^{1/2} + \left(h^2 + \left(x + \frac{d}{2}\right)^2\right)^{1/2}$$

Then it is a simple matter to see that a single point target traces out an approximately hyperbolic path in an image constructed by putting range profiles taken at different along track positions beside one another. Linear interpolations are then performed to determine the complex value of the time domain signal at the points where the hyperbola crosses each of the range profiles. The SAR algorithm then finds the mean of all these interpolated values. The length of the hyperbola over which this calculation is conducted is adjusted such that the hyperbola subtends a constant angle at the centre point between the antennas.

Where part of the hyperbola extends past the boundary of the data set the calculation of the mean is limited to that section of the hyperbola within the data set.

The algorithm could be further improved by performing a weighted mean (rather than an unweighted mean) along the hyperbola to account for spreading losses and to window the synthetic aperture.

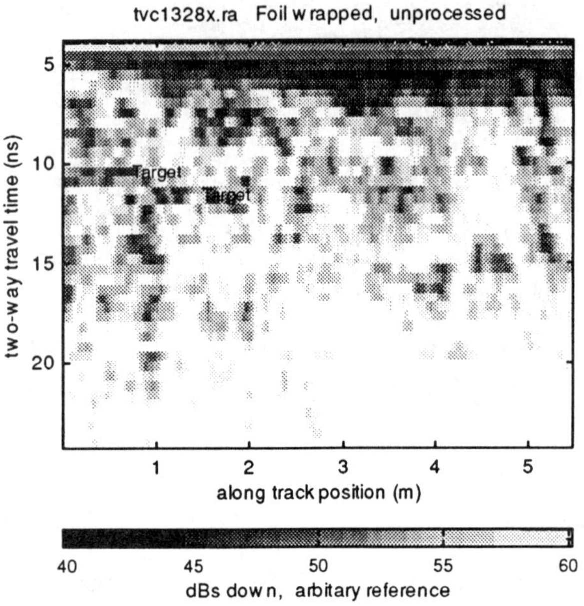

Figure 3- Unprocessed image of buried spheres wrapped in aluminium foil. The 24cm and 21cm diameter spheres which are buried closest to the antenna track are marked in the image.

RESULTS

In the unprocessed data the largest and closest spheres can just be seen when wrapped in aluminium foil (Figure 3). The data in all such images presented in this paper has been weighted by a range-gain function (Figure 4).

After SAR processing the two targets are quite easily distinguishable from the clutter (Figure 5). The 24cm diameter sphere 2.00m from the antenna track is also tentatively identified.

Similar results were also obtained when the antennas were lowered 30cm and the experiment repeated. Previous experiments with the antennas approximately 1.2m above the ground clearly detected cylindrical metal targets.

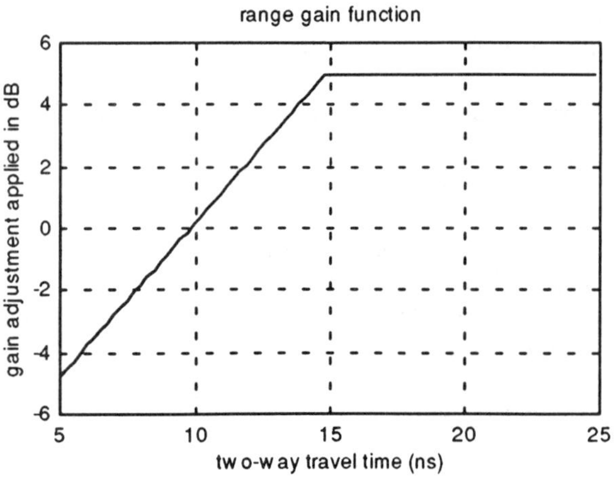

Figure 4- The range-gain function applied to the data. This function partially compensates for increased spreading losses for more distant targets.

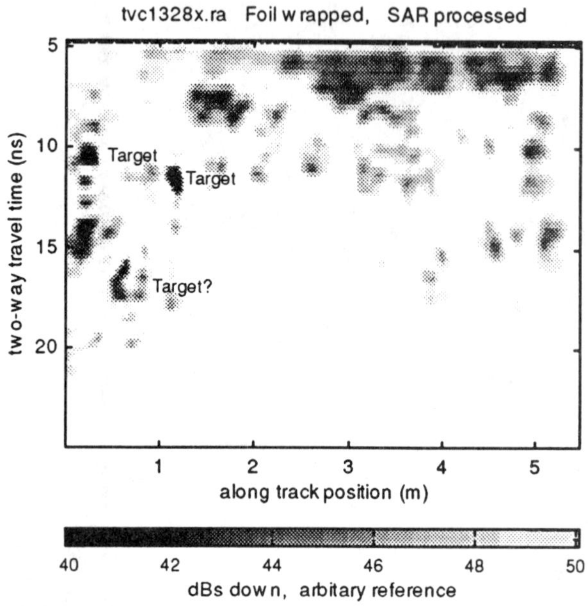

Figure 5- SAR processed image of buried spheres wrapped in aluminium foil.

After removal of the aluminium foil the spheres become much more difficult to detect. In the unprocessed image it is questionable whether even the return from the largest and closest sphere can be distinguished from the clutter (Figure 6).

After processing however the return from this sphere can be distinguished with a fair degree of certainty (Figure 7).

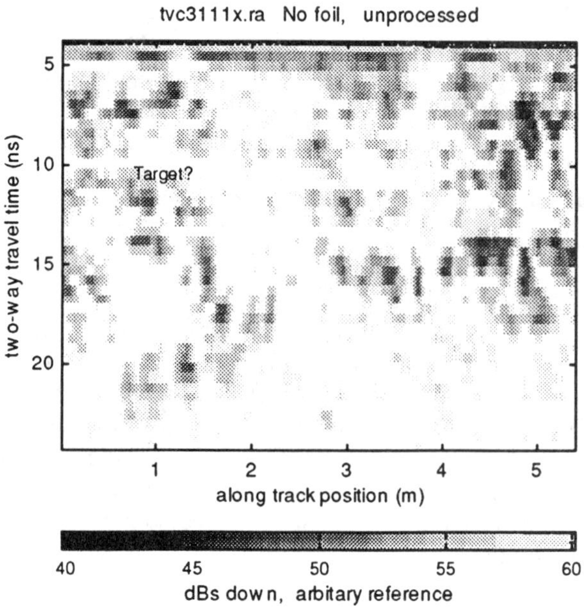

Figure 6- Unprocessed image of air filled spheres buried in dry sand.

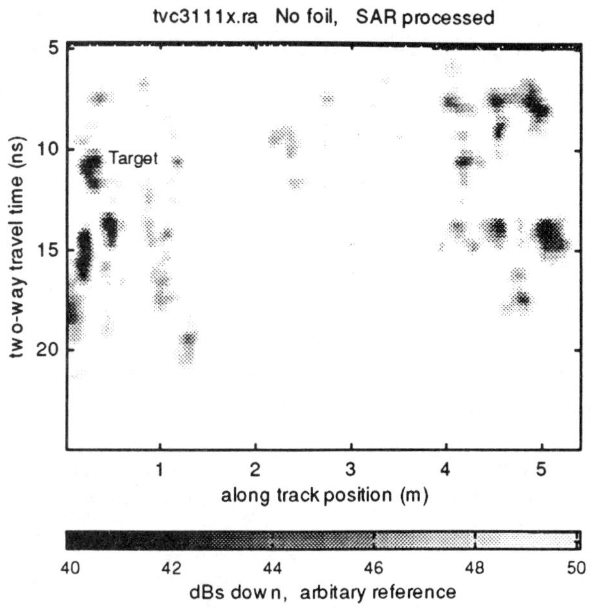

Figure 7- SAR processed image of air-filled spheres. The largest and closest sphere can be distinguished.

DISCUSSION AND CONCLUSIONS

In all experiments presented here the return signal from clutter far exceeds the receiver noise or the level of radio frequency interference. Hence refinements to this technique should focus on better clutter discrimination not better radar system performance.

A CFAR (Constant False Alarm Rate) analysis has not been conducted on the images obtained here however it is worth noting that the signal to clutter ratio which from Figure 7 can be estimated to be of the order of 7 to 10 dB is probably not high enough to result in an acceptably low false alarm rate for practical operation. It might be possible to augment a broad sweep approach such as side-looking SAR with a technique involving closer inspection of likely targets to limit false alarms.

The experiments reported here were conducted in dry sand which was raked flat. Such conditions are obviously very favourable for this technique. It is reasonable to expect clutter levels to rise substantially should the surface environment include large rocks, tree roots, water puddles or other inhomogeneities.

As with all GPR systems this radar, which currently has only an octave bandwidth, would perform better with a wider fractional bandwidth.

With the frequency band currently used (1 to 2 GHz) objects of a similar size to anti-tank mines have been detected. We expect that if the highest frequency of the radar was increased then smaller objects such as anti-personnel mines could be detected. For such objects the highest frequency of operation may need to be increased to between 4 and 6 GHz.

It is possible that some of the clutter imaged by this system may be coherent speckle. The effect of which could be reduced by using multi-look techniques.

In conclusion the technique described in this paper may, with further refinement, have a contribution to make to the problem of detection of abandoned land mines.

REFERENCES

1. Noon, A.D., 1996. "Stepped-frequency radar design enhances ground penetrating radar performance", PhD thesis, Uni. of Qld.

POTENTIALS OF MICROWAVE TOMOGRAPHIC IMAGING FOR ON LINE DETECTION OF LAND MINES

Ph. Garreau*, G. Cottard*, P. Berthaud*, E. Beaumont*, J. Ch. Bolomey**

*SATIMO, France and **SUPELEC, France

ABSTRACT

An Imaging Radar equipment devoted to the detection of buried objects has been developed for the French National Police (CREL DGPN). This Radar so called « Microwave Tomographic Imaging Tool » (MTIT) offers quasi-real time capabilities for analysing ground cross-sections of 1 meter depth by 1 meter width (1). In comparison with all existing GPR equipments, the MTIT is composed of an array of sensors. An adequate ground discriminator data processing is used to prepare properly data before applying Spectral Diffraction Tomographic Algorithms. Such algorithms (2), (3), have been already assessed on a previous equipment designed by SUPELEC and realized by SATIMO for the detection of metallic bars in concrete structures (4). This paper presents results obtained from in situ measurements (5). The robustness of the results with respect to parasitic movements and vibrations has been assessed.

INTRODUCTION

For a long time, microwaves have been considered as a possible sensing agent for remote detection. The detection of metallic or non-metallic targets *via* a mechanically scanned probe has proven to offer a convenient but time consuming way to measure local reflection or transmission coefficients. High speed measurements are now possible by using arrays of fixed probes, resulting in attractive imaging equipments. Such a probe array prototype has been designed for the French National Police to detect buried objects such as drugs, explosives, weapons or buried body. During one year, outdoor tests have been conducted and promising results have been obtained. The interest of Ministry of Defence (DGA) and of some private companies in using microwaves for the depollution of land mines has led to perform new outdoor tests. Although the MTIT prototype was not designed for land mines detection, since frequency range, permittivity, conductivity, span and depth of investigation were not optimised for such an application, an assessment campaign has been realized with this prototype to demonstrate the feasibility of a Tomographic Microwave Radar for land mines detection.

The MTIT equipment is mounted on a motorized vehicle for providing real-time detection capabilities at a speed of about 1 mph. The detection of a target is achieved from a PC display. A colour spot is clearly visible when the MTIT equipment goes over a target and allows to localize it in the ground. Such images are easy to interpret whereas most GPR demand skilled operator and time consuming treatments, such as Synthetic Aperture Radar processing.

The images recorded during the assessment campaign have been used to provide 3D reconstruction of the buried targets.

DYNAMIC ASSESSMENTS

The Test Area and the Targets

The test area, as described in figure 1, consists in a natural field overlaid with grass. The dimensions of the test area are 1 m x 4 m. Part of the ground has been removed and replaced without any degradations of its structure in order to bury the targets. Typically, the bottom of the targets lays at a depth of 50 cm.

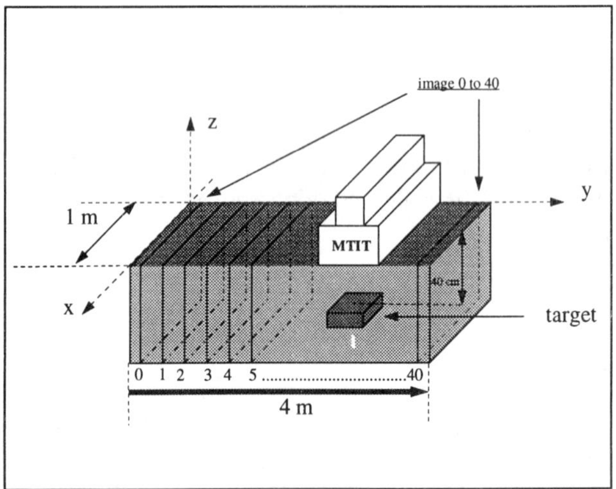

Figure 1: Presentation of the test area

Three types of targets (see figure 2) have been mainly tested, an empty plastic box (called target 1), a full metal cylinder (called target 2) and a plastic box (called target 3).

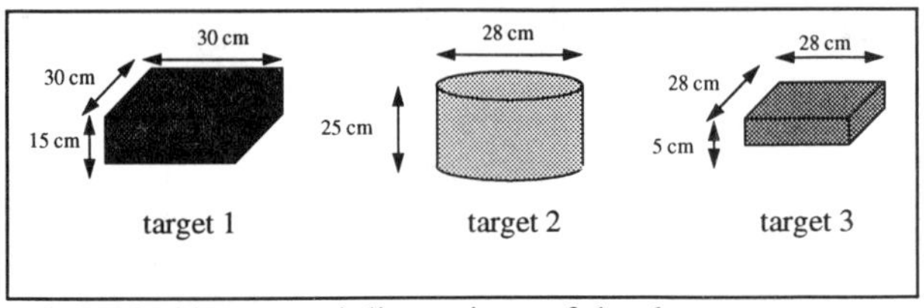

Figure 2: Shapes and dimensions of the three targets

Calibration of the MTIT

Before the beginning of the measurements, the MTIT must be calibrated. Three parameters are adjusted by the operator.

The first parameter is the permittivity of the soil ε_r. This parameter modifies the transmitting and receiving time gating and the Tomographic Algorithms. It can be either approximated or more accurately determined by using a reference target buried at a known depth. ε_r is adjusted in order to obtain the reconstructed target at the desired depth.

The last parameters are the upper and lower threshold levels, ρ_{max} and ρ_{min}, which determine the practical system dynamic range. The best results are obtained when ρ_{min} is determined to cut out the residual noise over a target-free soil. ρ_{max} is determined from the reference target.

All parameters (ε_r, ρ_{max} and ρ_{min}) are determined before any assessment campaigns and are **not modified** during the tests.

Before any assessment campaigns, the MTIT equipment is placed over a target-free ground. The target-free image is then subtracted to any other acquisitions in order to achieve background removal.

Assessment Campaign

The MTIT provides from 2 images/s to 12 images/s depending on the desired resolution. To be sure to detect the response of a 30 cm diameter target with the highest resolution, the speed of the vehicle can not exceed 1 km/h (0.6 mph, 28 cm/s). The MTIT is translated by the vehicle in the y-direction (see figure 1) over the target 1. During the translation, the MTIT equipment provides on the screen 2D **real time** tomographic reconstruction of the subsurface cross-sections (see figure 3).

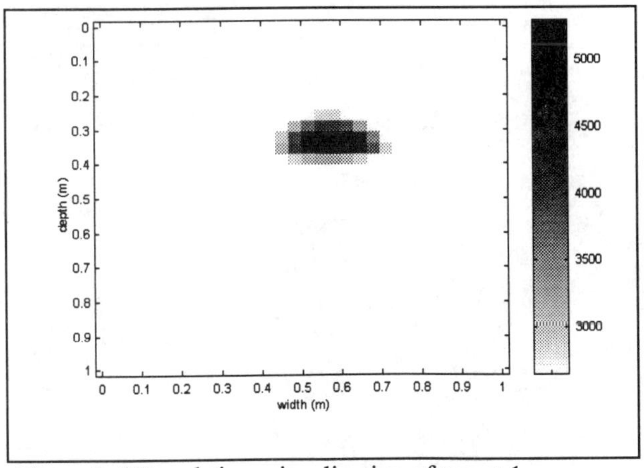

Figure 3: 2D real-time visualization of target 1

The signature of the target is maximum when the radar is just above it. Before and after, the spot increases and vanishes rapidly, so that the target can be detected during a few images. Comparable results have been obtained with target 2 and 3.

3D RECONSTRUCTION OF THE TARGETS

During the translation along the y-axis, 40 images have been stored, corresponding to 40 cross sections of the soil (see figure 1). These images can be used to visualize the 3D shape of each target (see figure 4). The shadows (*ie* projections of the reconstruction on each orthogonal plane) have been drawn in order to improve the visualization and the localization of the target.

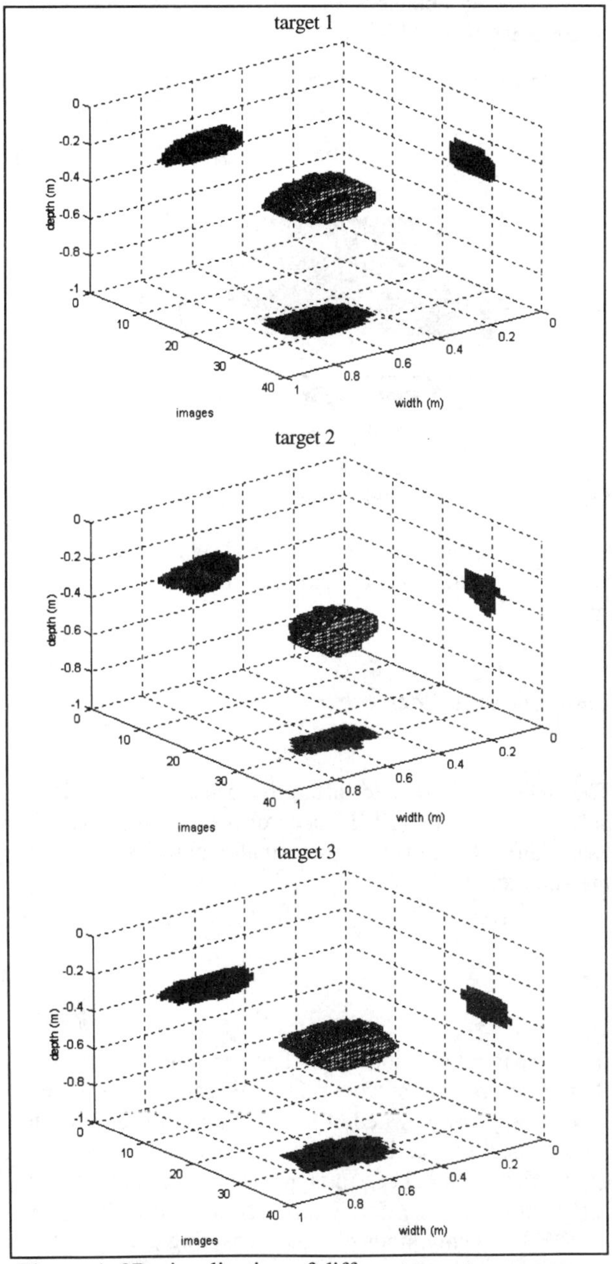

Figure 4: 3D visualization of different target responses

EFFECTS OF PARASITIC MOVEMENTS AND VIBRATIONS ON THE MTIT

The impact of parasitic movements and vibrations on the tomographic reconstruction have been more thoroughly investigated.

Impact of Parasitic Movements

Four types of movements have been tested, vertically (± 4 cm), horizontally (± 4 cm), rotation around the x-axis (± 4 degrees) and rotation around the z-axis(± 4 degrees), as shown in figure 6. The frequency of these movements is about 2 Hz. The impact of these deviations has been shown to be acceptable for an optimized sensor to ground height of the order of 8 cm.

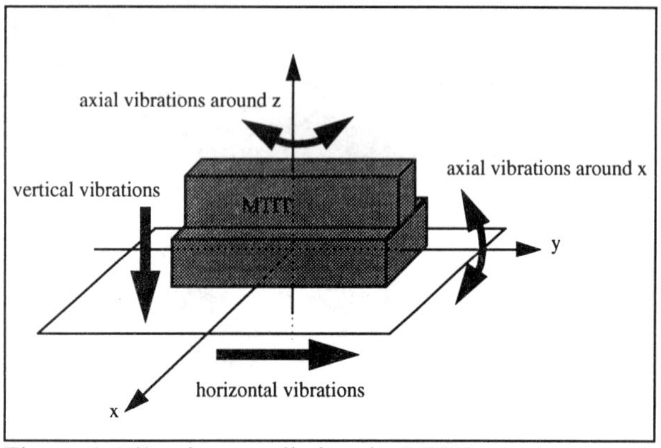

Figure 5: Vibrations applied to the MTIT

Such a robustness provides a good indication of the behaviour of the MTIT equipment on a rough ground surface.

Impact of Vibrations

The motorized vehicle naturally generates vibrations between 40 Hz and 60 Hz, according to its motor (2500 turns/min). The impact of such small vibrations were not measurable.

CONCLUSION

In this paper, the ability of Tomographic Microwave Radar to be used for mines detection has been successfully investigated. The depollution of an uneven field could be done in real time at a speed of at least 1 km/h. The same MTIT prototype has been used to detect real anti-tank mines during an assessment campaign for the DGA. Twenty kinds of mines, covering a large range of permittivities have been tested and promising results have been obtained. The next step could be the construction of a MTIT equipment with optimised parameters and dual-polarized antennas. The software could also be improved to allow automatic detection and thresholding and finally the implementation of a real time 3D reconstruction algorithm could be considered. The civil engineering applications are not forgotten since an assessment campaign with the same equipment for the detection of buried pipes is planned.

REFERENCES

1. Ph.Garreau, G.Cottard, P.Berthaud, J.Ch.Bolomey, EUROEM 1994, Bordeaux, France, in « Electromagnetic Environments and Consequences », D. Serafin et al, part II, 1664-1670

2. L.Chommeloux, Ch.Pichot, J.Ch.Bolomey, 1986, IEEE Trans.MTT,34, 1064-1076

3. J.Ch.Bolomey, Ch.Pichot, 1990, Intern. Journal Of Imaging Systems And Technology,2, 144-156

4. C.Pichot, A.Trouillet, Février 1987, « Cartographie des aciers dans le béton armé », Aide Fonds Valorisation CNRS Ní94-0063

5. Ph.Garreau, G.Cottard, P.Berthaud, J.Ch.Bolomey, J.L. Politano, E. Beaumont, « Real Time Microwave Tomographic Inspection Tool for the Detection of Buried Objects: Assessment Campaign », to be published in the proceedings of the AMEREM 96, Albuquerque, U.S.A.

DATA FUSION FOR VEHICLE-BORNE MINE DETECTION

Daniel W. McMichael

The Centre for Sensor Signal and Information Processing, Adelaide, Australia.

ABSTRACT

This paper introduces a statistically well-founded evaluation of the fusion of dual-band infrared and tri-band visual imagery for detecting buried land mines using real-data. A unified Bayesian framework is presented for representing and inferring from data captured by multiple sensors of scenes that may contain buried land mines.

1 INTRODUCTION

It is likely that electro-optical imagery will be one of the major sources of data for the vehicle-borne mine detection and clearance systems. Infrared imagery has proved invaluable in detecting the tell-tale thermal signatures of buried land mines [1, 2]. Dual-band approaches appear to offer the prospect of more reliable operation [3, 4, 5]. In this regard, DelGrande has shown, in an analysis based on heat transfer theory, that dual-band infrared systems can extract accurate temperature maps by compensating for the spatial variation of emissivity. Not surprisingly, visual imagery has not been used successfully for detecting buried mines.

Even though they are encouraging, these studies, which have attempted to detect or classify land mines, suffer from two major methodological faults: (i) the samples are too small to give statistically significant results, and (ii) no effort has been made to estimate confidence intervals on the estimated values of P_D and P_{FA}. The appropriate calculations show these intervals to be large.

These early studies, while they hold out the possibility of ultimate success, do not satisfy the requirements of military mine detection or civilian mine clearance. If smart signal processing is going to contribute to mine detection, it is likely that other sensors, including thermal nuclear activation, ground penetrating radar and metal detectors will also be necessary. However, mine detection is difficult, not just because mines are well hidden, but because the conditions are so variable. One can expect changes of soil type and humidity, weather, mine type, time of day, surface clutter and geometry. The methods reported in the studies cited above cannot be expected to perform well under a reasonable range of conditions.

The aim of this paper is two-fold: firstly, to present the experimental design calculations for estimating the necessary minimum number of sample for statistical proof of hypotheses on P_D and P_{FA}; and secondly, to propose an evolvable model structure capable of representing the necessary variation of targets and conditions.

2 A CLASSICAL STUDY

A short initial study is being mounted to produce methodologically sound evidence that (i) fusion of dual-band infrared imagery enhances detection/false alarm rate trade-off and (ii) that three-band visible imagery can be used to reduce the probability of false alarm by identifying surface clutter. The approach to detection is conventional; the imagery is registered, regions of interest (ROIs) are extracted, features of the ROIs are generated, and then classified as "mine" or "not mine".

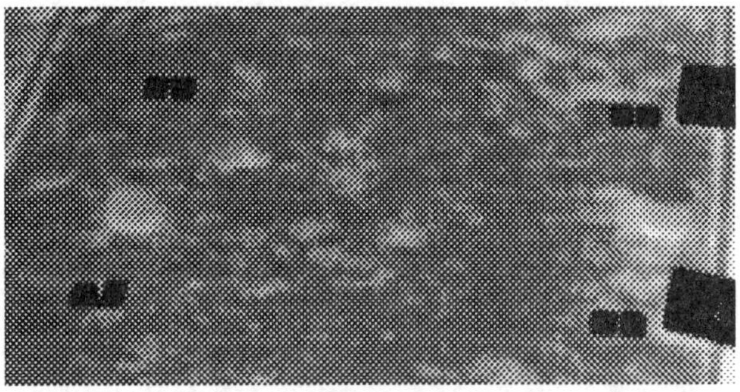

Figure 1: An example of a long wave band IR image from the test site. There are three buried mines.

The key methodological questions arise at the experimental design stage, in classifier training, and in interpreting the results. The discussion here focusses on experimental design. On the basis of Clark *et al.*s' results [6] sand is a difficult soil type, for which dual-band infrared imagery may give significant benefits. Comparison of single-band and dual-band detection results from a non-vegetated sandy soil test site should confirm or disprove the value of dual-band fusion under conditions such as these, which are so common in Australia (figure 1). The ability of the additional visual imagery to reduce clutter-induced false alarms will be tested by comparing these results with the outcome of similar experiments, in which the test site is partially cluttered by dry leaves. The required experimental conditions are shown in table 1 below.

	Experimental variables			
Class	time	soil type	clutter	# samples
class I	daylight	sand	yes	100
class II	daylight	sand	no	200

Table 1: The data classes, sample numbers and conditions for experiments.

2.1 Estimating Required Sample Sizes

The detection test results are either (i) a number of missed detections in a set of examples in which mines are present, or (ii) a number of false alarms in a set of examples where they are not. The ratios of these numbers of errors to the total number of examples in each set are estimates of the underlying error rates. However, these estimates are uncertain and have confidence intervals. The purpose of the study is to demonstrate statistically significant improvements in algorithm performance. Specifically, we seek to detect differences in underlying error rates. In the approach presented here, the significance level is stated and the required sample number is calculated by inverting a generalised likelihood ratio test of error rate difference. The working detailed by Roughan and McMichael [7] shows that the minimum number of data to generate a test with significance α for distinguishing underlying error rates e_1 and e_2 is

$$N = \frac{\chi^2(\alpha)(e_1 + e_2 - (e_1^2 + e_2^2))}{(e_1 - e_2)^2}. \quad (1)$$

Prior estimates of the error rates under the different treatments were obtained from [6], and the required number of samples was estimated to be 100 for class I and 200 for class II. This is significantly more data than has been used in previous studies. A diagram showing graphs of minimum sample numbers for $e_2 = 0.2$ and variable e_1 is shown in figure 2.

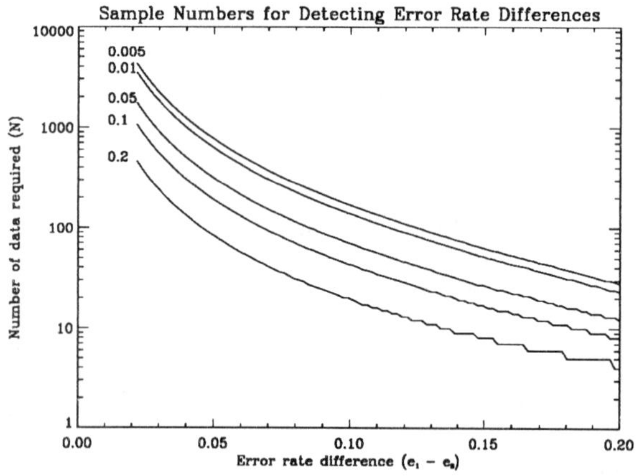

Figure 2: Plot of number of data required versus $e_1 - e_2$ for various test sizes α; $e_1 = 0.2$.

As the error rate difference, $e_1 - e_2$, becomes smaller the number of samples required increases. As the size α of the test increases, the test becomes less significant, and the required number of samples decreases.

3 A BAYESIAN FRAMEWORK

It is possible, that by muddling through using the type of *ad hoc* approach, characteristic of this and earlier work, a practical multisensor mine detection system could be developed. However, as we have seen, the problem is difficult, and requires error rates far below those provided by most automatic target detection systems. The remainder of this paper provides a unifying model structure for representing the problem, and an optimisation route for developing efficient algorithms.

3.1 Semantic Models

The problem of detecting mines involves providing answers to two questions: "where are the likely locations of mines?" and "how likely is a mine to be there?". To cover territory quickly, vehicle-borne sensors need to view considerable areas, which may contain many objects, including soil, roads, vegetation, surface mines, buried mines and rocks. The scene model needs to represent sufficient detail for mines to be recognised and non-mine objects to be discarded. This *semantic* representation would attribute meaning and properties to scene objects.

Earlier attempts to construct such structures, reported in the vision literature, have extracted features from visual scenes, and then modelled objects and their interrelationships with rules [8, 9, 10, 11, 12, 13, 14]. In logic-based and rule-based systems objects are abstract entities that satisfy or break the rules. In satisfying a rule an entity or group of entities is conferred with the property associated with the rule. Relationships can be represented in many ways besides rules, such as semantic nets, and propositional and predicate logic. In statistics, objects occur as components of mixture distributions. Uncertain relationships have been represented by fuzzy logic, Dempster-Shafer theory and conventional probability.

3.2 Good Model Structures

In drawing from these approaches we seek frameworks that can coherently represent the problem without adding artifactual constraints. A good framework should have at least some of the following properties:

1. allow consistent inference;
2. be a declarative problem statement;
3. be able to incorporate "deep" knowledge;
4. be able to model rule-like constraints;
5. be able to represent rule bindings;
6. contain objects directly driven by data;
7. contain meta-objects driven by other objects;
8. be able to infer from statistical data;
9. be able to infer from images and feature fields;
10. be able to fuse data from several sensors;
11. facilitate good algorithms;
12. enable algorithm optimisation.

Property 1 and the requirement to represent uncertain states force the use of Bayesian inference, which also provides a declarative problem statement. At least some of the objects in the representation should be directly constrained by the data, while others should be available to model higher level relationships between the data driven objects. Substantial reductions in the amount of data required to train the system can be achieved by constraining the inter-object relationships using expert knowledge, and this is facilitated by representing relationships as intelligible rules. The detection system should be able to infer the structure of its model and parameters from statistical data including imagery and feature fields (e.g. edge maps and wavelet transforms) and other sensor inputs. Most important of all, the framework must be capable of generating effective algorithms.

Bayesian inference seeks to infer the relative degree of consistency of each member of a portfolio of models $\{M\}$ with data X about an unknown reality. The relative degree of consistency of a model with the data is determined by its absolute degree of consistency with the data, as measured by the *likelihood function* $P(X|M)$. The prior prejudice of the modeller is represented by the *prior probability* $P(M)$. The relative degree of consistency of model with data is quantified by its *posterior probability* $P(M|X)$, calculated using Bayes' rule:

$$P(M|X) = \frac{P(X|M)\,P(M)}{\sum_{\{M\}} P(X|M)\,P(M)}. \quad (2)$$

This simple method for coherent reasoning under uncertainty will be progressively developed to infer the content of complicated scenes.

4 A SINGLE-SENSOR MODEL

In the Bayesian approach, the first step is to define the likelihood function that links the data to the space of models. The models are generally parameterised by large numbers of real and discrete variables that enable accurate representation of varied scenes at the necessary level of detail. In advanced problems, such as this, these functions are complicated, and the process of inference (applying Bayes rule, or approximations thereto) is made much simpler if they factorise into terms in discrete subsets of data and parameters. This form of "divide-and-rule" can be achieved here by segmenting the image. Let X be a list of pixel values in an image arranged by location and indexed by a variable k; i.e. $X = \{(x_k, k)\}_{k \in K}$. Let the *segment* $S_j(X)$ be a subset of X; i.e. $S_j(X) = \{(x_k, k)\}_{k \in K_j}$. The set of index variable sets K_j of the segments is termed the *segmentation* S and is equal to the partition $\{K_j\}_{j=1}^{n}$. We consider likelihoods of the form

$$P(X|B,S) = \prod_{j=1}^{n} P(S_j(X)|O_j), \quad (3)$$

where the model B is defined by the set of *objects* $\{O_j\}_{j=1}^{n}$ which themselves contain parameters, at least one defining the object's *type* t_j and possibly others defining other properties of the object; the latter are grouped for each object into a set θ_j. The posterior of the object set B given a fixed segmentation S is

$$p(B|X,S) = \frac{P(X|B,S)\,p(B|S)}{\int P(X|B,S)\,p(B|S)\,dB}, \quad (4)$$

where $p(B|S)$ is the prior probability density of the object set B given a fixed segmentation S. This term contains both subjective prior knowledge about the objects' parameters and about relationships between objects. The segmentation itself may be marginalised out, giving the posterior of the object set conditional only on the data,

$$p(B|X) = \int p(B|X,S)\,p(S)\,dS. \quad (5)$$

4.1 Associating Objects with Features

In all multi-object scene analysis problems, such as tracking, blind deconvolution and visual scene analysis, objects have to be associated with subsets of the data. These association problems are, in general, NP-hard; however effective procedures have been designed in each application domain by incorporating sufficient prior information. We swap the trivial problem of associating n identical objects with n segments for a NP-hard one, in which there are several object types, in order to reduce the complexity of the object representations and to allow expert knowledge to be built in to each type of object. For example, specific objects can be constructed to model areas of rock, vegetation etc..

The form of the relationships between pairs of objects is determined by the type parameters of each. Given the types of all the objects a particular *graphical model* or *Bayes net* can be constructed. Object type estimation is equivalent to estimating the structure of the Bayes net.

The object set parameters B may be divided into the *type matrix* $T = \{t_j\}_{j=1}^{n}$ and the *object properties* $\Theta = \{\theta_j\}_{j=1}^{n}$. The posterior of the type matrix is

$$P(T|X) = \int p(B|X)\,d\Theta, \quad (6)$$

and if the segmentation is fixed:

$$P(T|X,S) = \int p(B|X,S)\,d\Theta. \quad (7)$$

4.2 Inferring the Segmentation

Segmentation is a commonly used process in scene analysis; its purpose is to divide up the image in to regions that can be modelled individually using simple representations. It is a tool which simplifies both scene inference and its implementation. Although interesting algorithms, such as relaxation labelling and simulated annealing have been used to solve segmentation problems, the starting point is the objective of the segmentation, here defined by $P(X|B)$ and $p(B|S)$. Under the model structure proposed above, the posterior of a segmentation S is

$$P(S|X) = \int \frac{P(X|B,S)\,P(S)}{\sum_{\{S\}} P(X|B,S)\,P(S)}\,dB. \quad (8)$$

This posterior favours segmentations that fit the object types included in B, and is the basis of a maximum *a posteriori* probability (MAP) estimation algorithm for segmentations.

4.3 Inferring Higher Level Objects

To simplify the low level object association and estimation problem, the low level objects in B should have relatively few types. If so, probable image segments are likely to contain small fragments of larger objects of interest, and the model structure needs to be augmented to represent high level relationships between the low level objects. High level objects can be introduced which relate with the data indirectly via the low level objects. Again, their types determine the connectivity of the Bayes net representation of their interactions. Let the set of high level objects be $A = \{t'_i, \theta'_i\}_{i=1}^{m}$, and their relationship to the low level objects is determined by the conditional probability density $p(B|A)$. Their posterior is

$$p(A|X) = \int \frac{P(X|B)\,p(B|A)\,p(A)}{\int P(X|B)\,p(B|A)\,p(A)\,dA}\,dB. \quad (9)$$

An example of this model structure is presented in figure 3. It shows the Bayes net formed by image segments, low level objects and high level objects for a particular instance of the object type matrices and segmentation. The presence of higher level objects affects the probabilities of low level objects as follows:

$$p(B|X) = \frac{\int p(B|A)\,P(X|B)\,p(A)\,dA}{\int p(B|A)\,P(X|B)\,p(A)\,dB\,dA}. \quad (10)$$

This approach is readily extended to handle segmentations of stacked planes of feature fields.

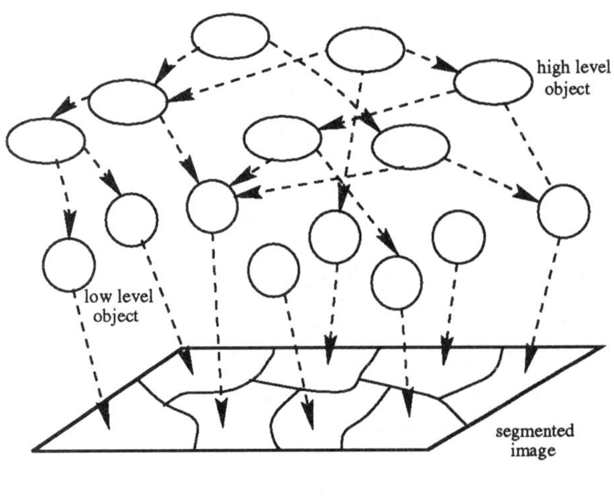

Figure 3: An instantiation of the object/segmentation model.

4.4 Decision Making

Ultimately, we want to test whether objects identified by this process are mines with sufficient probability to warrant action. Decision theory can be applied so that the costs of false alarms and failures to detect can be incorporated. Let one of the allowable object types be "mine"; action is taken when its expected utility is greater than that of inaction. The utility calculation is based on the marginal probability of the "mine" type for each object.

5 A MULTISENSOR MODEL

The concept of image segmentation is one example of a decomposition process in which data is processed into discrete components each of which is attributable to a different object. Sharp images are naturally decomposed into discrete segments, however ground penetrating radar images are better understood in terms of overlapping geometrical features. Never-the-less, low level objects can be inferred, from which physical causes can be estimated; networks of objects are plausible semantic representations of many forms of sensor data.

Sensors can support their own semantic models, or objects can be shared. This enables data fusion, cross sensor cuing, and, depending on the granularity of the decomposition, signal level decomposition. Figure 4 illustrates a multisensor Bayes net showing three registered images being fused with a non-imaging sensor. This network corresponds to a single instance of the type matrices and segmentation.

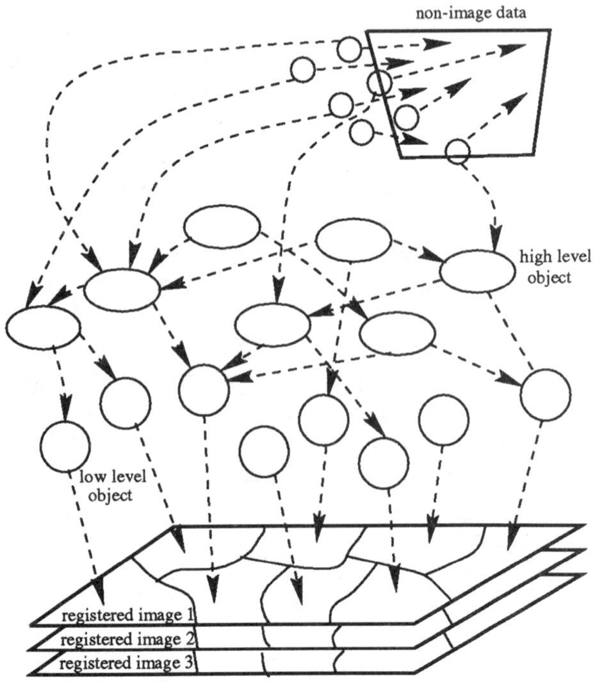

Figure 4: An instantiation of a multisensor model.

6 IMPLEMENTATION

This consistent declarative statement of the semantic scene analysis problem is a useful step; however, the ultimate aim is to design good algorithms. The integrals of the Bayesian solution cannot be calculated by any conceivable computational device, and approximation is necessary. MAP algorithms force the drastic simplification of selecting only one model from the portfolio. However, intermediate approaches, that involve retaining groups of likely models and approximate local modelling of the posterior distributions are likely to offer very large reductions in effort, provide sufficient information about the scene, and represent the uncertainty about objects' types required for reliable decision making.

7 CONCLUSION

This paper has sought to make two contributions, firstly, to provide a technique for setting minimum sample numbers to detect differences in error rates, and secondly, to set out a declarative framework for semantic analysis of scenes like those provided by forward looking vehicle-borne sensors for detecting land mines.

Recalling the properties of good scene models suggested in section 3.2, we see that the framework satisfies properties 1-7, 9 and 10. Property 8 would satisfied if the object representations were tuned in the light of historical data, and this is readily accommodated within the framework. However, despite clear lines of approach, further work is necessary to establish whether efficient algorithms can be derived from this declarative formulation. It is our aim to apply this, or similar models, to a mine detection problem using real data, in the near future.

8 References

1. J. Hackett, E. Gold, D. Long, E. Cloud, and H. Duvoisin, "Model-based approach to real-time target detection," in *Automatic Object Recognition II*, vol. 1700 of *Proceedings of the SPIE*, pp. 269–276, SPIE, 1992. Orlando, FL, USA.

2. P. Ngan, S. A. Garcia, R. L. Barnard, M. Bower, E. Cloud, and H. Duvoisin, "Development of automatic target recognition for infrared sensor-based close range land mine detector," 1994.

3. N. DelGrande, "Temperture evaluated mine position survey (temps) application of dual-band infrared methodology," in *Proceedings of the SPIE*, vol. 2073, pp. 365–384, 1990.

4. G. Clark, J. Hernandez, S. Sengupta, R. Sherwood, P. Schaich, M. Buhl, R. Kane, M. Barth, and N. DelGrande, "Computer vision and sensor fusion for detecting buried objects," in *Twenty-Sixth Asilomar Conference on Signals, Systems and Computers*, vol. 1, pp. 466–471, 1992.

5. G. A. Clark, S. K. Sengupta, W. D. Aimonetti, J. Roeske, J. G. Donetti, D. J. Fields, R. J. Sherwood, and P. C. Schlaich, "Multispectral image fusion for detectin land mines," in *Detection Technologies for Mines and Minelike Targets*, vol. SPIE vol. 2496, (Orlando, Florida, USA), SPIE, 1995.

6. G. Clark, S. Sengupta, M. Buhl, R. Sherwood, P. Schaich, N. Bull, R. Kane, M. Barth, D. Fields, and M. Carter, "Detecting buried objects by fusing dual-band infrared images," in *Twenty-Seventh Asilomar Conference on Signals, Systems and Computers*, vol. 1, pp. 135–143, 1993.

7. M. Roughan and D. W. McMichael, "Data fusion for land mine detection: experimental design," Tech. Rep. 10/96, Cooperative Research Centre for Sensor, Signal and Information Processing, Adelaide, Australia, 5095, June 1996.

8. R. A. Brooks, "Model-based 3-D interpretations of 2-D images," *IEEE PAMI*, vol. 5, pp. 140–150, March 1983.

9. B. V. Funt, "Problem solving with diagrammatic representations," *Artificial Intelligence*, vol. 13, pp. 201–230, 1980.

10. C. Goad, "Special purpose automatic programming for 3D model-based vision," in *Proc. DARPA Image Understanding Workshop*, pp. 94–104, June 1983.

11. D. T. Lawton, T. S. Levitt, C. McConnell, and J. Glicksman, "Terrain models for an autonomous land vehicle," in *Proc. IEEE Int. Conf. on Robotics and Automation*, (San Francisco), pp. 2043–2051, April 1986.

12. M. Herman and T. Kanade, "The 3D mosaic scene understanding system: incremental reconstruction of 3D scenes from complex images," in *Proc. DARPA Image Understanding Workshop*, (New Orleans), pp. 137–148, DARPA, October 1984.

13. D. M. McKeown, Jr., W. A. Harvey, and J. McDermott, "Rule-based interpretation of aerial imagery," *IEEE PAMI*, vol. 7, pp. 570–585, September 1985.

14. J. K. Tsotsos, "Knowledge organization and its role in representation and interpretation for time-varying data: the ALVEN system," *Computational Intelligence*, vol. 1, pp. 16–32, 1985.

Acknowledgement

The author is grateful to Dr. Matthew Roughan for interesting comments, and to the Defence Science and Technology Organisation of Australia for sponsoring this work; particular thanks go to Dr. Robert Seymour and Mr. Karl Fueloep.

U.S. ARMY'S NEW HANDHELD STANDOFF MINE DETECTION SYSTEM (HSTAMIDS)

C A Amazeen
M C Locke

U.S. Army Communications and Electronics Command
Night Vision and Electronic Sensors Directorate
USA

ABSTRACT

The U.S. Army recently launched a new program, called HSTAMIDS, to develop a handheld standoff mine detection capability for the dismounted soldier. This system will be comprised of active and passive sensor technology. Sensor integration will be accomplished using a high speed digital parallel processor; providing near real-time mine cueing and mine location data transmission to the operator's audio and visual display. The envisioned system will allow "plug-and-play" operation, permitting use in a variety of environmental conditions and operational scenarios.
The HSTAMIDS program is the next developmental phase of the U.S. Army's Close-In Man Portable Mine Detector (CIMMD) Advanced Technology Demonstration (ATD) Program. During extensive field testing conducted by the U.S. Army's Test and Evaluation Command (TECOM) in 1995, CIMMD technologies were evaluated against real metallic and non-metallic anti-personnel and anti-tank mines. Results of the field tests showed that the CIMMD technologies met user requirements for continued development.

BACKGROUND

Since the Second World War the U.S. Army has made numerous attempts to provide the dismounted soldier with portable nonmetallic and metallic mine detection capability. The challenge of reliably detecting mines of diverse physical properties in nearly every type of terrain in which the Army is expected to conduct operations has until recently been insurmountable. The mine detection tools the dismounted soldier presently uses are a handheld metal detector and a nonmetallic mine probe-- virtually the same approach as was employed 50 years ago (Reference 1, 2, and 3). During the past half century, while the detection approach has changed little (only changing from tube/transistor technology to integrated circuit/microprocessor technology) mines have become more sophisticated and far less metallic, to having nearly no metal content.

Though the currently fielded metal detector, the U.S. Army's AN/PSS-12, is very sensitive to small amounts of metal, the false alarm rate (FAR) when operating in a metallic cluttered environment can slow detection operations to a crawl; and shallow buried nonmetallic antipersonnel mines more often are undetectable.
With little work done by the U.S. Army during the late 1970s and most of the 1980s towards development of a fieldable handheld metallic/nonmetallic mine detector, a 1987 Defense Science Board study recognized the growing nonmetallic mine threat and recommended work commence to address this problem. The Defense Advanced Research Projects Agency (DARPA) and the Countermine Directorate of the U.S. Army's Belvoir Research, Development, and Engineering Center began a joint program in 1988/89 to pursue technologies that would eventually yield a portable mine detection system acceptable to the U.S. Army's user community. The Close-In Man-Portable Mine Detector (CIMMD) Advanced Technology Demonstration (ATD) Program was the result of this collaboration.

CIMMD ATD PROGRAM

The purpose of the CIMMD ATD was to evaluate and demonstrate the maturity of technology approaches, or combinations of technologies, to reliably detect metallic and non-metallic anti-personnel and anti-tank mines at a rate-of-advance commensurate with dismounted mobility on the battlefield, and with a low FAR. The principal technologies investigated included an Infrared (IR) Thermal Imaging approach, Ground Penetrating Radar (GPR) techniques, and a Pulsed Induction system integrated with two of the GPRs. These systems were portable, with the GPR-based sensors mounted on a wand swept directly in-front of the operator indicating via the operator audio/visual interface the location of a mine when a detection occurred.
The IR Thermal Imaging system (Figure 1) was helmet- and backpack-mounted and featured a Forward Looking Infrared (FLIR) sensor operating in the 3-5 micron spectral range and a miniaturized high-speed parallel processor which implemented an Automatic

Target Recognition (ATR) algorithm, based upon neural network techniques, to discriminate mines from clutter. FLIR imagery and ATR-outputted visual indications, using the thermal difference between the mine and the surrounding soil, were transmitted to the operator's Heads-Up Display (HUD) indicating likely mine locations (Reference 4). This ATR/FLIR sensor combination provided the operator standoff cueing as to the presence of mines.

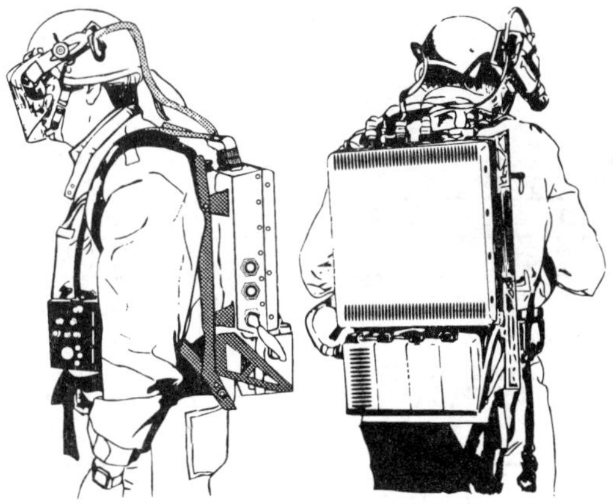

Figure 1. Thermal Infrared Mine Detection System

GPR and FLIR combinations were also evaluated for the potential of high probability of detection, lower FAR, and greater advance speed (Figure 2).

Figure 2. Combined GPR and FLIR System

The GPRs featured broad bandwidth (operating from 0.5 to 4.2 GHz), advanced signal processing which included clutter suppressing algorithms, sensor position feedback, operating heights of from 3 to 6 inches above the ground surface, self-contained power supplies, and improved audio/visual displays. Three GPR techniques were evaluated: Balanced Bridge, Separated Aperture, and Synthetic Pulse. The Balanced Bridge and Separated Aperture techniques were integrated with Pulsed Induction technology.

The Balanced Bridge Ground Penetrating Radar (GPR) (Figure 3) utilized a coplanar and linearly configured symmetrical set of antennas arranged in a receive/transmit/receive orientation. The transmitted signal is a stepped frequency and extends across a frequency bandwidth from 500 MHz to 3,000 MHz. When the sensor is swept over the ground surface, the direct coupled signals and ground reflection signals are nearly identical and cancel when the receiver outputs are combined through a 180 degree hardware coupler. As it passes over objects that scatter energy nonuniformly between the two receiver channels, a detection signal is produced. The system included a Liquid Crystal Display (LCD) operator interface, audio headset, and two algorithm suites to assist the operator in detecting and discriminating mine signals (Reference 5).

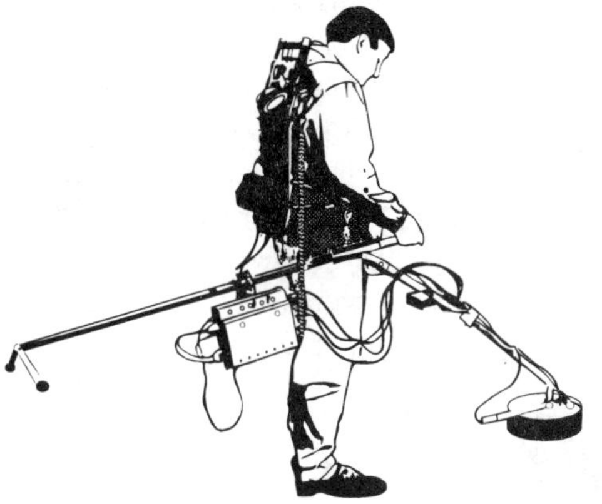

Figure 3. Balanced Bridge Ground Penetrating Radar

The Separated Aperture GPR (Figure 4) utilized two antennas each residing in a metal corner reflector, with a metal septum separating the corner reflectors. The transmitted signal is a stepped frequency with a transmitted bandwidth from 500 MHz to 4,000 MHz. When the sensor is swept at approximately one-half wavelength or less above the ground surface, the direct coupled and ground reflection signals are destructively canceled. The mine signal becomes the dominant signal observed at the receiver. The system included positional sensors, an HUD with an integrated audio headset, and algorithm suites to assist the operator in detecting and discriminating mine signals from background clutter (Reference 6,7,8, and 9).

Figure 4. Separated Aperture Ground Penetrating Radar

The Synthetic Pulse GPR (Figure 5) utilized a bi-static antenna system which transmits a continuous waveform that steps across a frequency band from 700 MHz to 4,200 MHz. The received signal is captured and Fourier transformed to process the return signal in the time domain. Synthetic Aperture radar techniques are employed to enhance the mine signal in cluttered environments. The system likewise included an HUD, audio headset, and algorithm suites to assist the operator in detecting and discriminating mine signals (Reference 10).

Figure 5. Synthetic Pulse Ground Penetrating Radar

Field testing of these systems was performed by TECOM at both arid and temperate sites during the summer of 1995. The test sites were composed of minelanes 1.5 meters wide and approximately 20 meters long with a variety of real (booster-less) metallic/nonmetallic anti-personnel and anti-tank mines. Mines were buried from flush with the ground surface to 15 cm below the surface. Preparation of the site was done well in advance of the tests to eliminate visual cues to the operator. Each system was walked through the test lanes, scanning all lane area within a test lane. The number of detections versus the total number of mines per each lane was measured, as was the number of false detections and scan rate. The statistics were further broken down by performance against mine types. From this information, probabilities of detection (Pd), FAR, and advance rate were calculated for each of the systems. TABLE 1 shows the resulting Pd's obtained for combined GPR and IR performance.

TABLE 1-ATD Test Results

Mine Type	Pd
Metallic	
Anti-tank (AT)	0.95
Anti-personnel (AP)	0.95
Non-metallic (<2 grams metal content)	
Anti-tank (AT)	0.95
Anti-personnel (AP)	0.70

False alarms were generally low to moderate with higher performance obtained by using a combination of GPR and FLIR sensors. Overall CIMMD ATD results showed improved performance over previously investigated mine detection technologies (Figure 6). The subsequent HSTAMIDS Program is expected to further improve performance. The CIMMD program completed the Advanced Technology Demonstration Phase in September 1995.

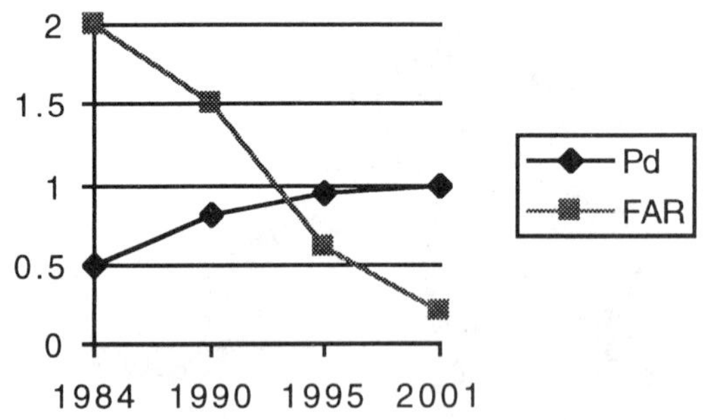

Pd-probability of detection FAR-false alarm rate (#/m^2)

Figure 6. Performance Growth

HSTAMIDS

The CIMMD ATD program transitioned to the HSTAMIDS Demonstration & Validation (Dem/Val) Phase in November 1995. The objective of HSTAMIDS is to use two competing contractors for development and demonstration of prototype portable mine detectors that combine the best technology approaches from the CIMMD Program into an integrated sensor system. User testing will be conducted several months before the conclusion of the Dem/Val phase. The principal criteria for acceptance by the U.S. Army for production of a fieldable system are:

(1) a Pd of no less than 90% against all mine types;
(2) a FAR less than 0.6 false alarms per square meter;
(3) a signal mine alert at a distance of three meters; and,
(4) accurately locate mines to within 0.15 meters of their actual location.

A synopsis of the program schedule is shown in TABLE 2.

TABLE 2-HSTAMIDS Program Schedule

Operational Requirements Document Approval	Apr 95
Milestone I In-Process Review (IPR)	Nov 95
Dem/Val Contract Award	May 95
Milestone II IPR	FY99
Milestone III IPR / Production	FY01

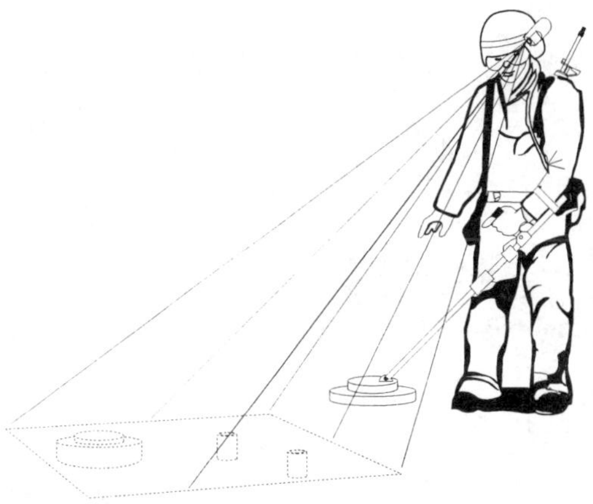

Figure 7. HSTAMIDS Concept

HSTAMIDS SYSTEM DESIGN

Key components of the HSTAMIDS (Figure 7) are the GPR, the metal detector, and the FLIR sensor used in each of the two competing system designs. The GPR sensors use lightweight RF hardware, unique antennas, and mature detection algorithms to find all metallic and non-metallic mines. A metal detector will be used to find mines with metal content. The metal detector search head and GPR antenna are collocated at the end of a lightweight wand, one meter from the operator, to form the GPR/metal detector search head. A position sensor is used to determine search head position for use in GPR algorithms. The IR sensor uses a militarized commercially available camera and modified CIMMD-phase detection algorithms to alert the operator as to the presence of non-metallic and metallic mines at a stand-off distance. Confirmation of mine location is achieved with the wand-mounted search head. Associated with the IR camera is a range finder to enhance algorithm performance.
A single board processor performs system control, sensor specific processing, and multisensor fusion. The system outputs processed detection data to a helmet mounted display and, for enhanced operator comprehension, presents complementary audio data through an earphone.

CONCLUSION

Recent mine detection research and development programs have resulted in detection techniques surpassing prior mine detection state-of-the-art. The technologies resulting from the Close-In Man Portable Mine Detection Program now enable reliable detection of metallic and non-metallic anti-tank mines and metallic anti-personnel mines. The detection rate for non-metallic anti-personnel mines has increased from virtually no capability prior to 1990 to near 70% probability of detection. During the next two to three years, portable, handheld mine detection efforts within the HSTAMIDS Program will focus on multi-sensor integration, detection algorithm development, and enhancement of the user interface. Multi-sensor integration will increase the number of mine properties detectable by the sensor suite, reducing false alarms and increasing rate of advance. The passive IR sensor will provide the capability of standoff mine detection to improve soldier survivability. Sensor integration will be accomplished using a single, high speed parallel processor and novel audio/visual display concepts. "Plug-and-Play" flexibility will provide for a lightweight, versatile design adaptable to limitations imposed by terrain type, ground conditions, weather, and operational scenario. Algorithm development will encompass two areas: target discrimination and automatic target recognition. Discrimination algorithms will improve clutter or false alarm rejection, separating mines from numerous other mine-like

objects present in the environment. Automatic Target Recognition (ATR) will sort through and correlate information from the sensor suite as well as recognize low signal-to-clutter responses from anti-personnel landmines, both approaches augmenting the operator's training and judgment.
The HSTAMIDS will provide a reliable, versatile, and robust mine detection solution that far surpasses any previous system, leading the way for man-portable Countermine systems in the 21st century.

ACKNOWLEDGEMENTS

Recognition is given to GDE Systems, Inc. of San Diego, California and Coleman Research Corporation of Orlando, Florida who will be developing the HSTAMIDS prototypes.

REFERENCES

1. "Basic Studies - Detecting, Destroying or Inactivating Mines", Final Report, Contract No. DA-44-009 Eng- 1773, December 1953, U.S. Army Engineer Center, Fort Belvoir, VA.

2. Stewart, C., May 1960, "Summary of Mine Detection Research, Volumes 1 and 2", Technical Report 1636, U.S. Army Engineer Research and Development Laboratories, Fort Belvoir, VA (AD 320124 and 320125).

3. Nolan, R.C., Egghart, H.C., Mittleman, L., Brooke, R.L., Roder, F.L., and Gravitte, D.L., March 1980, "MERADCOM Mine Detection Program: 1960-1980", Report 2294, U.S. Army MERADCOM, Fort Belvoir, VA (AD C021731).

4. Ngan, P., et al, April 1995, "Development of automatic target recognition for infrared sensor-based close range land mine detector", Proc. SPIE Conf. Detection Technologies for Mines and Minelike Targets, vol. 2496, pp. 881-889.

5. Sherbondy, K.D., April 1995, "Developmental GPR mine detection technology known as the balanced bridge", Proc. SPIE Conf. Detection Technologies for Mines and Minelike Targets, vol. 2496, pp. 31-41.

6. Clague, F.R., December 1989, "Summary of Experiments With the Separated Aperture Technique of Dielectric Anomaly Detection", Technical Report SR-723-29-89, U.S. Department of Commerce, National Institute of Standards and Technology.

7. Riggs, L.S., and Amazeen, C.A., August 1990, "Research With the Waveguide Beyond Cutoff or Separated Aperture Dielectric Anomaly Detection Scheme", Report 2497, U.S. Army BRDEC, Fort Belvoir, VA.

8. Riggs, L.S., and Amazeen, C.A., June 1991, "Measurements in Moist and Wet Soils With the Waveguide Beyond Cutoff or Separated Aperture Dielectric Anomaly Detection Technique", Report 2509, U.S. Army BRDEC, Fort Belvoir, VA.

9. Amazeen, C.A., April 1995, "Wideband ground penetrating radar portable data acquisition system", Proc. SPIE Conf. Detection Technologies for Mines and Minelike Targets, vol. 2496, pp. 84-91.

10. Barrett, C.R., Locke, M.C., et al, April 1995, "Measurement results from the technology assessment for Close-In Man Portable Mine Detection (CIMMD) program", Proc. SPIE Conf. Detection Technologies for Mines and Minelike Targets, vol. 2496, pp. 76-83.

TWO CHANNEL DETECTOR, COMBINING NEUTRON BACKSCATTERING AND ACTIVE ELECTRO-MAGNETIC METAL DETECTION

J.W. Leonhardt[1], A. Küster[1], H. Neff[2]

[1] IUT Institut für Umwelttechnologien GmbH, Berlin, Germany
[2] TZN Forschungs- und Entwicklungs-zentrum Unterlüß GmbH, Unterlüß, Germany

INTRODUCTION

The detection of personal mines is one of the most discussed problems today. Have these mines any metallic components a sensitive metal detector can give some indication. The false alarm rate is high. Combining the signal of this method with that of a neutron backscatter detector, which reflects the possible presence of explosives, the information is improved and false alarms should be reduced. A 2 channel detector is discussed. If there are personal mines without metallic parts than only a nbs-detector has a chance to give an information.

METHODICAL

(1) Neutron backscatter detector

The idea is to use the slow down of fast neutrons by collisions with hydrogen nucleids, which are a part of the chemical content of explosives. Being in a surrounding of material, which has a lower hydrogen concentration, these mines can be detected by means of the backscattered thermalized neutrons.

As neutron sources are available isotopic neutron sources as ^{252}Cf and ^{241}Am/Be-types neutron sources with a neutron flux of 10^6 neutron per second. The mean energies of the emitted neutrons are 2 MeV and 8 MeV respectively. A much more intensive neutron source is a neutron generator, giving a mean flux of 10^8 neutrons per second. The used equipment is schematically shown in fig. 1.

The used detectors for thermalized neutrons are boron coated gauges, in which the n,α-reaction is used to produce the signal. This is proportional to the flux of incoming thermalized neutrons.

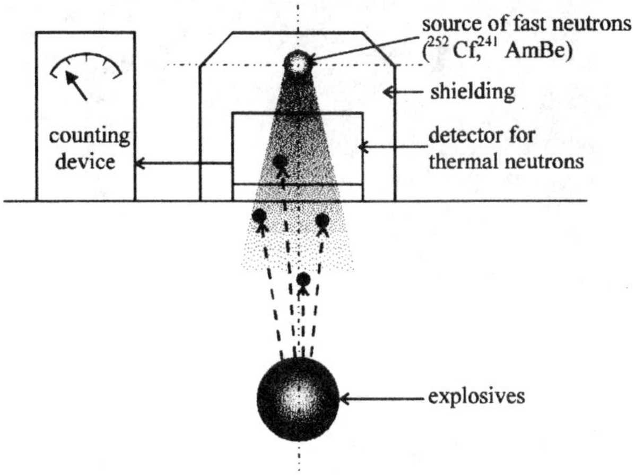

Fig. 1: Scheme of the neutron backscatter arrangement for explosive detection

(2) Metal detector

The pulse remance [1] principle is used for the detection. A dc-current is fed to the transmitter coil via an electronic swith (fig. 2). By this current a magnetic field is built up around the transmitter coil. When the current has reached the steady state, the switch is opened, and the magnetic field decays. By the decaying magnetic field eddy currents are induced in metallic bodies inside the field. These eddy currents generate secondary magnetic fields decaying slower than the original magnetic field. The secondary fields are detected by a receiving coil. It's time dependend output voltages contain information on the coordinates, the volume, the receptivity and - to some extend - the shape of the metallic objects. In order to reach a good sensitivity of such a device it is necessary to have a high field strength of the original magnetic field and a very rapid decay of the field after swith off the current.

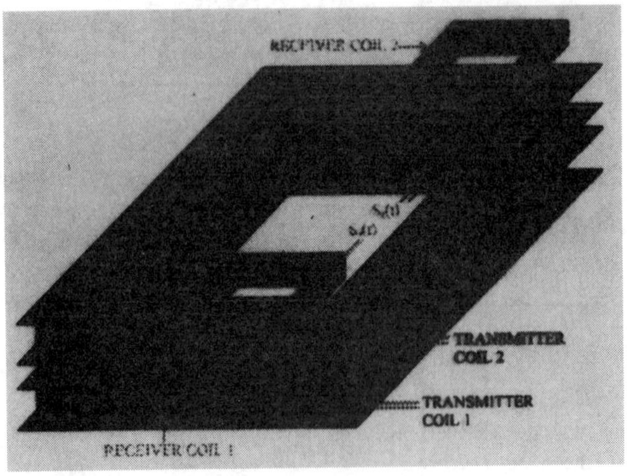

Fig. 2: Scheme of the used remanence system: transmitter coil and 2 receiver coils

RESULTS

(1) Neutronbackscatter

Fig. 3 shows the number (a.u.) of backscattered neutrons produced by an 100 ml water flask covered by 5 cm sand. The maximum flux is registered in the perpendicular direction above the object. There was done some comparision between water and explosives, the hydrogen content of them is relatively lower. The minimal detectable amount of explosives is about 100 g TNT, covered by 5 cm dry sand.

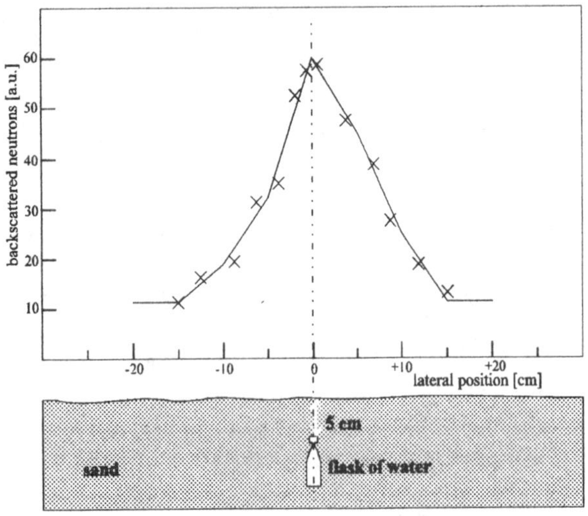

Fig. 3: Number (a.u.)of backscattered neutrons produced by an 100 ml water flask covered by 5 cm sand

The fig. 4 shows 5 letters filled with different amounts of TNT equivalent. The backscattered neutrons are able to identify the explosive down to 20 g TNT. 5 letters are as follows: paper only, paper + plastic bag, plastic bag, paper + plastic bag + 15 g TNT, and finally paper + plastic bag + 250 g TNT.

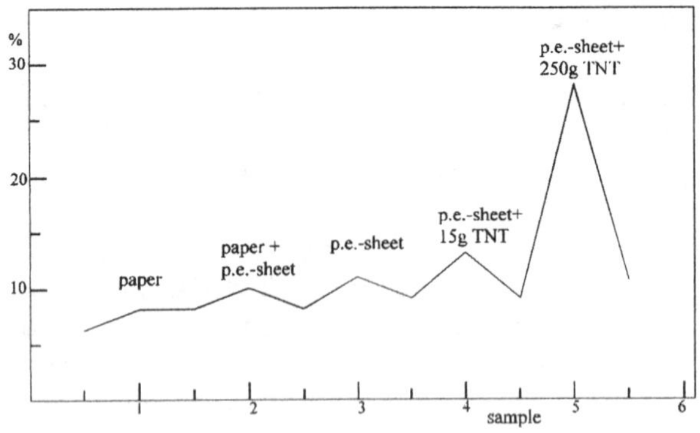

Fig. 4: Letters with TNT - detected by means of backscattered neutrons

(2) Metaldetector

The time dependence of the signal, S (t), is given in fig. 5 for different objects shapes: steel ring, shell in 3 positions and also a sphere. The conclusion is that the decay time T of the signal reflects some information about the shape of the object. Also it is shown that the sphere is a good approach to describe shells in the horizontal positions.

An other result is the correlation between the local half width of the received signal with the depth of the object. The agreement between theory and experiment is better than ± 10 %.

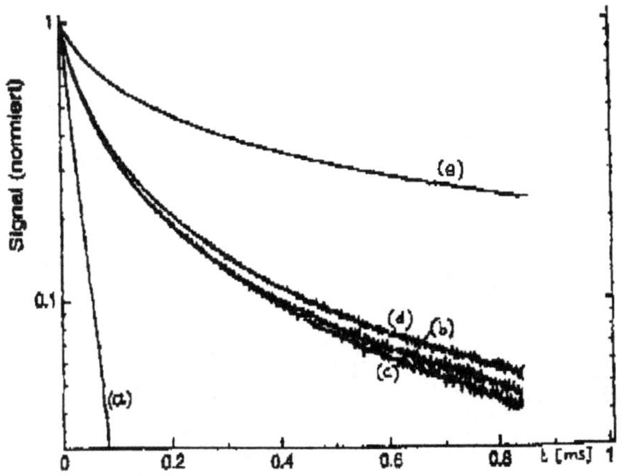

Fig. 5: Dependence of the signals of objetcs with different shapes on the fime: a) steel ring, b) steel sphere (10 cm), c) shell in horizontal position along the motion, d) shell in horizontal position - perpendicular to the motion, direction, e) shell in a vertical position.

CONCLUSIONS

The method is well applicable in many cases as follows:

(1) for the monitoring of personal mines on dry sand surfaces - especially in deserts - and may be in luggage.

(2) for the monitoring of personal mines in incoming letters, parcels, etc.

(3) the combination with the active electro magnetic detector allow the install 2 channel detectors, which are detecting both - metallic and non-metal objects.

The practical use of this 2 channel methods is limited by some restrictions:

- Explosives without metallic components can't be detected in a wet surrounding
- The user of equipment must have the permission to handle neutron sources.

DEVELOPMENTS IN MINE DETECTION

J J Flynn and P Trinkhaus
Foerster UK Limited and Institut Dr Förster

MINE DETECTORS

Due to the proliferation of mines with ever reducing metal content, search equipment manufacturers have been under pressure to design equipments to meet the increasing sensitivity requirements. The required increase in sensitivity has been difficult to achieve with standard detectors.

Continuous Wave Detectors (CW)

The continuous wave mine detector is the search industry standard. Increasing the sensitivity of these units without effecting their stability and ability to work in magnetic soil conditions and within the surf zone on beaches proved difficult.

Pulsed Detectors (PD)

The design of pulsed detectors made a major improvement in search sensitivity, with the ability to provide high power in a single pulse rather than a continuous wave, and still maintain a good working duration with a set of batteries.

When inspecting large areas teams of searchers are used, one major hazard associated with some instruments is a problem of interference between units which can cause them to "go blind" without the operator knowing. Developments have been made enabling up to 7 instruments to work in close proximity by synchronisation. One instrument acting as the master and others as slaves.

Twin Frequency Detectors (2FD)

The development of a twin frequency detector (2FD) has enabled detection sensitivity to be increased with additional benefits such as selectivity and operation with various ground conditions. Table 1 compares sensitivities of mines and metallic objects for the 3 types of detector. A clear distinction is shown between the continuous wave and the pulsed plus 2 frequency units' detection capabilities. The two frequency unit clearly indicates a higher sensitivity over the pulsed equipment for all objects, which is more pronounced with the smaller object sizes.

Selectivity, the ability to detect antipersonnel mines in close proximity to antitank mines is an important design consideration. Table 2 shows a clear superiority off 2 frequency units over the pulse detectors.

Frequency of operation plays an important roll in detection sensitivity, as high frequencies are better for detecting non ferrous metals and low frequencies better for ferrous metals. The 2 frequency detector together with microprocessor control is able to use signal phase-control techniques extensively used in eddy current instruments in Non Destructive Testing and developed by Dr Förster in the 1940s. This enables ground conditions to be balanced without adverse effects on sensitivity.

Sensitivity

The sensitivity of instruments is now very high and as a result has reduced the quality of detection because of the increase in false call rate. The speed of searching has, therefore, been reduced. Further increases in instrument sensitivity may not produce the gains required in detection quality.

Current Developments

Developments have begun in a number of areas to enhance the capabilities of mine detection and include pattern recognition techniques, mapping, use of new technologies and a combination of technologies.

Pattern Recognition

The 2 frequency detector has a printed circuit board probe head which has two sensing areas and products 2 different signals as the head passes over an object. This ability enables not only pin pointing of objects but different signal patterns dependent upon object shape and how the probe traverses the object. One system developed in

Germany utilises two probes, superimposed and mounted on a scanning frame with positional feedback. The resulting test information and positional data can be used to provide position, size and shape of objects detected. The ability to differentiate between mines and other objects greatly increases the quality of detection.

Combined Technologies

The combination of fluxgate magnetometers and mine detectors have proved successful in battle area clearance trials and the ability to differentiate between surface and subsurface objects.

Ground probing radar has proved to be invaluable in many search scenarios including police, military and civilian uses, but still has a number of limitations, which maybe overcome with future developments. The combination of GPR and mine detectors is considered to be a potential solution and requires further investigation to prove its viability. The test data produced by both types of equipment lend themselves to pattern recognition techniques. Portable computers now combine small size with high processing speeds and data storage to enable speedy processing and evaluation of data. Developments could lead to automatic or operator aided decision making.

Vehicle Mounted Systems

The size and weight of any unit is important when considering a portable system. However, if the worldwide problem of vast areas of mine infested land is to be overcome, portable units will not answer all the requirements in a reasonable time frame. Vehicle mounting lends itself to the ability to combine two or more technologies without the same limitations to weight and size governing portable equipment.

Details and photographs of vehicle mounted systems and developments discussed in the paper will be shown during the conference.

TABLE 1. SENSIVITITY COMPARISON TESTS (CLEAN GROUND)			
OBJECTS	2 FREQUENCY	PULSE	CONTINUOUS WAVE
A/Pers Mine SB 33 (MISAR)	10.50	6.00	2.00
A/Pers Mine PMN 6	23.00	19.50	23.00
Nut & Bolt 7.62 cm x 13 mm dia.	33.00	26.00	29.50
<u>Cube 1 = 12.7mm dia x 15.2mm height</u>			
Aluminium	21.50	18.00	11.20
Brass	21.00	19.00	21.00
Copper	19.00	16.00	16.00
Stainless Steel	23.00	20.00	18.50
Mild Steel	26.50	23.50	22.50
<u>Cube 2 = 4.75mm dia. x 4.75mm height</u>			
Aluminium	12.50	8.50	4.50
Brass	12.50	9.00	5.50
Copper	13.5	8.00	5.00
Stainless Steel	5.50	*	*
Mild Steel	13.50	6.50	4.50
Note: * = not possible to find			

TABLE 2. SELECTIVITY OF ANTITANK AND ANTIPERSONNEL MINES		
MINIMUM LOCATION DISTANCE		
MINES	2 FREQUENCY	PULSED
AT - PMN6	25 cm	60 cm
AT - PPM2	21 cm	52 cm
AT - SB33	50 cm	66 cm

AIRBORNE MULTISENSOR SYSTEM FOR THE AUTONOMOUS DETECTION OF LANDMINES

Klaus Scheerer
Bodenseewerk Gerätetechnik GmbH, Germany

SUMMARY

A concept of a modular multisensor system for use on an airborne platform is presented. The sensor system comprises two high resolution IR sensors working in the 3 μ - 5 μ and 8 μ - 12 μ spectral regions, a RGB videocamera with its sensitivity extended to the near IR in connection with a laser illuminator, and a 10 GHz radar. The sensor concept emerged from the evaluation of comprehensive static and airborne measurements on numerous unburied and buried mines. The system has an on-board real time image processing capability and is intended to operate autonomously with a data link to a mobile ground station. Integration of the sensor system into a manned airborne platform is also considered. It is planned to realize an experimental system by 1999.

INTRODUCTION

Considering the various scenarios for abandoned landmines, an airborne mine detection system offers obvious advantages. Concerning the sensors which could provide a solution to the mine detection problem, especially the problem of buried or otherwise hidden mines, no brilliant idea is in sight and one is forced to fall back upon conventional solutions.

Clearly, the requirements for mine and minefield detection are not identical for military and humanitarian purposes. On the one hand, military requirements are more stringent, especially concerning e. g. round the clock and all weather capabilities, which are irrelevant for the humanitarian mine clearance problem. On the other hand, the detection and clearance of anti-personnel mines as used in civil war regions is a matter of major humanitarian concern, but is of minor relevance in military considerations. Nevertheless, a mine detection system designed according to military specifications will also give a useful contribution to the solution of the general mine clearance problem.

In what follows a short description of several measurement campaigns of mine signatures with various electro-optical and radar sensors is presented. The operational sensor concept is described and explanations of the real time image processing and the sensor fusion concept are given.

MEASUREMENTS

Goal of the measurements was to investigate the potential of available electro-optical, millimeterwave and radar state-of-the-art sensors for the mine detection problem. The measurements were performed on 15 different mine types of domestic and foreign origin.

Several measurement campaigns were carried out from September 1993 through spring 1996 at the tower measurement facility of the Diehl company at Röthenbach/Germany. Further measurements were carried out from the measurement tower at Meppen during spring 1994. Airborne trials were performed during summer 1995 at Manching. At the measurement site in Röthenbach, the ground is sandy and the vegetation sparse, at the sites in Meppen and Manching, the ground is grass-covered.

The following sensors were used in the trials:

- Imaging IR sensor 8 - 12 μ
- Imaging IR sensor 3 - 5 μ
- CO_2 Laser Radar (Ladar), 10.6 μ
- Sony RGB video camera
- 94 GHz radiometer
- 400 MHz pulsed radar
- 94 GHz fm-cw radar
- 94 GHz bistatic imaging radar
- 10 GHz pulse doppler radar

The IR sensors and the Ladar were designed and built by BGT, the radar and mmw sensors were designed and built by Diehl. Additional measurements were carried out with a video camera with extended sensitivity in the red channel up to a wavelength of 1.1 μ in combination with a YLF-Laser working at a wavelength of 1047 nm.

The measurements were performed on single mines and on minefields, consisting of buried and unburied mines, and layed down according to military requirements. For the tower measurements, the altitude of the sensors was about 50 m, the depression angles between 50 ° and 90 °.

RESULTS AND CONCLUSIONS

The evaluation of the measurement results led to the following conclusions:

1. Under fair weather conditions, the video and IR channels are capable of detecting almost all mines layed out by the military based on tactical considerations. Buried mines will become visible in the IR a certain time after they have been hidden, provided the thermal resolution of the sensor is sufficient. Buried mines were visible remarkably clear at midnight during rainfall, conditions which are usually regarded as very unfavorable for an IR-system. Figure 1 shows a photo and IR images of a mine scene in the 3 μ - 5 μ spectral range taken at successive times during the field measurements in February 1994.

   The scene contains a buried antitank mine and two further mines (1 a). 1 b shows the IR image taken at midnight during rainfall from a distance of 50 m. 1 c shows the same scene at 8 h next morning. The detectibility of buried mines may be further enhanced by low level data fusion, Del Grande (1).

   The video system with the extended red channel in connection with a laser illuminator is useful when mines with camouflage colours are used in background with a rich vegetation under unsufficient natural illumination conditions.

1 a) Photo

1 b) midnight, rain

1c) next morning, 8 h, dry

Figure 1: IR image of a mine scene

Green vegetation has a large reflectivity in the near infrared which mines have not, resulting in a large negative contrast.

2. Passive radiometry at 94 GHz showed promising results. A high resolution imaging radiometer would be a useful sensor on its own, but the development risk would be high and the benefits in combination with a passive IR system doubtful.

3. The measurement results show, that an imaging radar is capable of providing excellent scene information. As an active system, it would provide an additional all weather and night capability. For an operational system the high spatial resolution requirement can only be realized by a focused synthetic aperture radar, which is difficult to synchronize with other imaging sensors. The development risk has to be regarded as very high, and cost is almost prohibitive.

4. The measurement results with the 10 GHz pulse doppler radar showed, that such a system with a resolution adapted to the mine size provides a certain capability to detect mines which are hidden by grass, bushes etc. Additionally the radar provides range information, which can be used by the image processing to decide wether a detected object is a mine or can be ruled out. The development risk for such a system with sufficient resolution is reasonably low.

SYSTEM CONCEPT

Figure 2 shows a block diagram of the operational system

The system will be integrated into a cylindrical structure of about 40 cm diameter. This may be a drone or simply a tube which can be mounted on any carrier whatever.

The sensor unit comprises:

- RGB video sensor with extended sensitivity to the near IR
- Near IR laser illuminator
- Imaging 8 µ - 12 µ IR sensor
- Imaging 3 µ - 5 µ IR sensor
- 10 GHz pulse-doppler radar with doppler beam sharpening

The electro-optical sensors use linear detector arrays with identical instantaneous fields of view. They are synchronized to each other and to the laser by the use of a common polygon scanner with roll compensation. Scanning is done perpendicular to the flight direction. The total field of view is $\pm$ 26.5 ° x 19 °, the resolution less than 0.5 mrad.

The 10 GHz pulse doppler radar uses several microwave patch antennas, which are integrated into the hull of the structure. The size of a resolution cell corresponds to typical mine sizes, resolution in down range is achieved by the use of a high bandwidth, resolution in cross range by synthetic aperture processing. The radar field of view coincides with the field of view of the electro-optical sensors.

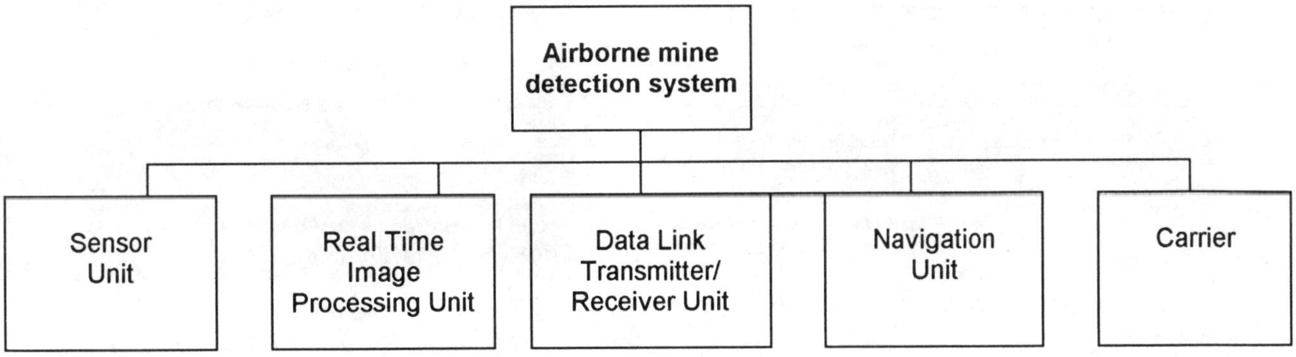

Figure 2: Airborne mine detection system block diagram

The output of the radar sensor is a reflectivity map in range-doppler coordinates.

The real time image processing unit consists of a massively parallel computer network and several high performance digital signal processors for the symbolic image processing.

When operating autonomously, the system uses a bi-directional data link to a mobile ground station working in S- and L-band, respectively. In case there is no line of sight connection between ground station and platform, a balloon is used as a relay station. The data link is used to transmit commands to the platform and to transmit processed image data for human inspection from the platform to the ground station.

The carrier is intended to fly a preprogrammed flight path. In case mines are detected, the navigation unit determines the mine position in geodetic coordinates.
This is achieved by means of an inertial navigation sensor, which is updated by GPS and image aided navigation.

SENSOR FUSION AND IMAGE PROCESSING

Figure 3 shows the architecture of the multisensor image processing.

The realtime pre-processing is done by a BGT developed massive parallel computer based on several Systolic Array Processor (SAP) -chips. A SAP chip consists of 64 processing elements operating in parallel in a single instruction multiple data architecture. The number of SAP chips required depends on the sensor data output stream. The SAP chips are programmable in the higher programming language LISA. In this processing stage, image conditioning, filtering and pixel classification is performed. Furthermore, the images of the IR-sensors are fused on a pixel related basis to separate temperature and emissivity effects /1/. The output of this processing stage are pixels belonging to potential objects. In the object related processing stage, features are extracted from the potential objects and are fused into a single feature vector for each possible object.

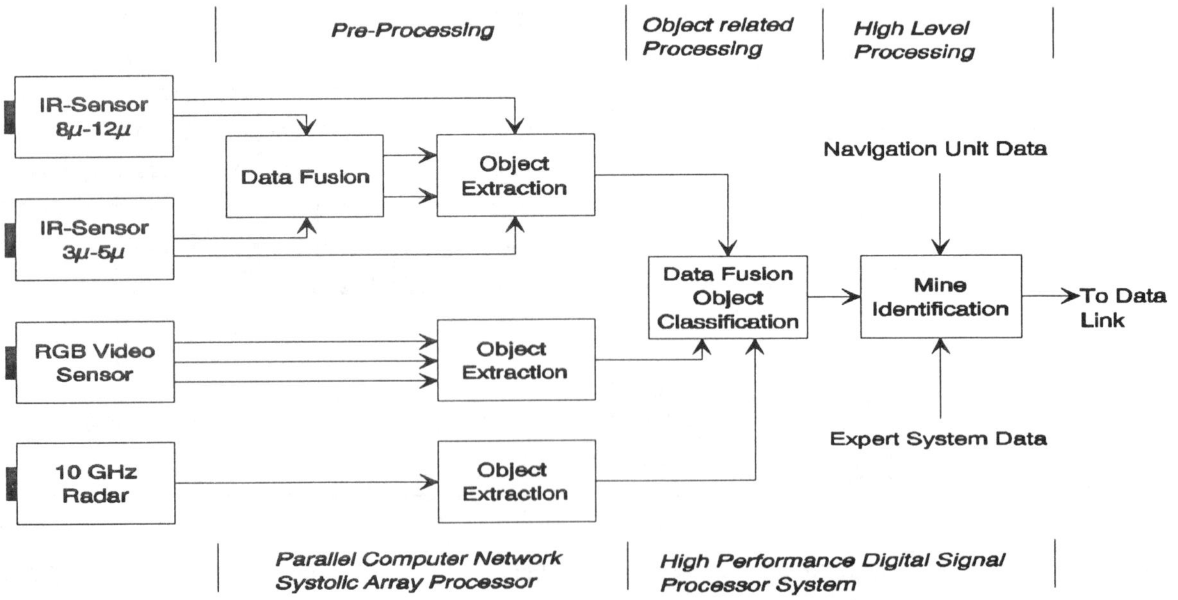

Figure 3: Multisensor image processing architecture

In the high level processing stage, a knowledge based mine identification and classification takes place. The identification and classification process for single mines is essentially based on reflectivity features in the visible and near IR spectrum, on radar backscatter characteristics, on the emissivity and temperature distribution, and on size and shape. Furthermore, in case of minefields, the detection capability of mines and minefields can be improved by taking environmental and structural information into account, Bargel et al (2). The location of identified mines is transformed into a geodetic coordinate system with help of data from the navigation unit.

References

1. Del Grande, N. 1990. Sensor Fusion Methodology For Remote Detection of Buried Landmines, Lawrence Livermore National Laboratory, UCRL-IC-103626, 1990

2. Bargel, B., Bers, K.-H., Stein, G., Traeger, G., 1995, SPIE Proceedings Vol. 2496, 509-518

This work was performed under a contract of the Bundsamt für Wehrtechnik und Beschaffung (BWB) for an airborne minefield reconnaissance system (AAMIS, Aufklärungsausstattung Minensperren).

THE IMPROVED LANDMINE DETECTION SYSTEM

B. M. Cain, T. V. Meidinger

Computing Devices Canada, Defence Research Establishment Suffield, Canada

INTRODUCTION

This paper is a product of the Improved Landmine Detection Project (ILDP) initiated by the Defence Research Establishment Suffield (DRES) as part of its research and development mandate, and in direct support of an urgent Canadian Forces operational requirement. A significant portion of the Canadian Forces' involvement in international military operations has been directed toward peacekeeping assignments. The objective of the ILDP is to develop a landmine detection and marking system for use in peacekeeping operations and other low intensity conflicts. The Improved Landmine Detection System (ILDS) is being developed in conjunction with Computing Devices Canada.

A solution to the landmine detection problem has been sought for over 40 years by various countries. Although the quality of research in this discipline has been excellent, an operationally satisfactory solution has not been developed. This is largely due to the conflict between system requirements and the capabilities of available detection technologies. For example, previous efforts have attempted to address the requirement for a single system capable of detecting all landmine types and sizes, buried and surface laid, which could be used for all types of terrain. In addition, this single solution was to operate at high speeds (no stopping), be simple to operate and maintain, work in all weather conditions, and be usable in high, medium, and low intensity conflicts. However, current detection technologies cannot satisfy this combined set of detectability and operational requirements.

The current Canadian approach is to focus R&D efforts on the development of an improved landmine detection capability for use in peacekeeping operations. This will address the most urgent operational need, and allows system requirements to be compatible with currently available technologies.

ROLE OF THE ILDS IN A COMPLETE LANDMINE CLEARANCE SYSTEM

Landmine clearance personnel currently use manual methods and equipment such as handheld metal detectors, prodders, and protective clothing. A landmine clearance system will allow soldiers to be removed from the immediate danger area. It will also enable a high percentage of the landmines encountered in recent conflicts to be detected, at a rate which is much faster than existing manual methods. The complete landmine clearance system consists of preparation, detection, and neutralization elements (in that order).

Route preparation is necessary to remove immediate threats to the detection system. This function is performed by a lead vehicle which removes surface-laid Anti-Personnel (AP) landmines, and detonates tilt rod and magnetic influence Anti-Tank (AT) landmines. Development or procurement of a route preparation vehicle is outside the scope of the ILDP.

The detection function is performed by the ILDS and includes both landmine detection and marking. The ILDS concept is shown in Figure 1, and consists of a teleoperated Remote Detection Vehicle (RDV) platform and a Command Vehicle (CV). The RDV contains the detectors, navigation system, marking system, and all other equipment required for teleoperation. Passive protection from AT landmines is achieved by controlling RDV ground force. The CV follows behind the RDV by up to 500 m. The control stations, including all communication equipment required to control and monitor teleoperated RDV and detector functions, are installed within the CV.

The neutralization function involves temporarily or permanently rendering detected landmines harmless. Methods such as destroying, removing, or isolating landmines may be used. Development of neutralization methods and procedures is outside the scope of the ILDP.

SYSTEM REQUIREMENTS

The primary threat being addressed by the ILDS is buried metal and low-metal content Anti-Tank (AT) landmines. Advances in landmine technology have included the development of plastic landmines which cannot be detected using metal detection techniques; however, more conventional metal and low-metal content landmines are still most commonly used. Also, new landmine technologies such as scatterable landmines and off-route landmines have not been encountered regularly in peacekeeping operations to date. The ability to detect AP landmines will be

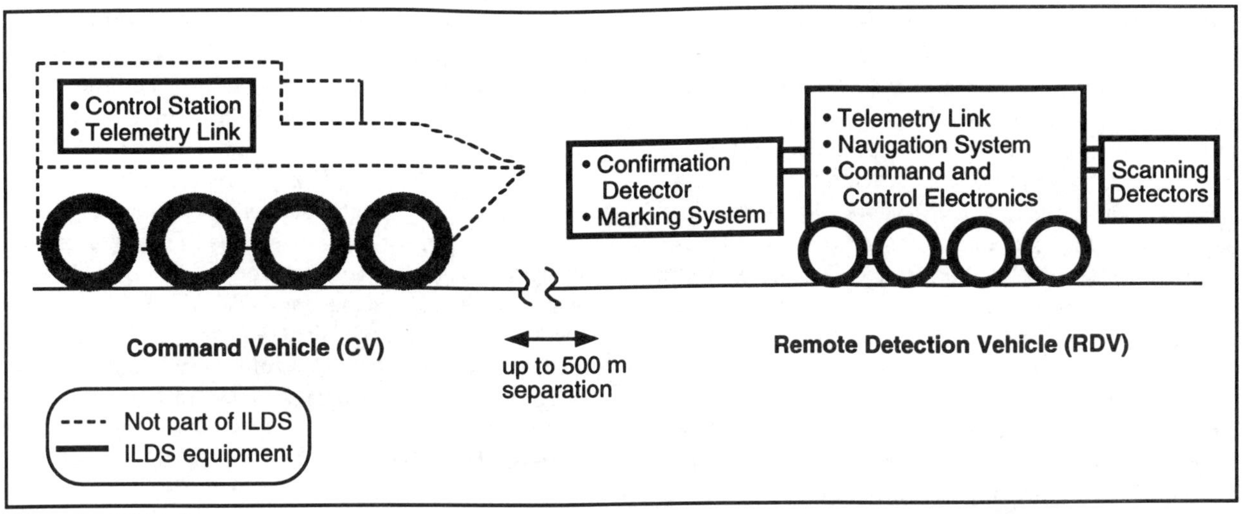

Figure 1: ILDS Concept

somewhat less than for AT landmines primarily due to the limitations of the currently available detector suite.

The ILDS is designed to detect landmines buried in roads. Typical peacekeeping missions include detection and clearance of landmines on supply and patrol routes, as well as around logistic installations and base camps. Proofing of roads that have been previously swept for landmines by other means would also be performed by the system.

Landmine clearance tasks which are part of peacekeeping missions focus on the detection and removal of all landmines from the area being cleared. Safety of landmine clearance personnel is essential, and except for emergency situations, speed is not an overriding factor. Detector technologies which were previously discarded, largely due to their inability to satisfy combined detectability and system operating speed requirements, may be used on the ILDS. Low speed operation also enables data fusion techniques to be employed, which reduces the false alarm rates associated with individual detectors and provides a satisfactory system false alarm rate.

To summarize, a detection system whose performance characteristics were once categorized as unsatisfactory for a fieldable system in an active combat situation is acceptable for use in peacekeeping roles. This enables an improved detection capability to be pursued immediately using currently available detection technologies. Since the safety of mine clearance personnel is vital, it is critical that all system limitations be defined and understood by the end user. A significant part of the ILDS development will be to test and quantify system capabilities and limitations under a variety of operating conditions.

ILDS OVERVIEW

Operating Modes

The ILDS will operate in one of two fundamental operating modes: scanning and confirmation.

In scanning mode, the RDV will move forward at a constant speed with the two system operators interpreting data from the scanning detectors, identifying possible targets, and looking for correspondence between the various scanning detector outputs. Once a possible target has been identified, the RDV will be stopped to check the location with a confirmation detector.

In confirmation mode, the RDV is stopped and the confirmation detector is placed over the potential landmine location, and the detection process initiated. If the target is confirmed to be a landmine, a physical marker will be placed at a known distance and orientation with respect to the landmine location. If the target is not confirmed, the operator may discard the possible detection and proceed, or reposition the detector and repeat the confirmation process.

Detector Suite

To date, no single detection technology is able to detect all landmines under all conditions. Instead, a multiple detector system will be used to attain the required level of landmine detection for a variety of operating conditions. The ILDS detector suite consists of four primary detectors; three for area scanning while the RDV is in motion, and one for confirmation of scanned detections while the RDV is stationary. Selection of this detector suite resulted from ongoing studies in the areas of landmine and ordnance detection conducted by the DRES [1], [2].

A data fusion process will compensate for individual detector limitations and provide the system operators with spatially-resolved detector data displays. It will enable the reduction of relatively high false alarm rates associated with each of the individual detectors since each detector "observes" a different, uncorrelated landmine characteristic. The combination of multiple scanning detectors and data fusion processing will also enable the system to function (at reduced performance levels) even when one or more detector cannot be used due to failure or ineffectiveness resulting from the specific operating environment.

System probability of detection is determined largely by the performance of the scanning detectors and data fusion process. System speed is reduced by nuisance alarms (false alarms from the scanning detectors), but the system false alarm rate is a function of confirmatory detector performance and the ability of the system to accurately track possible landmine locations from the scanning detectors back to the confirmatory detector.

Passive Infrared (IR) Detection. A commercial IR camera is one of three scanning detectors and is used to observe heat flow disturbances in the road. One of the two system operators interpret the IR imagery and designate possible landmine locations during scanning operations.

Two types of detection are used during the IR image interpretation process:

1. bulk thermal effect of a buried landmine - detection of a thermal anomaly due to the difference in thermal resistance between the buried landmine and the surrounding soil layers; and
2. surface thermal effect of the burial site - detection of a thermal anomaly caused by the perturbation of the soil surrounding the burial site which was created by the burial process.

Minimum Metal Detector (MMD). The MMD is the second scanning detector and is capable of detecting the subsurface metal content of both metal and low-metal content landmines. This detector consists of a multiple-coil array providing 3 m wide coverage, with performance characteristics which are similar to handheld metal detectors. The MMD array is mounted in front of the RDV.

Ground Penetrating Radar (GPR). The final scanning detector is a GPR which detects dielectric anomalies associated with buried landmines. This detector consists of a fixed antenna array mounted behind the MMD, and provides 3 m wide coverage.

Thermal Neutron Activation (TNA) Detector. The TNA detector has a high sensitivity to bulk explosives and a low false alarm rate. Specifically, it detects nitrogen content in soil; nitrogen levels are abnormally high when explosives are present. It is a point detector which requires the RDV to stop so that the sensor head can be brought to bear and held stationary in close proximity to the suspected landmine location. The TNA detector is restricted to a confirmation role, since the detection process requires stationary exposure to a small patch of ground for several seconds in order to achieve acceptable signal-to-noise performance. The detector head is mounted behind the RDV which provides the necessary separation between the scanning and confirmation detectors to enable maximum time for image interpretation and data fusion tasks to occur.

Detector Data Fusion

Data fusion is an essential element of the ILDS and represents one of the most challenging aspects of the system development. Data fusion involves the correlation of targets from the multiple scanning detectors and is critical to the operation of the ILDS. The use of multiple detectors is necessary to achieve the required probability of detection; data fusion is necessary to reduce the high system false alarm rate associated with the use of multiple detectors. If the nuisance and false alarm rates are not controlled in this way, the overall speed may be reduced such that use of the system is not practical.

An automatic position estimation system is used to track the positions of various detector targets, and enables the control station computer to provide the operator with a spatially-corrected display of the targets. A semi-automated process is used to evaluate the correlation of these targets.

Remote Detection Vehicle

The RDV is designed specifically to provide mobility for the landmine detection system. The multi-wheeled, hydraulically-driven platform uses differential skid steering and has wheel-pair bogies to ensure multiple independent wheel-to-ground contacts. The total RDV weight is distributed over a sufficiently large area for traveling over bumps, obstacles, or on otherwise rough road. The RDV also includes the electrical power source for all RDV-mounted equipment.

Marking System

The marking system is capable of depositing a landmine location indicator which is visible during the day as well as at night, when illuminated. These markers are placed near the center of the confirmed landmine location to minimize the total error between the marker's location and the surface-projected centroid of the detected landmine.

Control Station

The control station is located within the CV and comprises the controls, indicators, and displays required to control the RDV and to monitor detector data outputs. Only two system operators are required to operate the system via two workstation computers.

RDV ARCHITECTURE

The major RDV-mounted detection system components and their interconnections are shown in Figure 2. Each of these components is controlled and/or monitored from the control station.

At the center of this system is the Command and Control (C^2) Electronics subsystem, which is based on the Ancaeus Control System previously developed by DRES. This subsystem performs both video and data multiplexing functions to minimize telemetry link requirements. This is important, not only to reduce system costs, but also to address the end user's requirement to minimize the system's transmission bandwidth.

Video signals from the IR camera, RDV steering camera, and TNA positioning camera are switched within the C^2 Electronics subsystem and connected directly to video transmitters within the Telemetry Link subsystem.

The C^2 Electronics subsystem also performs the necessary serial data multiplexing and demultiplexing functions for control and monitoring of MMD, GPR, TNA detector, navigation system, and marking system data outputs from the control station using a single RF data link. This subsystem is expandable to accommodate the addition of new detectors or upgrades to existing system components.

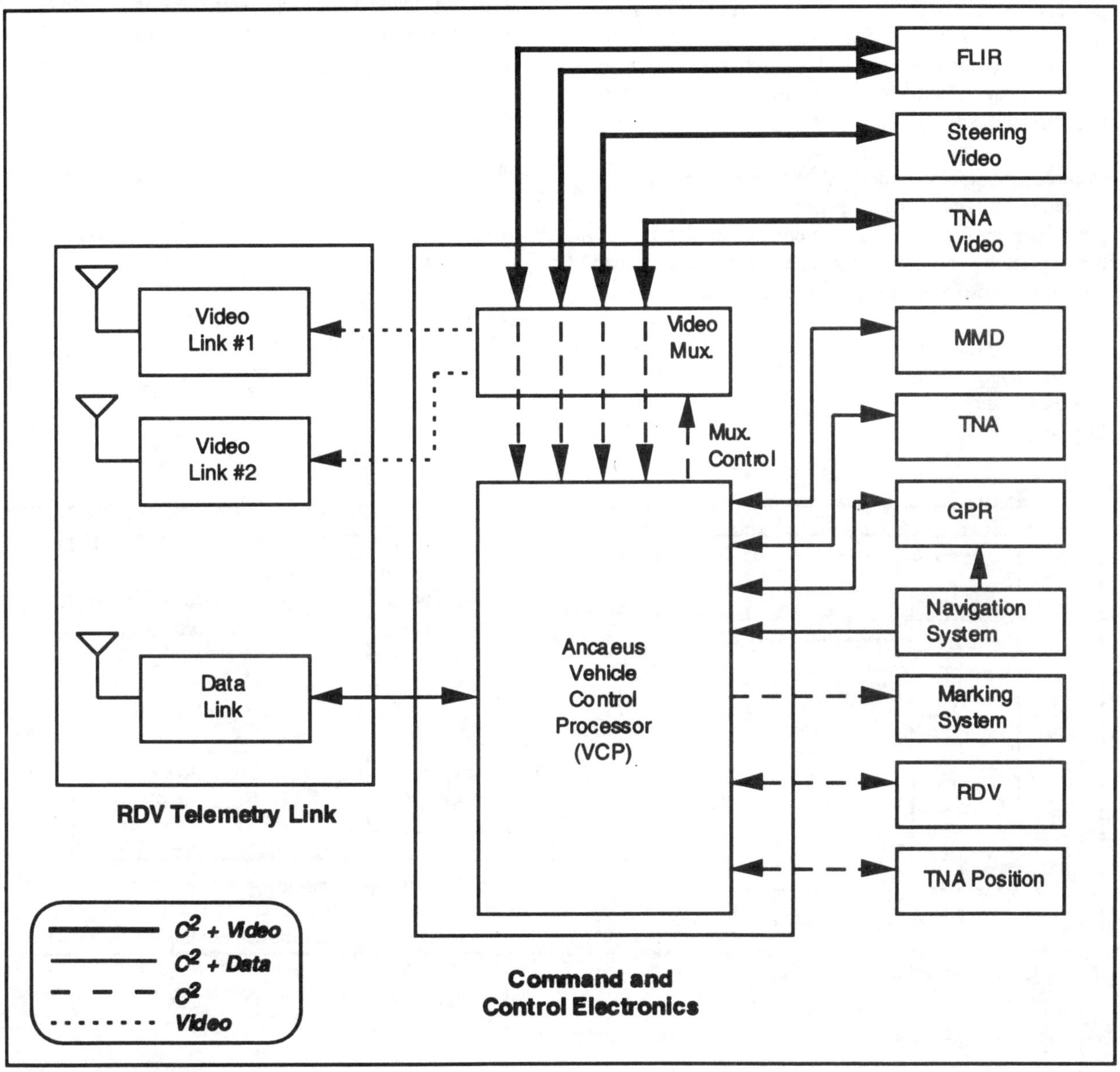

Figure 2: RDV Detection System Components Block Diagram.

The C^2 Electronics (Ancaeus) system also performs all vehicle control and monitoring functions. The modular design enables the detector package to be installed on an alternative vehicle by changing a single module which interfaces to the vehicle's monitoring instrumentation and control actuators. Built-in safety devices alert the system operators when the remotely-operated RDV is about to move, and stop the RDV if communication between the RDV and CV is interrupted.

COMMAND VEHICLE ARCHITECTURE

Control stations and telemetry link equipment within the CV enable the two system operators to control the teleoperated RDV detection platform, monitor detector outputs, and make decisions concerning the presence of landmines. The major CV-mounted system components are shown in Figure 3.

The control consoles are identical and use programmable controls to enable operators to switch roles without physically changing position. This feature allows the operators to quickly and frequently change tasks thereby maintaining a high level of concentration and performance.

Control computers perform complementary data multiplexing and demultiplexing operations to those performed on the RDV. These computers also perform target tracking and data fusion calculations, and provide meaningful displays for the system operators.

Video signals are fed directly from the video receivers to the control station computers where they are processed and displayed in the appropriate operator display screens.

CONCLUSION

This paper has outlined the approach being taken for the development of an improved landmine detection and marking system for peacekeeping roles. An incremental approach is being used which enables an urgent operational requirement to be addressed expeditiously in spite of the considerable difficulty of the fundamental detection problem. It also enables significant progress to be made in a relatively short time period with very limited budgets and R&D resources.

The main features of the ILDS solution have also been described. By limiting the scope of the total landmine detection problem, currently available detection technologies with known capabilities may be used in conjunction with data fusion techniques to produce an improved landmine detection capability. This system will provide landmine clearance personnel with a system which is much faster and safer than the currently available manual methods, and represents a significant step towards a solution to the overall landmine detection problem.

REFERENCES

[1] McFee, J. et al, 1994, DRES-SSP-174, 19-36, 65-85.

[2] McFee, J., and Das, Y., 1991, DRES-SR-548, 3-37.

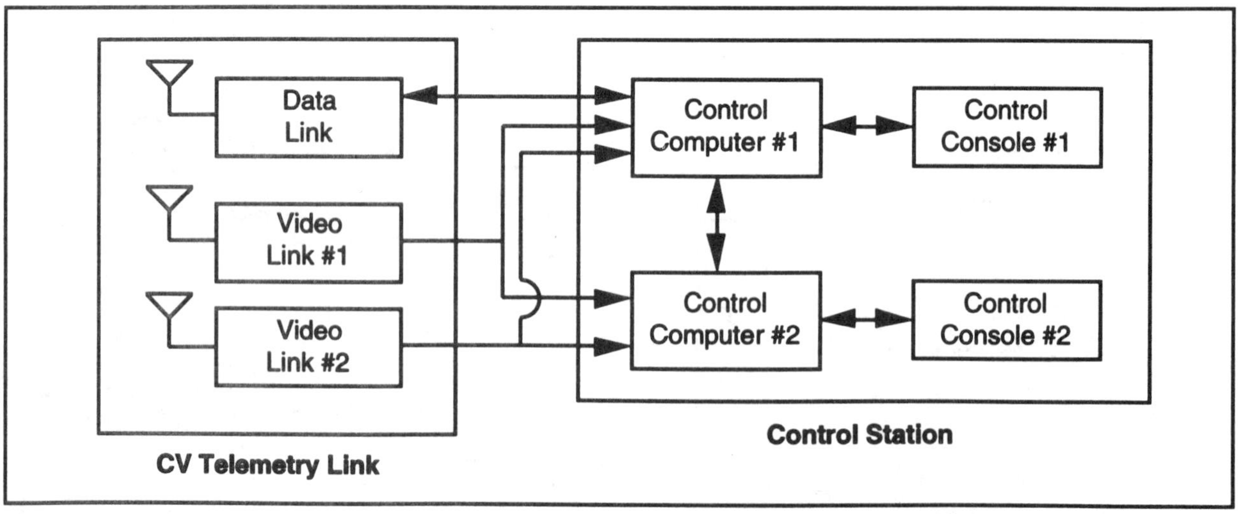

Figure 3: CV Block Diagram.